T0100732

Birkhäuser

Modeling and Simulation in Science, Engineering and Technology

For further volumes:
http://www.springer.com/series/4960

Animesh Mukherjee • Monojit Choudhury
Fernando Peruani • Niloy Ganguly • Bivas Mitra
Editors

Dynamics On and Of Complex Networks, Volume 2

Applications to Time-Varying Dynamical Systems

Editors
Animesh Mukherjee
Department of Computer Science
and Engineering
Indian Institute of Technology
Kharagpur, India

Monojit Choudhury
Microsoft Research Lab India
Bangalore, India

Fernando Peruani
Laboratoire J.A. Dieudonné
Université de Nice-Sophia Antipolis
Nice, France

Niloy Ganguly
Department of Computer Science
and Engineering
Indian Institute of Technology
Kharagpur, India

Bivas Mitra
Department of Mathematical Engineering
(INMA)
Université Catholique de Louvain
Louvain-la-Neuve, Belgium

ISSN 2164-3679 ISSN 2164-3725 (electronic)
ISBN 978-1-4614-6728-1 ISBN 978-1-4614-6729-8 (eBook)
DOI 10.1007/978-1-4614-6729-8
Springer New York Heidelberg Dordrecht London

Library of Congress Control Number: 2009921285

Mathematics Subject Classification (2010): 82B31, 82B43, 90B15, 90B18, 90B40, 91D30, 92D30

Printed on acid-free paper

Springer is part of Springer Science+Business Media (www.birkhauser-science.com)

Preface

"Network science" has recently attracted the attention of a large number of researchers from various disciplines, mainly due to its ubiquitous applicability in modeling the structure and dynamics of large-scale complex systems (both natural and man-made). Examples of such systems, exhibiting complex interaction patterns among their constituent entities, range from genetic pathways and ecological networks to the WWW, peer-to-peer networks, and blogs and online web-social networks.

Dynamics on networks refers to the different types of so-called processes (e.g., proliferation and diffusion) that take place on networks. The functionality/efficiency of such processes is strongly affected by the topology as well as the dynamic behavior of the network. On the other hand, *dynamics of networks* mainly refers to various phenomena (e.g., self-organization) that go on in order to bring about numerous changes in the topology of the network.

A first systematic exploration of the above field materialized into an edited volume: *Dynamics on and of Complex Networks: Applications to Biology, Computer Science and the Social Sciences* published by Birkhäuser (ISBN: 978-0-8176-4750-6). With a rich collection of surveys and cutting-edge research contributions exploring the interdisciplinary relationship of dynamics on and of complex networks this volume served as one of the pioneers of research in this filed. The entire piece was thematically organized into three main sections: Part I studies the application of complex networks to biological problems; Part II focuses on social networks; and Part III is an overview of networks prevalent in the information sciences.

With the progress of this field, it is becoming more and more apparent that the current challenges lie in understanding and solving problems that arise in dynamically changing networks, i.e., whose structure as well as function can both change with time. Examples include dynamic trafficking networks, telephone/mobile communication networks, or web-social networks. The processes that essentially take place on top of such networks are also dynamic, i.e., processes like epidemic spreading or the opinion formation co-evolve over time along with the underlying network. This second volume aims to put forward burgeoning multidisciplinary research contributions that combine methods from computer science, statistical physics, econometrics, and social network theory towards modeling time-varying social, biological, and information systems. This volume deems to significantly contribute to the scientific advancement of the field of complex networks and, most importantly, dynamically changing networks that has also been continuously, for the past two years, the central theme for the workshop series Dynamics On and Of Complex Networks, held in conjunction with the European Conference on Complex Systems. A closer inspection would make it clear that the issues and the related problems in this area are still very loosely defined. Therefore, in summary, the primary objective of this volume has been to tie these loose ends through in-depth review chapters and concretize new problems that need to be urgently addressed in this area through cutting-edge contributory chapters.

This volume consists of three parts. The contributions in Part I center around the dynamical properties of online social networks, the Internet, and the WWW. This part consists of five chapters. The first chapter, "Dynamics in Online Social Networks" by Grabowicz et al., describes some of the results of research studies on the structure, dynamics, and social activity in online social networks. In the next chapter, "An Empirical Validation of Growth Models for Complex Networks," Mislove et al. use empirical growth data from four different networks (the Flickr and the YouTube online social networks, Wikipedia's content graph, and the Internet's AS-level graph) to validate proposed models for the generation of power-law networks. In the third chapter, "On the Routability of the Internet," Erola et al. present methodologies based on complex network theory to construct navigable maps of the Internet to address the problem of the scalability of the Internet routing protocols. The fourth chapter, "The Evolution of Layered Protocol Stacks Leads to an Hourglass-Shaped Architecture" by Akhshabi et al., studies the Internet protocol stack and their evolution in a rigorous and quantitative manner. They propose an "EvoArch" model which predicts the emergence of an hourglass architecture and the appearance of few stable nodes (not always of the highest quality) at the waist of the hourglass. In the last chapter, "Opinion Dynamics on Coevolving Networks," Federico Vazquez describes some of the most representative opinion dynamics models on evolving networks.

Part II is spread over seven chapters and focuses on community analysis which at this point in time constitutes as one of the "hottest" areas of research in this field. This part begins with the chapter "A Template for Parallelizing the Louvain Method for Modularity Maximization," where Bhowmick et al. introduce a shared memory implementation of the popular Louvain method for maximizing modularity. They discuss the challenges in parallelizing this algorithm as well as metrics for evaluating correctness while handling large-scale data. The next chapter, "Multi-scale Modularity and Dynamics in Complex Networks," is authored by Renaud Lambiotte and focuses on the detection of communities in multi-scale networks, namely networks made of different levels of organization and in which modules exist at different scales. The third chapter, "Evaluating the Performance of Clustering Algorithms in Networks" by Andrea Lancichinetti, introduces a new model of benchmark graphs whose features are close to those found in real networks for evaluation of clustering algorithms. The fourth chapter, "Communities in Evolving Networks: Definitions, Detection and Analysis Techniques" by Aynaud et al., exposes a survey of recent advances in the definition, the detection and the analysis of communities in the particular case of evolving networks. The fifth chapter, "Clustering Hypergraphs for Discovery of Overlapping Communities in Folksonomies" by Chakraborty et al., presents the overlapping-hypergraph clustering algorithm, which detects overlapping communities in folksonomies using the complete tripartite hypergraph structure. The sixth chapter, "The Stability of a Graph Partition: A Dynamics-Based Framework for Community Detection" by Delvenne et al., develops a dynamical perspective towards community detection by introducing a stability measure that provides a unifying framework enabling deeper understanding of what a community structure is. In the last chapter, "Algorithms for Finding Motifs in Large Labeled Networks," Khan et al. introduce different kinds of subgraph analysis problems and discuss some of the important parallel algorithmic techniques that have been developed for them.

Part III presents an overview on diffusion, spreading, mobility, and transport on networks. This part is laid out in four chapters. The first chapter in this part, "A Dynamical Network View of Lyon's Vélo'v Shared Bicycle System" by Borgnat et al., discusses the dynamical properties of Lyon's shared bicycle system called Vélo'v. In the second chapter, "Generalized Voter-Like Models on Heterogeneous Networks," Moretti et al. presents a generalized model of consensus formation that, which is able to encompass all previous formulations of copy/invasion processes inspired by variations on the voter model and the Moran process. The authors consider the implementation of such generalized dynamics on a heterogeneous contact pattern, represented by a complex network, and derive the theoretical predictions for the relevant dynamical quantities, within the assumptions of the heterogeneous mean-field theory. The next chapter, "Epidemics on a Stochastic Model of Temporal Network" by Rocha et al., presents a simple and intuitive stochastic model of a temporal network and investigates how a simulated infection coevolves with the temporal structures, focusing on the growth dynamics of the epidemics. The last chapter, "Network-Based Information Filtering Algorithms: Ranking and Recommendation" by Matúš Medo, gives an overview of applications of random walks to information filtering, focusing on the tasks of ranking and recommendation in particular.

The aforementioned contributions collectively demonstrate that dynamic networks indeed provide an intellectually deep research area relevant to a variety of scientific disciplines. The chapters are designed to serve as the state of the art not only for students and newcomers but also for experts who intend to pursue research in this field. All the chapters have been carefully peer-reviewed in terms of their scientific content as well as readability and self-consistency.

We would like to thank the authors for their contributions, constructive cooperation, and gracious acceptance of the editorial comments. We are also indebted to Paolo Bajardi, Anirban Banerjee, Ritwik Banerjee, Fabricio Benevenuto, Sanjukta Bhowmick, Chris Biemann, Anja Voss Boehme, Arnaud Browet, Joydeep Chandra, Vineet Chaoji, Munmun De Choudhury, Amit Deshpande, Young-Ho Eom, Fakhteh

Ghanbarnejad, Saptarshi Ghosh, Sudarshan Iyengar, Susanta Karmakar, Srivatsan Laxman, Matúš Medo, Pabitra Mitra, Mainack Mondal, Anirban Mukherjee, Subrata Nandi, José J. Ramasco, Parag Singla, Abhijeet Sonawane, Guillermo Terranova, Michele Tizzoni, Francesca Tria, and Federico Vazquez for their constructive criticisms, comments, and suggestions, which have significantly improved the quality of the chapters. In addition, we would also like to extend our gratitude to Suman Kalyan Maity and Swadhin Pradhan for their painstaking efforts in preparing the final draft of the volume. Finally, we are also grateful to the entire editorial team from Birkhäuser for all the help and support extended by them towards the publication of this volume.

Kharagpur, India	Animesh Mukherjee
Bangalore, India	Monojit Choudhury
Nice, France	Fernando Peruani
Kharagpur, India	Niloy Ganguly
Louvain-la-Neuve, Belgium	Bivas Mitra

Contents

Contributors

Patrice Abry Laboratoire de Physique, ENS de Lyon, CNRS, Lyon, France

Saamer Akhshabi College of Computing, Georgia Institute of Technology, Atlanta, GA, USA

V. S. Anil Kumar Department of Computer Science and Virginia Bioinformatics Institute, Virginia Tech, Blacksburg, VA, USA

Alex Arenas Departament d'Enginyeria Informàtica i Matemàtiques, Universitat Rovira i Virgili, Tarragona, Spain

Thomas Aynaud UPMC, CNRS, France

Mauricio Barahona Department of Mathematics, Imperial College London, London, UK

Andrea Baronchelli Department of Physics, College of Computer and Information Sciences, Bouvé College of Health Sciences, Northeastern University, Boston, MA, USA

Bobby Bhattacharjee University of Maryland, College Park, MD, USA

Sanjukta Bhowmick Department of Computer Science, University of Nebraska at Omaha, Omaha, NE, USA

Pierre Borgnat Laboratoire de Physique, ENS de Lyon, CNRS, Lyon, France

Vincent D. Blondel Department of Mathematical Engineering Université catholique de Louvain, Louvain-la-Neuve, Belgium

Abhijnan Chakraborty Microsoft Research India, Bangalore, India

Adeline Decuyper Department of Mathematical Engineering Université catholique de Louvain, Louvain-la-Neuve, Belgium

Jean-Charles Delvenne Institute of Information and Communication Technologies, Electronics and Applied Mathematics (ICTEAM) and Center for Operations Research and Optimisation (CORE), Université catholique de Louvain, Louvain-la-Neuve, Belgium

and
Namur Center for Complex Systems (naXys), Facultés Universitaires Notre-Dame de la Paix, Namur, Belgium

Constantine Dovrolis College of Computing, Georgia Institute of Technology, Atlanta, GA, USA

Peter Druschel Max Planck Institute for Software Systems, Kaiserslautern–Saarbruecken, Germany

Víctor M. Eguíluz Instituto de Física Interdisciplinar y Sistemas Complejos IFISC (CSIC-UIB), E07122 Palma de Mallorca, Spain

Pau Erola Departament d'Enginyeria Informàtica i Matemàtiques, Universitat Rovira i Virgili, Tarragona, Spain

Patrick Flandrin Laboratoire de Physique, ENS de Lyon, CNRS, Lyon, France

Eric Fleury ENS de Lyon (UMR CNRS – ENS de Lyon – UCB Lyon 1 – Inria), France

Saptarshi Ghosh Department of Computer Science and Engineering, Indian Institute of Technology Kharagpur, Kharagpur, India

Sergio Gómez Departament d'Enginyeria Informàtica i Matemàtiques, Universitat Rovira i Virgili, Tarragona, Spain

Przemyslaw A. Grabowicz Instituto de Física Interdisciplinar y Sistemas Complejos IFISC (CSIC-UIB), Palma de Mallorca, Spain

Jean-Loup Guillaume UPMC, CNRS (UMR), France

Krishna P. Gummadi Max Planck Institute for Software Systems, Kaiserslautern–Saarbruecken, Germany

Maleq Khan Department of Computer Science and Virginia Bioinformatics Institute, Virginia Tech, Blacksburg, VA, USA

Hema Swetha Koppula Cornell University, Ithaca, NY, USA

Renaud Lambiotte Department of Mathematics, University of Namur, Namur, Belgium

Andrea Lancichinetti Howard Hughes Medical Institute (HHMI), Northwestern University, Evanston, IL, USA
and
Department of Chemical and Biological Engineering, Northwestern University, Evanston, IL USA

Madhav V. Marathe Department of Computer Science and Virginia Bioinformatics Institute, Virginia Tech, Blacksburg, VA, USA

Matúš Medo Physics Department, University of Fribourg, Fribourg, Switzerland

Alan Mislove Northeastern University, Boston, MA, USA

Paolo Moretti Departamento de Electromagnetismo y Física de la Materia and Instituto "Carlos I" de Física Teórica y Computacional, Universidad de Granada, Facultad de Ciencias, Fuentenueva s/n, Granada, Spain

Romualdo Pastor-Satorras Departament de Física i Enginyeria Nuclear,

Universitat Politècnica de Catalunya, Barcelona, Spain

José J. Ramasco Instituto de Física Interdisciplinar y Sistemas Complejos IFISC (CSIC-UIB), Palma de Mallorca, Spain

Céline Robardet LIRIS, INSA-Lyon, CNRS, Bâtiment Blaise Pascal, Villeurbanne, France

Luis E.C. Rocha Department of Mathematical Engineering Université catholique de Louvain, Louvain-la-Neuve, Belgium

Jean-Baptiste Rouquier Eonos, Paris, France

Michael T. Schaub Department of Chemistry, Imperial College London, London, UK
and
Department of Mathematics, Imperial College London, London, UK

Sriram Srinivasan Department of Computer Science, University of Nebraska at Omaha, Omaha, NE, USA

Michele Starnini Departament de Física i Enginyeria Nuclear, Universitat Politècnica de Catalunya, Barcelona, Spain

Nicolas Tremblay Laboratoire de Physique, ENS de Lyon, CNRS, Lyon, France

Federico Vazquez Max Planck Institute for the Physics of Complex Systems, Dresden, Germany

Qinna Wang Inria (UMR CNRS – ENS de Lyon – UCB Lyon 1 – Inria), France

Sophia N. Yaliraki Department of Chemistry, Imperial College London, London, UK

Zhao Zhao Department of Computer Science and Virginia Bioinformatics Institute, Virginia Tech, Blacksburg, VA, USA

Part I
Online Social Media, Internet, and WWW

Dynamics in Online Social Networks

Przemyslaw A. Grabowicz, José J. Ramasco, and Víctor M. Eguíluz

1 Introduction

An increasing number of today's social interactions occur using online social media as communication channels. Some online social networks have become extremely popular in the last decade. They differ among themselves in the character of the service they provide to online users. For instance, Facebook can be seen mainly as a platform for keeping in touch with close friends and relatives, Twitter is used to propagate and receive news, LinkedIn facilitates the maintenance of professional contacts, and Flickr gathers amateurs and professionals of photography. Albeit different, all these online platforms share an ingredient that pervades all their applications. There exists an underlying social network that allows their users to keep in touch with each other and helps to engage them in common activities or interactions leading to a better fulfillment of the service's purposes. This is the reason why these platforms share a good number of functionalities, e.g., personal communication channels, broadcasted status updates, easy one-step information sharing, and news feeds exposing broadcasted content. As a result, online social networks are an interesting field to study an online social behavior that seems to be generic among the different online services. Since at the bottom of these services lays a network of declared relations and the basic interactions in these platforms tend to be pairwise, a natural methodology for studying these systems is provided by network science. In this chapter, we describe some of the results of research studies on the structure, dynamics, and social activity in online social networks. We present them in the interdisciplinary context of network science, sociological studies, and computer science.

P.A. Grabowicz (✉) • J.J. Ramasco • V.M. Eguíluz
Instituto de Física Interdisciplinar y Sistemas Complejos IFISC (CSIC-UIB),
E07122 Palma de Mallorca, Spain
e-mail: pms@ifisc.uib-csic.es

A. Mukherjee et al. (eds.), *Dynamics On and Of Complex Networks, Volume 2*,
Modeling and Simulation in Science, Engineering and Technology,
DOI 10.1007/978-1-4614-6729-8_1, © Springer Science+Business Media New York 2013

2 Structure of Social Networks

Social networks in general show a very rich internal structure [1] that in some aspects falls quite far from random graphs or even from artificial networks created by virtue of a preferential attachment mechanism. In this section we briefly review the most important features broadly found in social networks.

2.1 *Degree Distribution*

The most fundamental characteristic of a network is the distribution of degrees: a function that measures how many friends have the members of the network and what is the variability of this number among all the users. The degree distribution in social networks is usually broad. These distributions have been typically modeled as functions having a heavy tail such as a power-law or a lognormal combined with an exponential cutoff at large values of the number of friends [2–7]. This means that there is a large variability in the number of connections of the nodes, with many nodes having small or moderate number of friends and a small number of them maintaining large number of friends. Almost all users of online social networks are connected in a largest connected component [4, 7]. Some of the studies also point out that online social networks contain a densely connected core or cores [4, 5] consisting in groups of high-degree nodes that hold the network together. The existence of such cores provides paths for the connection between distinct parts of the network. A well-known aspect of the social networks is that the average shortest path distance is low [2, 4, 5, 7]. This characteristic is popularly known as six degrees of separation or small-world effect [8]. The importance of the shortcuts for reducing the network path length has been highlighted in [9].

2.2 *Triangles and Community Structure*

Possibly, the most important feature distinguishing social networks from other types of networks is their high level of clustering or transitivity [1,2,4–7,9]. The clustering coefficient measures the probability that two nodes sharing a common neighbor (a node to which they are both connected) are connected. This property is quantified with a global clustering coefficient C [6] which is defined as

$$C = \frac{\text{number of closed connected triples}}{\text{number of connected triples}}, \tag{1}$$

where a connected triple of nodes is a sequence of 3 nodes which have at least 2 connections between them and a closed triple is a triangle. One can also define a local clustering coefficient c_i as

$$c_i = \frac{\text{number of closed triples centered on node i}}{\text{number of triples centered on node i}}. \quad (2)$$

In this case a global value of clustering coefficient may be obtained averaging the local c_i over all the nodes of the network $\langle c \rangle$. One should note that $\langle c \rangle$ is in general different from the coefficient C and that the latter has a much worse scaling behavior. At the structural level, a high clustering coefficient indicates the presence of many triangles in the network. At the social level, this means that friends of an individual tend to be connected between themselves too. This is a well-known phenomena in sociology which is important for the formation of strong social ties [10, 11] and affects the emergence of positive and negative relations [12]. At the network macroscopic level, a high density of triangles can be related to the existence of community structure in social networks [13]. Furthermore the study [14] suggests that in real networks with high value of clustering coefficient community structure emerges without any additional ingredients included.

Existence of communities in social networks is considered to have high relevance both by sociologist [11, 13] and network scientists [15–17]. We give further arguments on this in the third section of this chapter. In online social networks, groups can be identified in several ways. One of them is searching for communities in the graph defined as more densely connected parts of the network compared with their neighborhood. This approach is usually taken in network science, and various community detection algorithms have been developed and continue to be under active development for detecting such clusters [15]. In addition, some online social networks allow their users to create explicit groups and to claim its membership. Although it seems straightforward to make use of such user-declared groups, one should be careful when interpreting them since incentives for creation of such groups may vary [18]. Nevertheless it has been found that the declared groups tend to have internally higher clustering coefficient [5] and therefore they may be correlated with the more densely connected parts of the network found by community detection algorithms.

2.3 *Assortativity and Homophily*

Another common feature of social networks is that connected users tend to be similar [19]. This effect is popularly known as *birds of a feather flock together* phenomena. It manifests itself in social networks through similarities in various properties of connected individuals. From pure network theory point of view the similarity may appear as a correlation of degrees between friends, which is called assortativity mixing, or as a rich-club effect [20]. In such assortative networks nodes of high degrees tend to be connected to other nodes of high degrees, and vice versa, nodes of low degree tend to be connected to other low-degree nodes. It has been found that offline social networks are assortative in contrast with networks of other types [1, 21]. However, this is not the only property in which friends are similar. This kind of phenomena is in general called homophily and is known to be present very broadly is social networks. People who are connected in online social networks tend to have similar age, live in close locations, and have similar interests [4,7,17,22]. It is also considered that people who belong to the same community, namely, the same well-connected group of people, talk about similar topics, which can have an important impact on information and innovation diffusion in social networks [11,23].

2.4 Differences Between Offline and Online Social Networks

As shown in the previous subsections many statistical properties of offline social networks are also found online. On the other hand creating links in an online social network is much less costly than developing offline social relations. These online connections can easily accumulate and pile up to large numbers [24]. If the number of connections increases to the millions, the amount of effort that a user can invest into a relation that each link represents must fall to near zero. An early illustration of the relevance of the definition of social tie in characterization of social networks was shown in the study of email networks: while the distribution of the number of contacts in address books is power law [25], it is exponential when the contacts are restricted to reciprocated emails [26]. Moreover disassortative mixing has been encountered in some online networks [2, 27] in contrast to the assortative mixing characteristic of offline social networks [6]. As a matter of fact there exists an open discussion on the validity of online interactions as indicators of real social activity [24,28–31]. In order to test the validity of online networks for social studies and to find its limitations, further investigation is needed. In this chapter we present some of the recent results of such studies.

3 Growth in Social Networks

3.1 Preferential Attachment

Many features of complex systems are characterized by heavy-tailed distributions [32, 33], e.g., frequency of words [34], the wealth of nations [35], and degree distribution of complex networks [36]. This property is typically perceived as a symptom of the rich-gets-richer principle, from which the so-called preferential growth models stem. The common concept of these models is that the elements of the system grow proportionally to their current size, what is referred to as preferential growth or preferential attachment rule. Typically, in these models, increments of the defining variables of the system occur in each time step. Such increments can involve the addition of new elements and/or to increase the sizes of the existing ones according to a preferential growth rule. Preferential models are usually the first approach to explain heavy-tailed distributions in many different systems [37–40]. In the case of networks, this kind of models got popularized a decade ago [36,41–44]. The first of these models in the context of complex networks was introduced by Barabási and Albert in [41]. To describe it shortly, in each time step, one node is introduced into the system with m edges. These edges are connected to existing nodes in the system with probability proportional to the degree of the present nodes. As a result a network with heavy-tailed (usually power-law) distribution of node degrees emerges. The rule of Barabási-Albert model yields high simplicity, which is typically a desirable feature, but that can be too rigid in some cases. In preferential-growth models, the time unit is directly coupled to the number of new arriving elements, which can complicate the comparison of the dynamics of these models with real data. Some other drawbacks include the lack of heterogeneity and strong correlation between age of elements and their size [45]. Because of these issues the basic preferential growth model is typically used as a simple model for generating networks with a power-law degree distribution. On the other hand, it is also successfully used as a component of models trying to simulate growth of real social networks [46,47].

3.2 Heterogeneity

In many real systems, especially in social systems, individuals or elements are very diverse. This factor is related to the heavy-tailed distributions that are so commonly found. In this direction, some models incorporating heterogeneity in the form of fitness, hidden variables, or ranking shuffling have been proposed [48–52]. In general this family of models determines growth of elements with some kind of intrinsic property. Whereas in preferential attachment models, the growth is proportional to the current size of the elements; in fitness models, it is usually proportional to the intrinsic fitness of each element. Typically the fitness is a random variable specific for each element drawn from a given distribution. This introduces high heterogeneity among the elements. A number of empirical works shows how this intrinsic fitness is distributed and what is its role in complex system growth [53–56]. We discuss in detail one of the models of this family in the next section when commenting on the growth of groups in online social networks.

3.3 Triadic Closure/Triangle Closing

Due to the fact that clustering coefficient is remarkably high in some networks (mainly social networks), other growth models have been introduced in order to reproduce high number of triangles in the network. One of the first models accounting for this was [9] in which regular network with initially high clustering had its connections rewired to make it more realistic and control clustering coefficient, as well as average shortest path length. A more sophisticated model used to simulate growth of social networks has been proposed in [47], and one of its main components is triangle closing. In this model new nodes appearing in the system connect to some node, usually using preferential attachment rule, and then start closing triangles with neighbors of this node. This simple triangle-closing mechanism exhibits much more realistic results in modeling online social networks [47].

3.4 Dynamics of Groups

As we have emphasized in the previous section the existence of communities plays an important role in functioning of social networks. In this section we present studies of the growth of such groups. Several aspects have been identified as positively influencing groups' growth and their persistence. It has been suggested that high internal connectivity helps declared groups' growth [57]. Other work argues that flexibility of big communities helps them stay alive longer, while small groups are more persistent if their composition stays unchanged [17].

From the macroscopic perspective growth of groups can be described and modeled using a version of preferential attachment model or a model with hidden variables/intrinsic heterogeneity. A comparison between these two approaches has been performed in [56] using real data from Flickr. The heterogeneous linear growth model suggested in this study assumes linear growth of groups with growth value (fitness) being drawn from heavy-tailed distribution (lognormal) and a number of new groups appearing in the systems growing linearly in time. As a comparison, a version of Simon model [37] has been used, which represents a model from preferential attachment family. As one can see in Fig. 1a, the average growth $\langle\alpha|g\rangle$ for groups of given size g is proportional to the size of the groups for high g. This commonly is interpreted as the consequence of preferential attachment. However, as it is shown in Fig. 1a, one obtains similar dependence using the heterogeneous linear growth model. This is the case because the average growth is an average over all groups of a given size, each of them growing linearly. Due to the heterogeneity and the linear growth, at a given time, larger groups consist of old groups that grow slowly and younger groups that grow faster. Thus, the observation of preferential growth for groups of the same size does not reflect in this case an underlying rich-gets-richer principle, but it is a consequence of the competition of groups with different growth values and ages. Both the heterogeneous linear growth model and the Simon model produce heavy-tailed distribution of group sizes. However, the former model performs better in other respects. First, in the Simon model the final size of groups is heavily determined by their initial size measured one year before (Fig. 1b); thus, there is little heterogeneity among the groups, in contrast to the heterogeneous linear growth model which displays a degree of heterogeneity similar to the one of real groups. Second, for the Simon model the correlation of size and age is strong, while it is weak for real groups and the heterogeneous

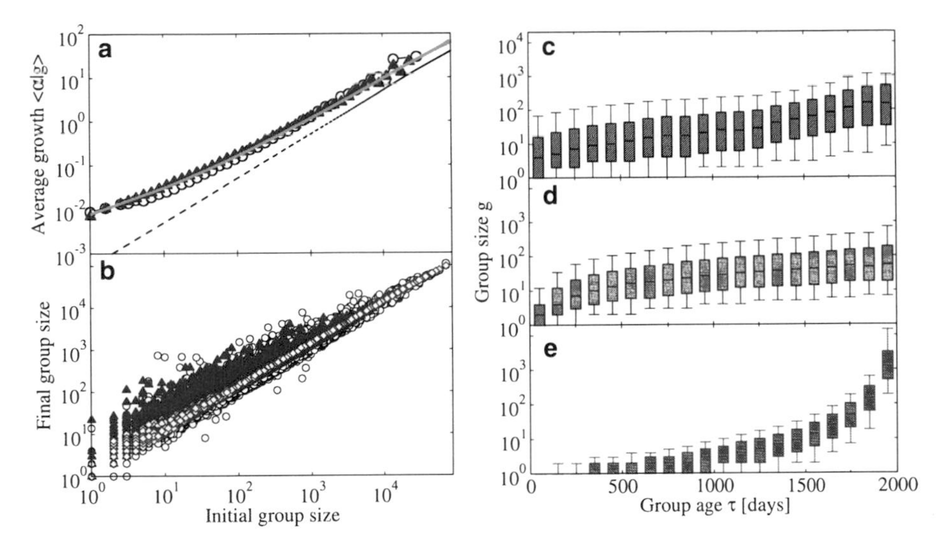

Fig. 1 Growth of groups in Flickr. (**a**) Average daily growth as a function of the initial size of the groups, estimated for the period of 6 weeks and averaged over 260,000 groups of a given initial size, for the real data from Flickr (*circles*), the heterogeneous linear growth model (*triangles*). The dashed line corresponds to the linear behavior $\langle \alpha | g \rangle \sim g$. (**b**) Initial and final group sizes over a period of 350 days for the real data (*circles*), the heterogeneous linear growth model (*filled triangles*) and Simon model (*diamonds*). Each point represents a single group, there are 9,503 points plotted for each set of points. (**c–e**) Box plots with whiskers at 9th/91st percentile of final size of groups as a function of their age at the time of the measurement for 260,000 groups for (**c**) the real data, (**d**) the heterogeneous linear growth model, and (**e**) the Simon model (Adapted from [56])

linear growth model (Fig. 1c–e). In the heterogeneous linear growth model the heavy-tailed distribution of final sizes of elements does not emerge from the growth process itself (e.g., rich-gets-richer principle), but from the intrinsic heterogeneity of elements which take part in this growth process. This certainly does not answer the question why some groups grow faster than the others, as we do not understand yet what factors influence the fitness of the groups. However it points that it does not have to be due to the fact that one group is bigger than the other as in preferential attachment models. The simplicity of this approach suggests that the characterization of the heterogeneity may play an important role in understanding the origin of broad distributions and the time evolution of many real systems.

4 Activity in Online Social Networks

In general a social network is a broad term, and it refers to a set of actors and a set of ties between them representing some kind of relation or interaction. In fact, however, there are many types of both relations and interactions, and usually they

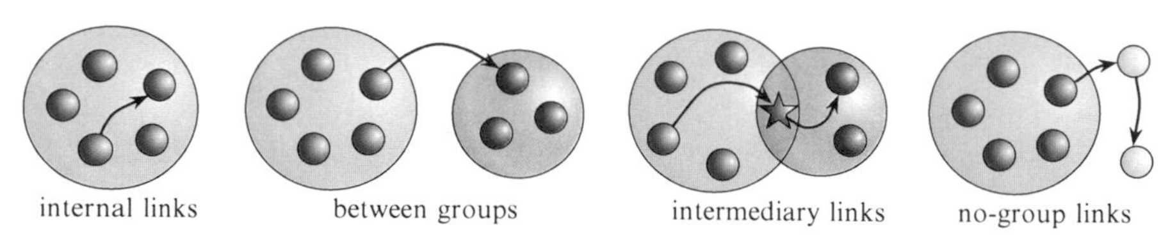

Fig. 2 Different types of links depending on their position with respect to the groups' structure: internal, between groups, intermediary links, and no-group links (Adapted from [16])

happen on top of each other. So far we mostly discussed social networks which represent a particular relation or interaction, e.g., coappearance in movies, boards of directors or coauthorship [1, 6, 21], network of online friendship [2, 5, 7, 47], and network of communication [4, 17, 58]. In online social networks, we can relate user activity with their declared relations with other users. In other words, one can relate pairwise (rarely one-to-many) interactions of users with their declared social network. We describe the studies which tackle this issue in this section.

4.1 Activity Networks Vs. Declared Social Network

The comparison of the network built from declared online friends and the network built from user interactions shows several differences at the structural level. First of all, the actors tend to interact with much less people than they declare as friends, what results in smaller degrees of nodes in the interactions network [59,60]. Moreover, the friends they interact with change rapidly, and only about 30% of pairwise interactions in one month continue over the next month [59]. Due to the fact that the degrees are lower, the properties related to small-world effect are also less evident, namely, average path lengths are higher [60] and there is less densely connected cores [61].

4.2 Theories on Social Ties and Information Diffusion

The theory known as *the strength of weak ties*, proposed by Granovetter [11], deals with the relation between structure, intensity of social ties, and diffusion of information in offline social networks. On one hand, a tie can be characterized by its strength, which is related to the time spent together, intimacy, and emotional intensity of a relation. Strong ties refer to relations with close friends or relatives, while weak ties represent links with distant acquaintances. On the other hand, a tie can be characterized by its position in the network. Social networks are usually composed of communities. A tie can thus be internal to a group or a bridge between groups, as in Fig. 2. Granovetter's theory predicts that weak ties act as bridges between groups and are important for the diffusion of new information across the network. Strong ties are predicted to be located at the interior of the groups between actors who have many friends in common. Burt's work [13] emphasizes the advantage of connecting different groups to access novel information due to the diversity in the sources.

Furthermore, more recent works point out that information propagation may be dependent on the type of content transmitted [62–64] and on a *diversity-bandwidth trade-off* [65]. The bandwidth of a tie is defined as the rate of information transmission per unit of time. Aral et al. [65] note that weak ties interact infrequently and, therefore, have low bandwidth, whereas strong ties interact more often and have high bandwidth. The authors claim that both diversity and bandwidth are relevant for the diffusion of novel information. Since these are anticorrelated, there has to be a trade-off to reach an optimal point in the propagation of new information. They also suggest that strong ties may be important to propagate information depending on the structural diversity, the number of topics, and the dynamic of the information. Due to the different nature of online and offline interactions, it is not clear whether online networks organize following the sociological theories. In the following subsection we present results of some works testing if these theories apply to online social networks.

4.3 Testing Social Theories in Online Social Networks

The predictions of the theory of *the strength of weak ties* have been checked in a mobile phone calls dataset [58] and, very recently, in online social networks [16, 66, 67]. Different predictors have been considered to estimate social tie strength [68] including, for instance, time spent together [68], the duration of phone calls [58], or number of messages exchanged [16, 66]. The two works [16, 58] have measured the dependence of strength of a tie on number of common friends shared by the two actors, showing that the more friends they share, the more likely it is that the tie is strong. This stays in agreement with homophily effect in social network described at the beginning of this chapter. Many shared friends of a pair of users coupled by a strong tie can be interpreted as high homophily between them in terms of acquired friends. Furthermore, large field experiment performed at Facebook [66] has isolated the effect of homophily and social impact on the probability of propagation of information in online social network. The study has shown that users are around 7 times more likely to rebroadcast a piece of information published by their friends if they are exposed to it, which is interpreted as 7 times higher chance of information propagation due to social influence than to homophily. Moreover, the work argues that the weaker is the tie for which information propagation is considered, the higher is the likelihood of information flow due to social influence. This corresponds to Granovetter's prediction that weak ties are important for information diffusion. In the following paragraphs we describe in more detail findings of a similar study in Twitter [16], a popular social microblogging platform.

Online networks are promising for social studies due to the wide availability of data and the fact that different types of interactions are explicitly separated, e.g., information diffusion events are distinguished from more personal communications. Diffusion events are implemented as a system option in the form of *share* or *repost* buttons with which it is enough to single-click on a piece of information to rebroadcast it to all the users' contacts. This is in contrast to personal communications for which more effort has to be invested to write a short message and to select the recipient(s). In Twitter such actions are called, respectively, *retweet* [69] and *mention/reply* [70]. The more mentions have been exchanged between two users, even more so if reciprocated, the stronger we consider the tie between them. On the other hand declared network does also exist in Twitter and is made of directed follower links. One, using clustering algorithms, can find communities of more densely connected users in such network. Specifically, in the study which we present, various clustering algorithms have been used (as shown in Supporting Information of [16]), and for brevity, we will focus only on results for OSLOM [71]. Granovetter's theory predicts that social ties should occur more often inside communities. This is what happens for links with mentions. We define the fraction f as the ratio between the number of links with specified type of interaction in a given position with respect to the groups of corresponding size and the total number of links with that interaction. The fraction f reveals an interesting pattern as function of the group size as can be seen in Fig. 3a. Links with mentions are more abundant inside communities than any other links. This effect is especially significant for groups of sizes from 10 to 150 members. In addition, the distribution of the number of times that a link is used (intensity) for mentions is wide, which allows for a systematic study of the dependence of intensity and position (see Fig. 3b). It turns out that the more intense (or reciprocated) a link with mentions is, the more likely it becomes to find this link as internal (Fig. 3c). This corresponds to Granovetter's expectation that the stronger the tie is, the higher the number of mutual contacts of both parties it has and the higher the chance that the parties belong to the same group.

The communication between groups can take place in two ways: the information can propagate by means of links between groups or by passing through an intermediary user belonging to more than one group; see Fig. 2. We have defined as intermediary the links connecting a pair of users sharing a common group and with at least one of the users belonging also to a different group. In order to estimate the efficiency of the different types of links as attractors of mentions and retweets, there was measured a ratio r defined as the number of links with specified interaction in a given position divided by the total number of links in that position was measured. The bar plot with the values of r is displayed in Fig. 4. The efficiency of the different type of links can thus be compared for the attraction of mentions (red bars) and retweets (green bars). Links internal to the groups attract more mentions and less retweets than links between groups in agreement with the predictions of the strength of weak ties theory. Intermediary links attract mentions as likely as internal links: the ratio of intermediary links with mentions is very close to the ratio of internal links with mentions. This is expected because intermediary links are also internal to the

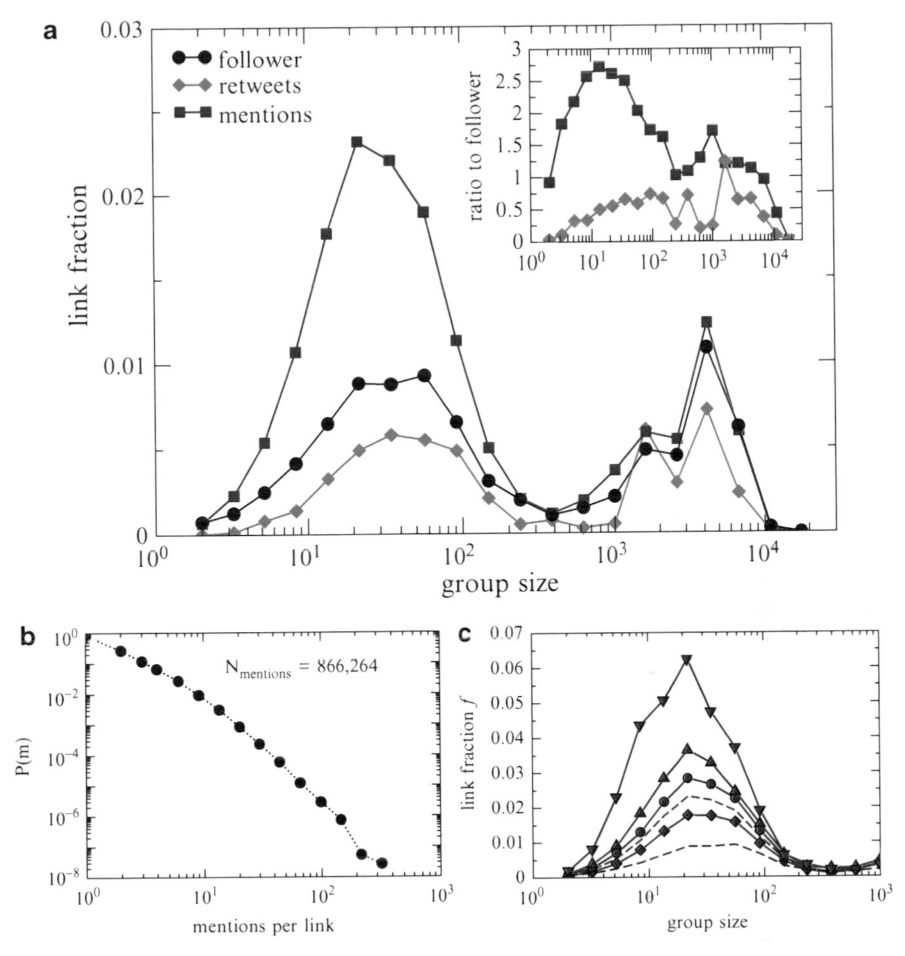

Fig. 3 Internal activity in Twitter. (**a**) Fraction f of internal links as a function of the group size in number of users. The curve for the follower network acts as baseline (*black*) for mentions (*red*) and retweets (*green*). Note that if mentions/retweets were randomly appearing over follower links, then the red/green curve should collapse with the black curve. *Inset*: links with mentions divided by the baseline (*red*) and links with retweets divided by the baseline (*green*). (**b**) Distribution of the number of mentions per link. (**c**) Fraction of links with mentions as a function of their intensity. The dashed curves are the total for the follower network (*black*) and for the links with mentions (*red*). While the other curves correspond (*from bottom to top*) to fractions of links with: 1 non reciprocated mention (*diamonds*), 3 mentions (*circles*), 6 mentions (*triangle up*), and more than 6 reciprocated mentions (*triangle down*) (From [16])

groups. However, the aspect that differentiates more intermediary links from other type of links is the way that they attract retweets. Intermediary links bear retweets with a higher likelihood than either internal or between-groups connections (see Fig. 4a). This fact can be interpreted within the framework of the trade-off between diversity and bandwidth [65]: strong ties are expected to be internal to the groups

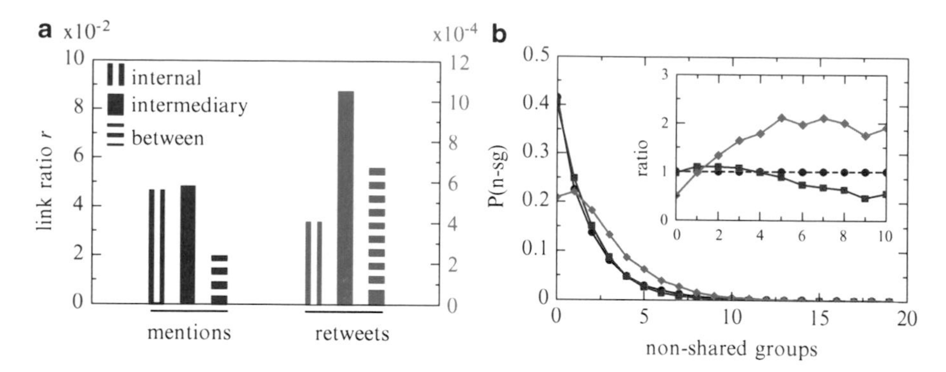

Fig. 4 Intermediary links. (**a**) Ratio r between the number of links with mentions or retweets and number of follower links. (**b**) Distribution of the links in the follower network (*black curve*), those with mentions (*red curve*) and retweets (*green curve*) as a function of the number of non-shared groups of the users connected by the link. Inset, ratios between these distributions and the follower network (From [16])

and to have high bandwidth, while ties connecting diverse environments or groups are more likely to propagate new information. High-bandwidth links in our case correspond to those with multiple mentions, while links providing large diversity are the ones between groups. Intermediary links exhibit these two features: they are internal to the groups and statistically bear more mentions and introduce diversity through the intermediary user membership in several groups. Moreover, in line with the theories [11, 13, 65], higher diversity increases the chances for a link to bear retweets as can be seen in Fig. 4b, which implies a more efficient information flow. In the inset of the figure it is shown that the number of non-shared groups assigned to the users connected by the link positively correlates with, up to twice higher than expected, number of retweets.

5 Summary

Research on online social networks is a rich and an active field of study. The availability of large amount of data allows for studies of both dynamics of social networks and user–user activity on the social network connections. Different growth models have been proposed to simulate the growth of the network, among which three main families are preferential growth models, fitness or hidden variable models, and triangle-closing models. The latter model is reported to yield most accurate results; however, it also incorporates mechanism of preferential attachment. The main advantage of triangle-closing model is that it directly produces network with enough clustering, which is reported to be a feature of social networks. Moreover, there are still open questions about the origin of these mechanisms and of some other phenomena observed during the growth process such as network densification [72]. While declared social network evolves, different types of interactions occur among its members, mostly among users already connected in the declared social network.

Recent studies have shown that different types of interactions happen according to the patterns predicted by the sociological theories. In general strong ties, which in online social networks are usually defined as the links with many messages exchanged between the pair of users, happen more often between users who have many friends in common or who belong to the same communities. On the other hand, weak ties appear more often between users who do not share friends and belong to different groups. It has been shown that weak ties are more efficient for the information spreading than strong ties. Closer study shows that trade-off between diversity and bandwidth may be crucial for diffusion of information.

In conclusion dynamics and activity in online social networks is remarkably rich and tells us much about our social behavior and confirms some of the known offline social theories. We expect that this field of research will be active and developing in the following years and that numerous further online observations and experiments will be undertaken to better understand and quantitatively describe social behaviors.

Acknowledgement We acknowledge partial support from the European Commission through PATRES project, the Spanish Ministry of Economy (MINECO) and FEDER (EU) through projects FISICOS (FIS2007-60327) and MODASS (FIS2011-247852). P. A. G. acknowledges support from the JAE program of the CSIC; J. J. R. acknowledges support from the Ramón y Cajal program of MINECO.

References

[1] M. Newman, J. Park, Phys. Rev. E **68**, 9 (2003)
[2] Y.Y. Ahn, S. Han, H. Kwak, S. Moon, H. Jeong, Analysis of topological characteristics of huge online social networking services, in *Proceedings of the 16th International Conference on World Wide Web - WWW '07* (ACM, Banff, Alberta, 2007), pp. 835–844
[3] S. Boccaletti, V. Latora, Y. Moreno, M. Chavez, D. Hwang, Phys. Rep. **424**, 175 (2006)
[4] J. Leskovec, E. Horvitz, Planetary-scale views on a large instant-messaging network, in *Proceeding of the 17th International Conference on World Wide Web* (ACM, San Diego, California, 2008), pp. 915–924
[5] A. Mislove, M. Marcon, K.P. Gummadi, P. Druschel, B. Bhattacharjee, Measurement and analysis of online social networks, in *Proceedings of the 7th ACM SIGCOMM Conference on Internet Measurement - IMC '07* (ACM, San Diego, California, 2007), pp. 29–42
[6] M.E.J. Newman, SIAM Rev. **45**, 167 (2003)
[7] J. Ugander, B. Karrer, L. Backstrom, C. Marlow, *Arxiv preprint arXiv:1111.4503* (2011)
[8] J. Travers, S. Milgram, Sociometry **1**, 125 (1969)
[9] D.J. Watts, S.H. Strogatz, Nature **393**, 440 (1998)
[10] D. Krackhardt, M. Handcock, Heider vs Simmel: emergent features in dynamic structures, in *Statistical Network Analysis: Models, Issues, and New Directions*, vol. 4503, ed. by E. Airoldi, D. Blei, S. Fienberg, A. Goldenberg, E. Xing, A. Zheng of *Lecture Notes in Computer Science* (Springer, Heidelberg, 2007), pp. 14–27
[11] M.S. Granovetter, Am. J. Sociol. **78**, 1360 (1973)
[12] J. Leskovec, D. Huttenlocher, J. Kleinberg, Predicting positive and negative links in online social networks, in *Proceedings of the 19th International Conference on World wide web* (ACM, New York, 2010), pp. 641–650
[13] R. Burt, *Brokerage and Closure: An Introduction to Social Capital* (Oxford University Press, Oxford, 2005)

[14] D. Foster, J. Foster, P. Grassberger, M. Paczuski, Phys. Rev. E **84**, 066117 (2011)
[15] S. Fortunato, Phys. Rep. **486**, 75 (2010)
[16] P.A. Grabowicz, J.J. Ramasco, E. Moro, J.M. Pujol, V.M. Eguiluz, PLoS ONE **7**, e29358 (2012)
[17] G. Palla, A.-L. Barabási, T. Vicsek, Nature **446**, 664 (2007)
[18] N. Pissard, C. Prieur, Thematic vs. social networks in web 2.0 communities: A case study on Flickr groups, in *Proc. of Algotel Conference*, 2007
[19] M. McPherson, L. Smith-Lovin, J.M. Cook, Ann. Rev. Sociol. **27**, 415 (2001)
[20] T. Opsahl, V. Colizza, P. Panzarasa, J. Ramasco, Phys. Rev. Lett. **101**, 168702 (2008)
[21] M. Newman, Phys. Rev. Lett. **89**, 208701 (2002)
[22] R. Schifanella, A. Barrat, C. Cattuto, B. Markines, F. Menczer, Folks in folksonomies, in *Proceedings of the Third ACM International Conference on Web Search and Data Mining - WSDM '10* (ACM, New York, 2010), p. 271
[23] D. Centola, Science **334**, 1269 (2011)
[24] A. Avnit, http://blog.pravdam.com/the-million-followers-fallacy-guest-post-by-adi-avnit (2009)
[25] H. Ebel, L.-I. Mielsch, S. Bornholdt, Phys. Rev. E **66**, 035103 (2002)
[26] R. Guimerà, L. Danon, A. Díaz-Guilera, F. Giralt, A. Arenas, Phys. Rev. E **68**, 065103 (2003)
[27] H.-B. Hu, X.-F. Wang, Europhys. Lett. **86**, 18003 (2009)
[28] J.N. Cummings, B. Butler, R. Kraut, Comm. ACM **45**, 103 (2002)
[29] D. Lazer, A. Pentland, L. Adamic, S. Aral, A.-L. Barabasi, D. Brewer, N. Christakis, N. Contractor, J. Fowler, M. Gutmann, T. Jebara, G. King, M. Macy, D. Roy, M. Van Alstyne, Science **323**, 721 (2009)
[30] A. Vespignani, Science **325**, 425 (2009)
[31] D.J. Watts, Nature **445**, 489 (2007)
[32] M. Newman, Contemp. Phys. **46**, 323 (2005)
[33] A. Saichev, Y. Malevergne, D. Sornette, *Theory of Zipf's Law and Beyond* (Springer, New York, 2009)
[34] G. Zipf, *Human Behaviour and the Principle of Least Effort: An Introduction to Human Ecology* (Addison-Wesley Press, Reading MA, 1949)
[35] V. Pareto, *Cours d'Économie Politique* (Librairie Droz, Geneve, 1964)
[36] A.-L. Barabási, R. Albert, Science **286**, 509 (1999)
[37] H.A. Simon, Biometrika **42**, 425 (1955)
[38] E. Eisenberg, E. Levanon, Phys. Rev. Lett. **91**, 138701 (2003)
[39] K. Yamasaki, K. Matia, S. Buldyrev, D. Fu, F. Pammolli, M. Riccaboni, H. Stanley, Phys. Rev. E **74**, 035103 (2006)
[40] Y.E. Maruvka, D.a. Kessler, N.M. Shnerb, PLoS ONE **6**, e26480 (2011)
[41] A.-L. Barabási, R. Albert, H. Jeong, Phys. A **272**, 173 (1999)
[42] B.A. Huberman, L.A. Adamic, Nature **401**, 131 (1999)
[43] S.N. Dorogovtsev, J.F.F. Mendes, A.N. Samukhin, Phys. Rev. Lett. **85**, 4633 (2000)
[44] S. Bornholdt, H. Ebel, Phys. Rev. E **64**, 035104 (2001)
[45] L.A. Adamic, B.A. Huberman, Science **287**, 2115 (2000)
[46] A. Mislove, H.S. Koppula, K.P. Gummadi, P. Druschel, B. Bhattacharjee, Growth of the flickr social network, in *Proceedings of the First Workshop on Online Social Networks - WOSP '08* (ACM, Seattle, WA, 2008), pp. 25–30
[47] J. Leskovec, L. Backstrom, R. Kumar, A. Tomkins, Microscopic evolution of social networks, in *Proceeding of the 14th ACM SIGKDD International Conference on Knowledge Discovery and Data Mining - KDD '08* (ACM, Las Vegas, Nevada, 2008), pp. 462–470
[48] G. Bianconi, A.-L. Barabási, Europhys. Lett. **54**, 436 (2001)
[49] G. Caldarelli, A. Capocci, P. De Los Rios, M.A. Muñoz, Phys. Rev. Lett. **89**, 258702 (2002)
[50] B. Söderberg, Phys. Rev. E **66**, 066121 (2002)
[51] M. Boguñá, R. Pastor-Satorras, Phys. Rev. E **68**, 036112 (2003)
[52] J. Ratkiewicz, S. Fortunato, A. Flammini, F. Menczer, A. Vespignani, Phys. Rev. Lett. **105**, 158701 (2010)

[53] D. Garlaschelli, M. Loffredo, Phys. Rev. Lett. **93**, 188701 (2004)
[54] G. De Masi, G. Iori, G. Caldarelli, Phys. Rev. E **74**, 066112 (2006)
[55] J.S. Kong, N. Sarshar, V.P. Roychowdhury, Proc. Natl. Acad. Sci. USA **105**, 13724 (2008)
[56] P.A. Grabowicz, V.M. Eguíluz, Europhys. Lett. **97**, 28002 (2012)
[57] D. Taraborelli, Viable web communities: two case studies, in *Viability and Resilience of Complex Systems*, ed. by G. Deffuant, N. Gilbert (Springer, Berlin, Heidelberg, 2011), pp. 75–105
[58] J.-P. Onnela, J. Saramäki, J. Hyvönen, G. Szabó, D. Lazer, K. Kaski, J. Kertész, A.-L. Barabási, Proc. Natl. Acad. Sci. USA **104**, 7332 (2007)
[59] B. Viswanath, A. Mislove, M. Cha, K.P. Gummadi, On the evolution of user interaction in facebook, in *Proceedings of the 2nd ACM Workshop on Online Social Networks - WOSN '09* (ACM, Barcelona, 2009), p. 37
[60] C. Wilson, B. Boe, A. Sala, K.P. Puttaswamy, B.Y. Zhao, User interactions in social networks and their implications, in *Proceedings of the Fourth ACM European Conference on Computer Systems - EuroSys '09* (ACM, Nuremberg, 2009), pp. 205–218
[61] H. Chun, H. Kwak, Y.H. Eom, Y.Y. Ahn, S. Moon, H. Jeong, Comparison of online social relations in volume vs interaction: a case study of cyworld, in *Proceedings of the 8th ACM SIGCOMM Conference on Internet Measurement Conference - IMC '08* (ACM, Vouliagmeni, 2008), pp. 57–70
[62] D. Centola, V.M. Eguíluz, M.W. Macy, Physica A **374**, 449 (2007)
[63] D. Centola, M. Macy, Am. J. Sociol. **113**, 702 (2007)
[64] D. Centola, Science **329**, 1194 (2010)
[65] S. Aral, M. Van Alstyne, Am. J. Sociol. **117**, 90 (2011)
[66] E. Bakshy, I. Rosenn, C. Marlow, L. Adamic, The role of social networks in information diffusion, in *Proceedings of the 21st International Conference on World Wide Web - WWW '12* (ACM, New York, 2012), p. 519
[67] J.L. Iribarren, E. Moro, Soc. Network **33**, 134 (2011)
[68] P.V. Marsden, K.E. Campbell, Soc. Forces **63**, 482 (1984)
[69] W. Galuba, K. Aberer, D. Chakraborty, Z. Despotovic, W. Kellerer, Outtweeting the twitterers-predicting information cascades in microblogs, in *Proceedings of the 3rd Conference on Online Social Networks* (USENIX Association, 2010)
[70] C. Honeycutt, S. Herring, Beyond microblogging: conversation and collaboration via twitter, in *42st Hawaii International Conference on Systems Science*, ed. by N. Fielding, R.M. Lee, G. Blank (IEEE, Waikoloa, Big Island, HI, 2009), pp. 1–10
[71] A. Lancichinetti, F. Radicchi, J.J. Ramasco, S. Fortunato, PLoS ONE **6**, e18961 (2011)
[72] J. Leskovec, J. Kleinberg, C. Faloutsos, ACM Trans. Knowl. Discov. Data **1**, Article 2 (2007)

An Empirical Validation of Growth Models for Complex Networks

Alan Mislove, Hema Swetha Koppula, Krishna P. Gummadi, Peter Druschel, and Bobby Bhattacharjee

1 Introduction

Complex networks arise in a variety of different domains, including social networks [2], Internet topologies [11], Web connections [3], electrical power grids [30], and networks of brain neurons [6]. Despite their disparate origins, these networks share a surprising number of common structural features such as a highly skewed (power-law) degree distribution, small diameter, and significant local clustering. The link structure of these networks has received significant research attention, and the resulting understanding of their structure has led to popular search algorithms like PageRank [28] and HITS [15].

In this paper, we wish to understand the dynamic processes that lead to the observed structures. A number of different network growth models have been proposed—e.g., the preferential attachment [5], the random walk model [32], and the common neighbors model [26]—which all lead to graphs with similar structural properties. Unfortunately, none of these growth models have been validated using large-scale data from real networks. It is not known if the models predict how real networks actually grow; that is, how, when, and where links are added to a network.

A. Mislove (✉)
Northeastern University, Boston, MA, USA
e-mail: amislove@ccs.neu.edu

H.S. Koppula
Cornell University, Ithaca, NY, USA
e-mail: hema@cs.cornell.edu

K.P. Gummadi • P. Druschel
Max Planck Institute for Software Systems, Kaiserslautern–Saarbruecken, Germany
e-mail: gummadi@mpi-sws.org; druschel@mpi-sws.org

B. Bhattacharjee
University of Maryland, College Park, MD, USA
e-mail: bobby@cs.umd.edu

A. Mukherjee et al. (eds.), *Dynamics On and Of Complex Networks, Volume 2*, Modeling and Simulation in Science, Engineering and Technology, DOI 10.1007/978-1-4614-6729-8_2,

In his editorial on the future of power-law networks research, Mitzenmacher [25] argues that power-law research must move from observing and modeling power-law behavior to the challenging problem of model validation.

In this paper, we apply growth data from four different real-world networks towards understanding the processes that lead to the observed structures. We have repeatedly crawled large social networks (Flickr and YouTube) daily for over four months. Our crawls have generated datasets that contain 10 million and 14 million new links in Flickr and YouTube, respectively. We have also downloaded and analyzed the link formation history between pages in the English language Wikipedia. This dataset covers over six years of growth history, encompassing almost 40 million links between almost 2 million pages. Finally, we use successive snapshots of the autonomous system (AS) level Internet topology graph to observe inter-AS links being created. This dataset covers over three years of AS topology evolution, representing over 75,000 new links.

Using these four datasets, we investigate the link formation processes in each of these networks. Our primary goal is to understand the underlying processes that lead to the observed static structural properties. We examine a number of previously proposed models, and test how well the properties of links created by the models match our observed data. In particular, if these properties match, then we have some assurance that the model might describe the underlying growth process in the "real" network (some other models may also describe the same process). However, if not, then we can assert that the model does *not* describe the dominant growth mechanism in the network under study.

Our analysis shows that links are created by nodes in direct proportion to their degree (as predicted by the preferential attachment model) and that nodes in directed networks tend to quickly respond to incoming links by creating a link in the reverse direction. However, we show that the preferential attachment mechanism alone is insufficient in explaining how a node picks the recipients for its links. Rather, our experiments point to a strong proximity bias: Nodes tend to connect to nearby nodes in the network much more often than would be expected by using preferential attachment alone. Additionally, we find that among the mechanisms which use local rules to form new links, those which select new links based on the indegree of the destination tend to have higher accuracy.

We believe our work is an important first step towards empirical validation of the processes underlying network formation and growth. It is likely that the simple proximity bias models we have explored will not capture the fine-grained dynamics of real systems since these depend on domain-specific parameters. However, our work is directly useful in constructing networks that reflect both global and local characteristics of real-world networks. Better structural and growth models are also useful for network analysis and planning. For appropriate networks, such models can be used in the design of search algorithms (e.g., by pre-identifying nodes that are likely to be hubs), in data mining (e.g., by identifying potential nodes for placing data monitors), and in system evaluation (e.g., by allowing networks to be simulated at arbitrary sizes).

2 Background and Related Work

In this section, we provide background on work related to complex information networks.

2.1 Static Structure of Complex Networks

A long thread of research examines the structure of various complex networks. Researchers have shown that many real-world networks are *power-law networks*, including Internet topologies [11], the Web [5, 20], social networks [2], neural networks [6], and power grids [30]. In such networks, the probability that a node has degree k is proportional to $k^{-\gamma}$. In addition to power-law degree distributions, these networks have been observed to share a number of common structural properties, such as a small diameter and significant local clustering. For more detail on these networks, we refer the reader to the survey by Newman [27].

2.2 Structural Growth Models

Researchers sought to explain the intriguing similarity in the high-level structural properties across networks of very different scales and types by hypothesizing that the networks are the result of a few common growth processes at work. Many models of these processes have been proposed and analyzed to explain the generation of complex networks.

The well-known preferential attachment model [5], where nodes acquire links in proportion to their current degree, has been shown to result in power-law networks. Preferential attachment, as proposed by Bárabási, is a global process whereby nodes create new links based only on the degree of the destination. Many extensions to the preferential attachment model have been proposed, such as to add a tunable level of clustering [13].

Another class of models that produce power-law networks is based on local rules, such as the random walk model [31, 32], where nodes select new neighbors by taking random walks; the common neighbors model [26], where nodes select new neighbors by picking nodes with whom they share many friends in common; and the finite memory model [16], where nodes eventually become inactive and stop receiving any new links.

Additionally, Eiron and McCurley [10] note that many complex information networks have a natural hierarchical structure (such as the hierarchical nature of URLs for the pages in the Web). They propose a new model for constructing such networks which takes into account this hierarchical structure, and they show that this approach more closely matches the observed networks' link structure. For a more detailed treatment of all of these models and others, we refer the reader to Mitzenmacher [24].

It is important to note here that these processes are, by and large, intuitive models that can explain the observed structural properties of the networks. But, they have not been validated using empirical data and they have not been shown to occur in practice. Mitzenmacher [25] poses this as one of the biggest challenges facing the future of power-law research. One of the contributions of this paper lies in evaluating how well these processes predict what actually occurs in different real-world networks at scale.

2.3 Empirical Validation of Growth Models

Some recent work compared snapshots of the same network at different points in time to verify the growth processes. Newman [26] examined the properties of two scientific collaboration networks and found evidence of preferential attachment in both. Peltomäki and Alava [29] examined a scientific collaboration network and a movie-actor network and found evidence of sublinear preferential attachment. Jeong et al. examined citation and coauthorship networks and found that nodes received links in proportion to their degree. Nowell et al. [22] investigated coauthorship networks in physics to test how well different graph proximity metrics can predict future collaborations.

Our work shares similar goals and methodology as the above studies. However, the datasets we use are orders of magnitude larger than the ones used before. Moreover, our data allows us to analyze network growth at very small time scales. We analyze daily snapshots of Flickr and YouTube networks and weekly snapshots of the Internet topology. For Wikipedia, we have sufficient data to create the snapshot of the network at the precise time a new link is established. Since the growth models rely solely on the current network structure to predict new link formation, having frequent snapshots of the network is crucial to validating the models with high accuracy.

Other work has studied the high-level properties of graph evolution, looking for evolution trends at the global level. For instance, Leskovec et al. [21] examined the evolution of a number of real-world graphs, including collaboration networks and recommendation networks. They found that the graphs tend to densify over time and that the average path length shrinks over time (instead of growing in proportion to the number of nodes). This line of work is largely complementary to our work, as we focus on the local link formation phenomena which might lead to these global observations.

Analysis of our detailed growth data, reveals that the preferential attachment model by itself cannot explain new link formation. We believe that the datasets we gathered (and plan to release publicly) represent a significant first step forward in the creation and validation of generative models for complex networks.

2.4 *Explanatory Growth Models*

Some recent studies, particularly on online social networks, have proposed explanatory models of the network growth. Unlike structural growth models, which try to model growth solely as a function of the network structure, explanatory models seek to account for the underlying sociological factors that cause the links to be established. For example, an explanatory growth model for Flickr, a photo-sharing social network, would be based on an understanding of how users behave when sharing pictures.

Examples of work on explanatory growth models include Kumar [19], who divided users into ones who are active and passive and presented a model describing their behavior. Kumar et al. also observed the early evolution of the blogosphere [18] and found that it is rapidly increasing in both scale and connectedness. Jin et al. [14] presented a model of social networks based on known human interactions. Backstrom et al. [4] looked at two snapshots of group membership in LiveJournal and presented a model for the growth of user groups over time. Kossinets and Watts [17] used an inferred social network from an email trace to show that new links in the network are more likely to be established between nodes close in the network. Finally, Chang et al. [9] proposed a model for the growth in connectivity of the Internet topology.

Compared to structural growth models, explanatory models are more detailed, but they also tend to be specific to the network being investigated. For example, the reasons why ISPs connect to each other in the Internet topology are very different from the reasons why users in Flickr connect to each other. By being agnostic to these factors, structural growth models are inherently less accurate. But they are far more general and can be compared across different types of networks. In this paper, we focus only on structural growth models.

3 Measurement Methodology

We now describe the data presented in this paper and the methodology we used to collect it. Whenever appropriate, we describe limitations of the measurement methodology.

3.1 *Flickr and YouTube*

We begin by describing in detail the methodology for collecting data on online social networks. For the two social networks we consider, we were unable to obtain data directly from the respective site operators. So, we chose to crawl the user graphs using the public Web interface provided by the sites. Below, we first describe the challenges and limitations of obtaining data in this manner, and then we describe the datasets we collected.

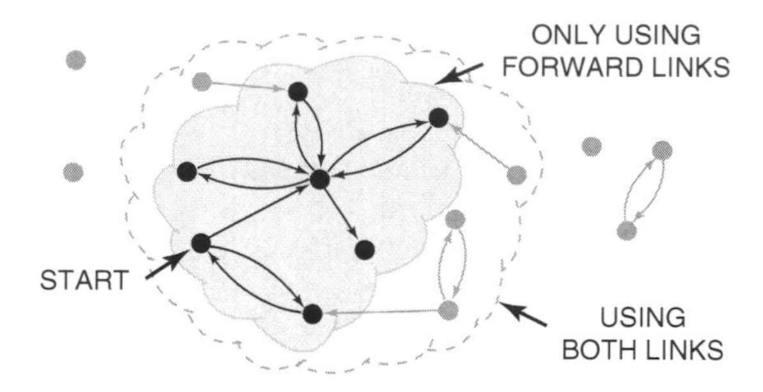

Fig. 1 Users reached using different links. Using only forward links crawls the inner cloud; using both forward and incoming links crawls the entire WCC (dashed cloud)

Crawling the Entire Graph

The primary challenge in crawling large graphs is covering the entire connected component. At each step, one can generally only obtain the set of links into or out of a specified node. In the case of online social networks, crawling the graph efficiently is important since the graphs are large and highly dynamic. Common algorithms for crawling graphs include breadth-first search (BFS) and depth-first search.

Crawling directed graphs, such as Flickr and YouTube, presents additional challenges over undirected graphs. In particular, most graphs can only be crawled by following links in the forward direction (i.e., one cannot directly determine the set of nodes which point *into* a specific node). Using only forward links does not necessarily crawl an entire weakly connected component (WCC); instead, it explores the connected component reachable from the set of seed users. This limitation is typical for studies that crawl online networks, including measurement studies of the Web [7].

Figure 1 shows an example of a directed graph crawl, where the users reached by using just forward links are shown in the inner cloud, and those discovered using both forward and incoming links are shown in the outer dashed cloud. Using both forward and incoming links allows us to crawl the entire WCC, while using only forward links results in a subset of the WCC. Both Flickr and YouTube can only be crawled using forward links.

Crawling Methodology

Using automated scripts on a cluster of 58 machines, we crawled the social network graphs of Flickr and YouTube once per day. We chose these sites because they represent different types of online social networking sites and because it is possible to crawl the entire network once per day. More details on our methodology and its limitations can be found in [23]. Here, we discuss the methodology and limitations that are relevant to the growth data.

We started each crawl by selecting a single known user as a seed. In each step, we retrieved the list of friends for a user we have not yet visited and added these users to the list of users to visit. We then continued until we exhausted the list, thereby performing a BFS of the social network graph, starting from the seed user. We performed these crawls once per day for each network. On each day, we revisited every user we had previously discovered, in addition to all nodes that were reachable from the seed node, and recorded any newly created or removed links and nodes.

Since the sites do not provide the time of creation for any node or link, our growth data for the social networks has a granularity of one day for the links we observed being created. As a result, we cannot determine the exact time of link creation or the order in which links were created within a single day. Moreover, new nodes cannot be observed until they become connected to one of the nodes we have already crawled. Additionally, in the rest of the paper, we only examine links that we observed being created. In other words, we may discover a new node that has a few established links, but we do not examine these previously established links in our growth analysis, as we did not observe them being created.

Flickr

Flickr (www.flickr.com) is a popular photo-sharing site based on a social network. Flickr's social network is directed, as users can create links to other users without any approval from the destination. Flickr exports an API that we used to conduct the crawl.

We crawled the Flickr network daily between November 2, 2006, and December 3, 2006, and again daily between February 3, 2007, and May 18, 2007, representing a total of 104 days of growth. During that period of daily growth observations, we observed over 10.7 million new links being formed and discovered over 680,000 new users. This represents, relative to the initial network snapshot, over 42% growth in the number of users and over 63% growth in the number of links.

YouTube

YouTube (www.youtube.com) is a popular video-sharing site that includes a social network. Similar to Flickr, YouTube exports an API, and we used this feature to conduct our crawls.

We crawled the YouTube network daily between December 10, 2006, and January 15, 2007, and again daily between February 8, 2007, and July 23, 2007, representing 201 days of growth. Between the two date ranges of our crawls, YouTube changed its policy to require confirmation from the destination of a link (previously, this approval was not required). Thus, YouTube changed from a directed network to an undirected network between our two observation periods. To properly deal with this significant change in policy, we treat the two YouTube networks separately—we denote the first set of growth data covering the directed graph as *YouTube-D* and the second set representing the undirected graph as *YouTube-U*.

The YouTube-D dataset represents the growth of a directed network over a period of 36 days. During that period of daily growth observations, we observed over 540,000 new links being formed and discovered over 130,000 new users. This represents, relative to the initial network snapshot, over 13% growth in the number of users and over 12% growth in the number of links.

The YouTube-U dataset represents the growth of an undirected network over a period of 165 days. We observed the network grow by over 11.7 million links and over 1.8 million users. This represents, relative to the initial network snapshot, over 129% growth in the number of users and over 173% growth in the number of links.

Crawling Limitations

There are two limitations to our crawls of Flickr and YouTube. First, we were only able to crawl using forward links, which does not necessarily result in an entire WCC. Second, we only crawled the single, large WCC; there may be users who are part of small clusters not connected to this WCC at all. In this section, we evaluate the number and characteristics of users who were missed by our crawl.

We performed the following experiment using Flickr. We used the fact that the vast majority of Flickr user identifiers take the form of *[randomly selected 8 -digit number]@N00*. We generated 100,000 random user identifiers of this form (from a possible pool of 90 million) and found that 6,902 (6.90%) of these were assigned usernames. These 6,902 nodes form a random sample of Flickr users.

Among these 6,902 users, 1,859 users (26.9%) had been discovered during our crawl. Focusing on the 5,043 users *not* previously discovered by our crawl, we conducted a BFS starting at each user to determine whether or not they could reach our set of previously crawled users. We found that only 250 (5.0%) of the missed users could reach our crawled set and were definitively in the WCC. While we cannot conclusively say that the remaining 4,793 (95.0%) missed users are not attached to the WCC (there could be some other user who points to them and to the WCC), the fact that 89.7% of these have no forward links suggests that many are not connected at all.

Thus, we believe that our crawl of the large WCC, although not complete, covers a large fraction of the users who are part of the WCC. Further, our experience with the randomly generated Flickr user identifiers indicates that (at least for Flickr) the nodes not in the largest WCC tend to have very low degree—in fact, almost 90% of them have no outgoing links at all.

3.2 Wikipedia

Wikipedia (www.wikipedia.org) is a popular online encyclopedia that allows any user to add or edit content. Wikipedia makes its entire edit history available on a monthly basis, and we downloaded the edit history of the English language Wikipedia as of April 6, 2007.

To extract the graph of links between Wikipedia pages, we use the following method: For each link in the current snapshot, we determine the time when this link was first created. We then construct a graph using these derived links and the associated timestamps. This method allows us to remove the effects of page vandalism, where malicious users sometimes overwrite entire pages, thereby temporarily removing all of the links from vandalized pages.

Since Wikipedia allows pages to redirect to other pages, we configured our tool to follow the redirects and treat a link to a redirect page as if it was a link to the destination page. Thus, if page A originally linked to B at time t, but later, B was set to redirect to C, we treat this like a link from A to C established at time t. This allows us to handle multiple layers of redirect pages, as well as large-scale naming convention changes.

Since the data represents the complete history of a complex network, we exclude startup effects by limiting our analysis to the recent history. This is similar to previous studies [8, 26]. In particular, we only consider links created between January 1, 2005, and April 6, 2007, a period of 826 days. During this period, we observed over 1.1 million new pages and over 33 million new links, representing 169% growth in the number of pages and 500% growth in the number of links relative to the snapshot on January 1, 2005.

3.3 Internet Topology

The Internet can be viewed as a collection of *autonomous systems* (AS), where each AS represents a single administrative domain (typically, an ISP). The inter-domain routing protocol of the Internet, BGP, uses unique AS numbers to allow ASes to advertise their connections to their neighbors. The union of these advertisements forms an undirected graph representing the AS-level connectivity of the Internet.

We used the AS topology graphs collected by CAIDA [1] to study the evolution of the AS network. CAIDA creates weekly (monthly for the first two years) snapshots of the AS topology using a number of BGP monitoring machines. We downloaded the entire history of their measurements, which covers the period from January 5, 2004, until July 9, 2007. The AS topology evolution data therefore covers 1,282 days of growth. During this period, the number of ASes in the network grew from 9,978 to 25,526, a growth of 155%. Similarly, the number of AS links grew from 29,504 to 104,824, a growth of 255%.

3.4 Summary

Table 1 shows the high-level statistics of the data we gathered. The network sizes vary by over three orders of magnitude. Similarly, other metrics, such as the average number of links per node and the yearly growth rate also vary greatly between the

Table 1 High-level statistics of the network growth data

	Flickr	Wikipedia	YouTube-D	YouTube-U	Internet
Network type	Directed	Directed	Directed	Undirected	Undirected
Days of observed growth	104	825	36	165	1,281
Resolution of link creation	Day	Second	Day	Day	Month/week
Fraction of links symmetric	62%	17%	79%	–	–
Initial number of nodes	1,620,392	695,353	1,003,975	1,402,949	9,978
Final number of nodes	2,570,535	1,892,691	1,137,638	3,218,658	25,526
Growth in number of nodes	42%	169%	13%	129%	155%
Norm. growth rate (nodes/year)	242%	54%	145%	525%	31%
Initial number of links	17,034,807	6,637,456	4,391,336	6,783,917	29,504
Final number of links	33,140,018	39,953,145	4,945,382	18,524,095	104,824
Growth in number of links	63%	500%	12%	173%	255%
Norm. growth rate (Links/Year)	455%	120%	215%	822%	43%

networks. Despite these differences, as our analysis later shows, the growth of these complex networks shows a number of commonalities.

4 Validation of Network Growth Models

Our goal in this section is to use our network growth data to validate existing models of network growth. In particular, we study how well the empirical data matches the predictions of growth models that have been proposed. Our findings can be summarized as follows. First, all of our data is consistent with the predictions of the popular preferential attachment model. Second, there are some properties in our datasets that cannot be explained by that model alone. Third, models that consider network proximity as a factor in link creation predict the empirical data better than preferential attachment. Fourth, no single proximity-based model best predicts link creation in all four of our datasets, but those which take into account the indegree of the destination tend to have higher accuracy. Fifth, reciprocation is a significant factor in the link creation of directed networks.

It is important to note that we can only study how well a particular model predicts the link creation that occurs in the empirical data. We fundamentally do not know why new links were established; we can only observe the source and destination of new links. Thus, we cannot ultimately prove or disprove any particular model; we can only examine the correlation between the observed data and what each model would predict. Nevertheless, knowing how well different models predict link creation in the data can improve our understanding of network evolution and can provide clues as to the actual underlying processes.

4.1 Growth Dominates Network Evolution

In all of the networks we examined, we found that link addition was significantly more frequent than link removal. In particular, we found that in Flickr, link additions exceeded link removals in our datasets at a rate of 2.43:1. Similar characteristics were observed in the other networks we studied: In YouTube-U, the ratio of link additions to removals was 3.71:1, and in the Internet, we found that the ratio was 2.06:1. Unfortunately , we did not record removed links for the YouTube-D dataset, and we are unable to estimate the fraction of removed links in Wikipedia due to the effects of page vandalism (i.e., vandalized pages often have their entire text, and therefore all of their outgoing links are replaced and then added back).

In summary, in the networks in which we were able to record link removals, we observed that link addition significantly exceeded link removal. Thus, in the rest of this paper, we focus only on how links are added to growing networks, and we leave examining link removal for future work.

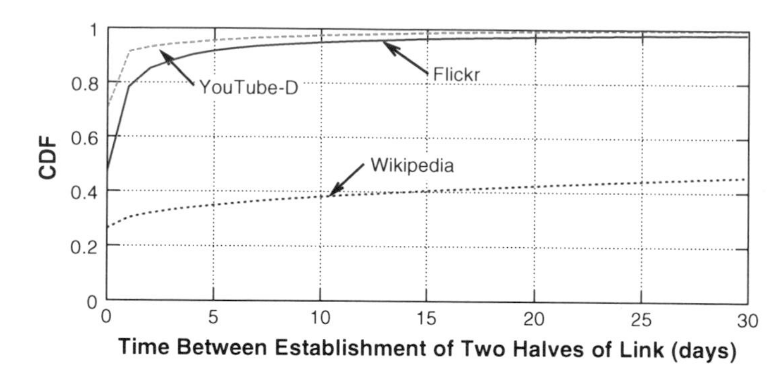

Fig. 2 *CDF of time between establishment of the two directed links of a symmetric link.* In both Flickr and Youtube, links are quickly reciprocated

All of the networks we observed showed a high growth rate: Normalizing for different observation periods across the networks reveals an average growth rate of between 31% and 525% per year in terms of nodes and a growth rate of between 43% and 822% per year in terms of links. These rapidly growing networks offer us a unique opportunity to observe new link creation.

4.2 Reciprocation

We begin by first examining *reciprocation*, a growth mechanism that exists only in directed graphs. Reciprocation occurs when the creation of a directed link between two nodes causes the reverse link to be established. Since undirected graphs are, by definition, symmetric, reciprocation does not make sense in the context of undirected graphs. Reciprocation has been proposed as an independent growth mechanism for large-scale directed graphs [12, 33].

Since we do not know why links were established, we rely on the timing between the creation of the two directed links of a symmetric link to guess whether the creation of the first causally affected the second. Figure 2 shows the distribution of the time between the establishment of the two links of a given symmetric link in the three directed graphs (Flickr, YouTube-D, and Wikipedia) that we studied.

From Fig. 2, it is clear that in the two social networks we observed, users often respond to incoming links by quickly establishing a reciprocal link back to the source node. In fact, over 83% of all symmetric links we observed in both Flickr and YouTube-D were established within 48 hours after the initial link creation. This suggests that users tend to quickly reciprocate links, if they reciprocate at all. Thus, it is highly likely that the establishment of the first link in these networks prompted the creation of the reciprocal link. The Wikipedia data, on the other hand, indicates a lower degree of reciprocation; only 30% of the symmetric links in Wikipedia had both halves of the link created within 48 hours of each other.

Our data suggests that reciprocation is an independent mechanism shaping the growth of directed networks. The degree of reciprocation is dependent on the network: the two social networks show significant reciprocation, while Wikipedia shows reciprocation, but to a less significant degree.

4.3 Preferential Attachment

Preferential attachment [5], colloquially referred to as the "rich get richer" phenomenon, is a growth model in which new links in a network are attached *preferentially* to nodes that already have a large number of links. Under preferential attachment, the probability that a new link attaches to a given node is proportional to the node's current degree.

To examine whether preferential attachment predicts the observed growth data, we calculated how the number of new links per day varies with the node degree. If preferential attachment is taking place, we would expect to see a linear correlation between the degree of a node and the number of new links it creates or receives. However, it is important to note that a linear correlation is a necessary but not sufficient condition for the validity of the preferential attachment mechanism, as other mechanisms could also result in such a linear correlation. For example, the "connecting nearest neighbors" model [32] has been shown to also exhibit such a linear correlation.

Figure 3 plots this distribution in log–log scale for each of the five networks we studied. For the three directed graphs, we separately plot the number of new links created and received, with respect to the node's current outdegree and indegree.

Undirected Networks

For the two undirected networks, YouTube-U and the AS-level Internet, we show how the degree of a node correlates with the number of new links per day. We find a strong linear correlation between the current degree and the number of newly created links in both of the networks.

Directed Networks

For the three directed networks, we separate the preferential attachment model into two aspects: *preferential creation* and *preferential reception*. Preferential creation describes the mechanism by which nodes *create* new links in proportion to their outdegree, and preferential reception describes the mechanism where nodes *receive* new links in proportion to their indegree. This distinction is consistent with previously proposed models of preferential attachment on directed graphs [8].

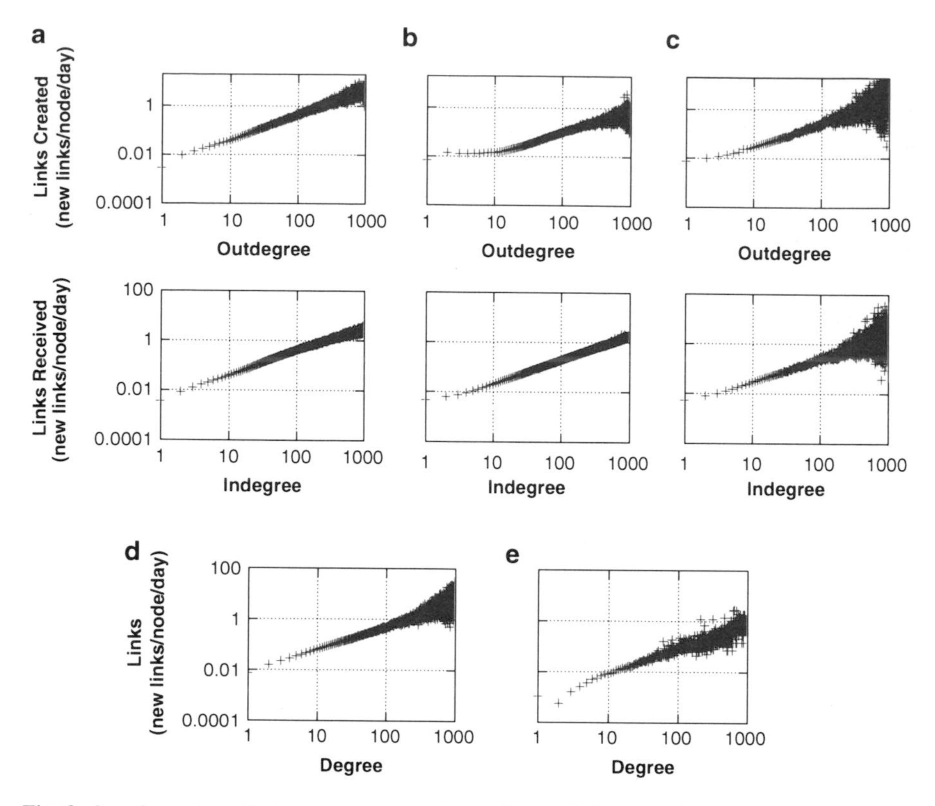

Fig. 3 *Log–log plot of degree versus number of new links per day.* For directed graphs (graphs a–c), separate plots are shown for outdegree (*top*) and indegree (*bottom*). All networks show strong evidence of preferential attachment. (**a**) Flickr, (**b**) Wikipedia, (**c**) YouTube-D, (**d**) YouTube-U, (**e**) Internet

It is important to understand why we separate preferential attachment into preferential creation and preferential reception for directed networks. Preferential attachment was originally defined for undirected graphs [5] and therefore does not distinguish between node indegree and outdegree. However, in the directed networks we study, link creation is very different from link reception. Nodes are in complete control over their outgoing links, since they decide who they link to, but they are not in control of their indegree, since it depends upon who they receive links from.

For the three directed networks, Flickr, Wikipedia, and YouTube-D, we separately examine how the current outdegree and indegree of a node is related to the number of newly created and received links per day. Figure 3 shows that the outdegree of nodes is linearly correlated with the number of new links created per node per day. This is a necessary, but not sufficient condition for the validity of the preferential creation mechanism. Figure 3 also shows, for the three directed networks, that the increase in node indegree is linearly correlated with the current indegree of the node. Similarly, this is a necessary condition for the validity of the preferential reception mechanism.

Discussion

Our data shows that a necessary condition for preferential attachment, a linear correlation between the degree of a node and the number of new links, is present in all five networks. However, this alone is not sufficient to claim that preferential attachment *is* the mechanism that is causing the growth, as other, different mechanisms could also result in this correlation. In the next section, we examine more closely the growth data to look for further evidence of preferential attachment.

4.4 Proximity Bias in Link Creation

In this section, we consider the distance in the network among the nodes connected by a new link and consider if preferential attachment can explain our observations. In particular, we examine the shortest path distance between the source and destination of newly created links, before a new link is created between them. If preferential attachment is the underlying mechanism, then the observed distance distribution between nodes should match that predicted by the model.

Over 50% of the links in all five networks are between nodes that have, a priori, some network path between them.[1] For these links, Fig. 4 shows the cumulative distribution of shortest-path hop distances between source and destination nodes for newly created links. It reveals a striking trend: Over 80% of such new links in Flickr connect nodes that were only two hops apart, meaning that the destination node was a friend of a friend of the source node. Similarly, this fraction is over 42% in YouTube-D, over 50% in Wikipedia, over 45% in YouTube-U, and over 57% in the Internet topology.

One might wonder whether in small diameter networks like the ones we observe, this high level of proximity in link establishment is simply a result of preferential attachment. This is plausible, since the high-degree nodes that preferential attachment prefers tend to be close to many nodes. To test this hypothesis, for each newly created link, we computed the expected distance from the source to the destination, if the destination is chosen using the preferential attachment mechanism (or preferential reception, for directed graphs). Figure 4 also plots this distribution for each network.

In all five networks that we study, the observed distances between the source and destination of links show a significant bias towards nearby nodes, relative to what preferential attachment or preferential reception would predict. In fact, in Flickr, Wikipedia, and YouTube-D, we found that the number of new links connecting

[1] For directed networks, we only count directed paths.

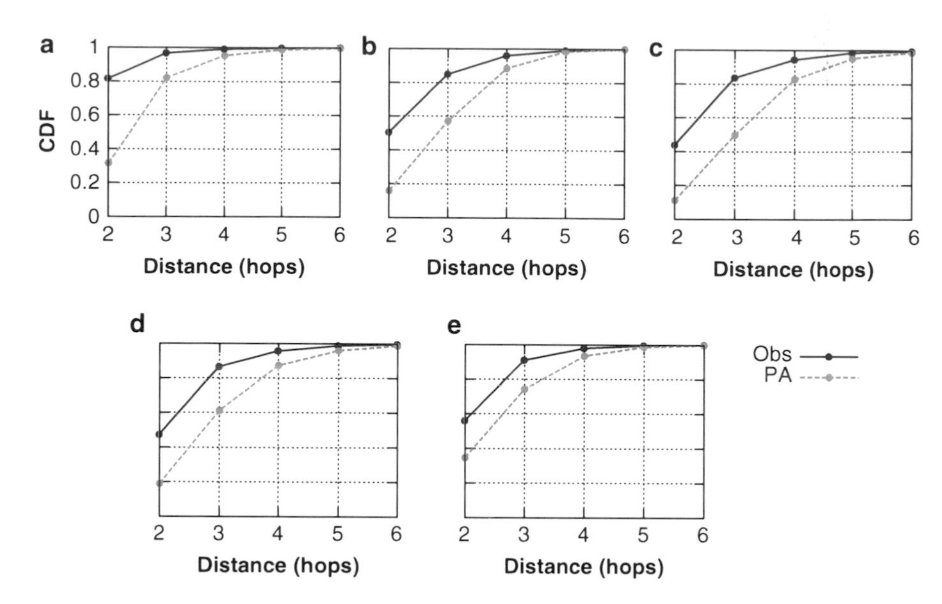

Fig. 4 *CDF of distance between source and destination of observed links (Obs).* Also shown is the expected CDF from preferential attachment (PA). The numbers in parenthesis are the fraction of all new links connecting nodes that had, a priori, some path between them. All networks show a proximity bias that is not predicted by preferential attachment. (**a**) Flickr (83%), (**b**) Wikipedia (58%), (**c**) YouTube-D (68%), (**d**) YouTube-U (82%), (**e**) Internet (54%)

2-hop neighbors in the empirical data exceeded that predicted by preferential reception by a factor of three.

This result shows that the new links created in the networks cannot be explained by a preferential attachment mechanism alone. Nodes are far more likely to link to nearby nodes than preferential attachment would suggest. This result is consistent with the previous observations on static networks which showed that the clustering coefficient was significantly higher than would be predicted by preferential attachment. In the next section, we focus on how nodes choose which nearby node to link to.

4.5 Mechanisms Causing Proximity Bias

Next, we examine network growth models that are known to have a stronger bias towards proximity than preferential attachment. To make the analysis tractable, we focus on new links that occur between nodes that are two hops apart. Such links account for over 45% of the links in all networks. We consider preferential creation, combined with five different proposed mechanisms for selecting the destination of a newly established link:

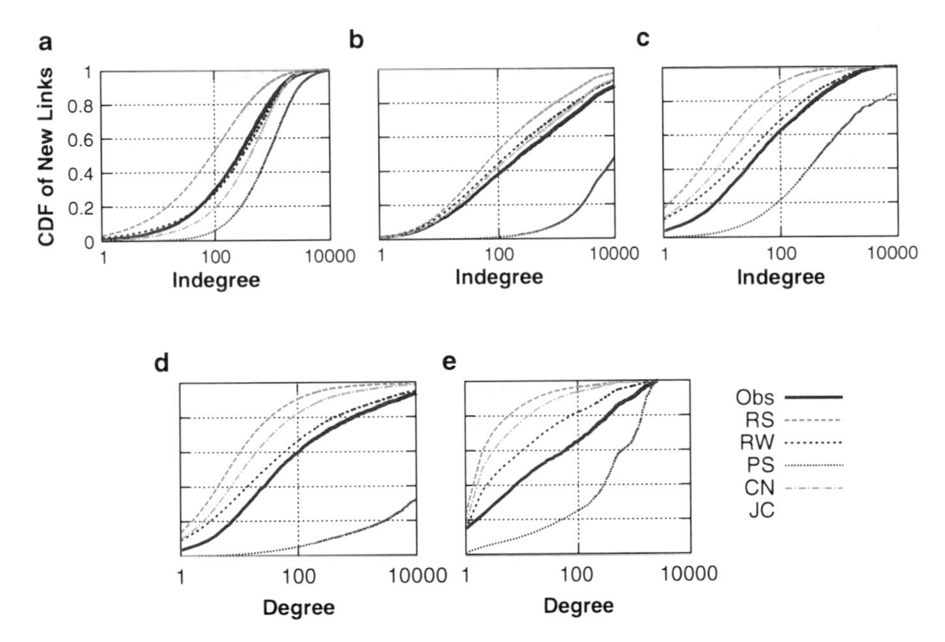

Fig. 5 *CDF of nodes receiving new links by indegree*. Plots are shown for observed data (Obs) and simulated mechanisms: random selection (RS), random 2-hop walk (RW), preferential selection (PS), common neighbors (CN), and Jaccard's coefficient (JC). The observed data does not match any one mechanism, suggesting that different mechanisms are at play in different networks. (**a**) Flickr (**b**) Wikipedia (**c**) YouTube-D (**d**) YouTube-U (**e**) Internet

- *Random selection (RS)*, where a node chooses the destination randomly from its set of two-hop neighbors. This mechanism serves as a baseline for evaluating the other mechanisms.
- *Random two-hop walk (RW)*, where a node performs a random two-hop walk to find the destination [32].
- *Preferential selection (PS)*, where a node chooses from its set of two-hop neighbors preferentially according to the nodes' indegrees. This is similar to preferential attachment, except that a node only considers its two-hop neighbors [22].
- *Common neighbors (CN)*, where a source makes a weighted random choice among its set of two-hop neighbors. The likelihood that a given candidate is chosen is proportional to the number of neighbors the source shares with the candidate [26].
- *Jaccard's coefficient (JC)*, where a source makes a weighted random choice among its set of two-hop neighbors. Here, the likelihood that a given candidate is chosen is proportional to the number of neighbors the source shares with the candidate divided by the candidate's indegree [22].

We examined newly established links in all networks that connect nodes that were previously two hops apart. We then calculated the expected indegree distribution of

Table 2 *Prediction accuracy of two-hop link creation mechanisms relative to the baseline random selection mechanism.* While no one mechanism appears to be the most accurate across all networks, random walk and preferential selection tend to have higher accuracy

	RS	RW	PS	CN	JC
Flickr	0.17%	2.0	1.1	1.2	1.2
Wikipedia	0.15%	2.9	2.9	1.3	0.7
YouTube-D	0.35%	1.6	1.5	1.1	1.0
YouTube-U	0.59%	1.7	1.3	1.1	1.4
Internet	0.53%	1.9	4.1	1.1	0.5

nodes that would have been selected using each of the five mechanisms above. We then compared the results to the distribution in the empirical data. Figure 5 plots these distributions for each network.

From Fig. 5, we can see that no one mechanism closely matches the empirical data in all networks. In fact, in two of the networks (Flickr and Wikipedia), the random walk mechanism most closely matches the observed data. However, in the other three networks, the results are less conclusive. To better quantify how well the various mechanisms predict the selected destination of new links, we calculated the accuracy of each mechanism, in the same manner as previous studies [22]. Thus, for each newly created link, we calculated the fraction of time each mechanism correctly predicted the selected destination. The results are shown in Table 2, relative to the random selection model.

The accuracy results in Table 2 show that no one model dominates in terms of accuracy across different networks. However, closely examining the results reveals that the two mechanisms which take into account the indegree of the destination (RW and PS) do tend to have higher accuracy. This suggests either that different mechanisms may be at play in different networks or that the actual mechanism driving link creation is not among the ones we evaluated or that the actual mechanism is a complex combination of some of the mechanisms we tested. This result is not surprising, though, as each of the networks represents a different system, and it is unlikely that one single mechanism would describe the link creation behavior in all of them.

4.6 Summary

In this section, we closely examined network growth data from five different networks and compared the empirical data to the predictions of previously proposed growth mechanisms. We found evidence of reciprocation as a mechanism in directed networks. We also found that nodes tend to create and receive links in proportion to the outdegree and indegree. However, we found that the preferential attachment mechanism did not accurately predict the proximity bias among nodes connected by new links in any of the empirical datasets. All networks showed a stronger bias towards proximity between new sources and destinations than would have been predicted by preferential attachment. Upon closer examination of these links, we found that no single proximity model we examined appears to accurately predict this proximity across all networks, suggesting that further research into growth mechanisms is necessary. In the next section, we discuss some of these future directions and describe the implications of our findings.

5 Discussion

In this paper, we have used empirical growth data from multiple large-scale complex networks to test if previously proposed growth models actually are at play in these networks. We have chosen to focus on the preferential attachment model because it is simple and has been suggested as the underlying growth mechanism in different contexts. Clearly, preferential attachment leads to global degree distributions of the type observed in many diverse networks, and absent other data, it is an attractive choice for researchers to explain static snapshots of crawled networks.

5.1 Is Proximity Fundamental?

We believe that some notion of proximity is inherent in the link creation processes underlying large networks. As a network grows larger, it is increasingly unlikely that nodes are influenced by knowledge of the global degree ranking when choosing their neighbors. In many networks, it may not even be possible to discover the global degree ranking of nodes, knowledge of which is required for pure preferential attachment. Other mechanisms that rely on global properties are equally unlikely because of technical and policy issues with computing global metrics.

In the networks we have examined, the bias towards proximity can be explained by considering the node discovery mechanisms available to users and the factors that constrain them. In the social networks (YouTube and Flickr), the primary mechanism available to users for exploring the network is to walk their neighborhood. This might explain our observation in Flickr and YouTube that there is a much stronger bias in link creation towards nearby nodes than would be predicted by preferential attachment alone, yet there still is a bias towards high-degree nodes (see Table 2). On Wikipedia, semantically closer pages are likely to be proximal in the network, leading to a proximity bias in link creation.

The Internet AS graph is fundamentally different because each AS consists of many different routers and there is a significant cost associated with creating new links. A model for AS link formation is given in [9], and our observations are consistent with the reasoning therein. The AS graph is naturally "tiered" with many small stub ASes interconnected by a few large backbone providers (who also tend to have high connectivity/degree). AS link creation is often constrained by financial, technical, and geographical factors: For most stub ASes, links to far away ASs tend to be costly (especially if the geographic distance is large) and are unlikely to be profitable since the upstream provider already provides transit to reach these ASes. Such links only make sense in specific cases where business relationships mandate a specific inter-AS peering. Thus, stub ASes tend to connect to their nearby backbone AS providers, and the resulting AS graph shows proximity bias coupled with strong preferential selection.

5.2 Proximity Mechanisms

While our growth data cannot *assert* which mechanism is at play when links are formed, it can be used to *disprove* existing hypotheses. Perhaps unsurprisingly, we find that the simplest mechanisms (such as global preferential attachment) are not sufficient to explain our observations. In particular, we have shown that to explain the empirical growth data, we must include some notion of proximity in the growth models. While proximity has been previously suggested as a factor in link creation in large networks, we believe we are the first to provide empirical date from multiple large-scale networks to support this conjecture.

The analysis in the previous section revealed some insights into how proximity affects the growth of complex networks. While our results are not conclusive, it appears that growth models that take into account the indegree of the destination (e.g., preferential selection and random walk) match the data more closely than other models. Moreover, preferential selection outperforms random walk only for the Internet AS graph.

6 Summary and Future Work

In this paper, we examined the link formation processes that govern the growth of large-scale information networks. We collected growth data from four different networks, including social networks, Wikipedia, and the Internet. We carefully analyzed this growth data, comparing the empirical observations to what would be predicted by the numerous proposed models. Our analysis shows that the link formation processes follow the well-known preferential attachment model but that this model alone is insufficient to explain the observed proximity between link sources and destinations. We then examined how well other models with local rules matched the observed data, and our findings suggest that different mechanisms are at play in different networks.

We believe that this work opens up many avenues for future research. In particular, the data we collected can be used to test other previously proposed growth models to see how well they match the observations. Similarly, the data we obtain could be used to guide the development of new models based on empirical data. While it is unlikely that a single model can capture all of the complexity of a large-scale real-world system, closely analyzing the data may reveal new insights into the link formation processes.

References

[1] http://as-rank.caida.org/data/
[2] L.A. Adamic, O. Buyukkokten, E. Adar, A social network caught in the Web. First Monday **8**(6) (2003)
[3] R. Albert, H. Jeong, A.-L. Bárabási, The diameter of the World Wide Web. Nature **401**, 130 (1999)
[4] L. Backstrom, D. Huttenlocher, J. Kleinberg, X. Lan, Group formation in large social networks: membership, growth, evolution, in *Proceedings of the 12th ACM SIGKDD International Conference on Knowledge Discovery and Data Mining (KDD'06)*, Philadelphia, PA, 2006
[5] A.-L. Bárabási, R. Albert, Emergence of scaling in random networks. Science **286**, 509–512 (1999)
[6] V. Braitenberg, A. Schz, *Anatomy of a Cortex: Statistics and Geometry* (Springer, Berlin, 1991)
[7] A. Broder, R. Kumar, F. Maghoul, P. Raghavan, S. Rajagopalan, R. Stata, A. Tomkins, J. Wiener, Graph structure in the web: experiments and models, in *Proceedings of the 9th International World Wide Web Conference (WWW'00)*, Amsterdam, 2000
[8] A. Capocci, V.D.P. Servedio, F. Colaiori, L.S. Buriol, D. Donato, S. Leonardi, G. Caldarelli, Preferential attachment in the growth of social networks: the internet encyclopedia Wikipedia. Phys. Rev. E, American Physical Society, College Park, MD **74**, 036116-1–0361166 (2006)
[9] H. Chang, S. Jamin, W. Willinger, To peer or not to peer: modeling the evolution of the internet's AS-level topology, in *Proceedings of the 25th Conference on Computer Communications (INFOCOM'06)*, Barcelona, Spain, 2006
[10] N. Eiron, K.S. McCurley, Link structure of hierarchical information networks, in *Proceedings of the Third Workshop on Algorithms and Models for the Web-Graph (WAW'04)*, Rome, Italy, 2004
[11] M. Faloutsos, P. Faloutsos, C. Faloutsos, On power-law relationships of the Internet topology, in *Proceedings of the Annual Conference of the ACM Special Interest Group on Data Communication (SIGCOMM'99)*, Cambridge, MA, 1999
[12] D. Garlaschelli, M. Loffredo, Patterns of link reciprocity in directed networks. Phys. Rev. Lett., American Physical Society, College Park, MD **93**, 268701-1–268701-4 (2004)
[13] P. Holme, B.J. Kim, Growing scale-free networks with tunable clustering. Phys. Rev. E, American Physical Society, College Park, MD **65**, 026107-1–026107-4 (2002)
[14] E.M. Jin, M. Grivan, M. Newman, The structure of growing social networks. Phys. Rev. E, American Physical Society, College Park, MD **64**, 046132-1–046132-8 (2001)
[15] J. Kleinberg, Authoritative sources in a hyperlinked environment, in *Proceedings of the 9th ACM-SIAM Symposium on Discrete Algorithms (SODA'98)*, San Francisco, CA, 1998
[16] K. Klemm, V.M. Eguiluz, Highly clustered scale-free networks. Phys. Rev. E, American Physical Society, College Park, MD **65**, 036123-1–036123-5 (2002)
[17] G. Kossinets, D.J. Watts, Empirical analysis of an evolving social network. Science **311**, 88–90 (2006)

[18] R. Kumar, J. Novak, P. Raghavan, A. Tomkins, On the bursty evolution of blogspace, in *Proceedings of the 12th International Conference on World Wide Web (WWW'03)*, Budapest, Hungary, 2003
[19] R. Kumar, J. Novak, A. Tomkins, Structure and evolution of online social networks, in *Proceedings of the 12th ACM SIGKDD International Conference on Knowledge Discovery and Data Mining (KDD'06)*, Philadelphia, PA, 2006
[20] R. Kumar, P. Raghavan, S. Rajagopalan, A. Tomkins, Trawling the web for emerging cyber-communities. Comput. Network **31**, 1481–1493 (1999)
[21] J. Leskovec, J. Kleinberg, C. Faloutsos, Graph evolution: densification and shrinking diameters. ACM Trans. Knowl. Discov. Data, Association for Computing Machinery, New York, NY **1**, 1–41 (2007)
[22] D. Liben-Nowell, J. Kleinberg, The link prediction problem for social networks, in *Proceedings of the 2003 ACM International Conference on Information and Knowledge Management (CIKM'03)*, New Orleans, LA, 2003
[23] A. Mislove, M. Marcon, K.P. Gummadi, P. Druschel, B. Bhattacharjee, Measurement and analysis of online social networks, in *Proceedings of the 5th ACM/USENIX Internet Measurement Conference (IMC'07)*, San Diego, CA, 2007
[24] M. Mitzenmacher, A brief history of generative models for power law and lognormal distributions. Internet Math. **1**(2), 226–251 (2004)
[25] M. Mitzenmacher, Editorial: the future of power law research. Internet Math. **2**(4), 525–534 (2006)
[26] M.E.J. Newman, Clustering and preferential attachment in growing networks. Phys. Rev. E, American Physical Society, College Park, MD **64**, 025102-1–025102-4 (2001)
[27] M.E.J. Newman, The structure and function of complex networks. SIAM Rev. **45**, 167–256 (2003)
[28] L. Page, S. Brin, R. Motwani, T. Winograd, The pagerank citation ranking: bringing order to the web. Technical Report, Stanford University, 1998
[29] M. Peltomäki, M. Alava, Correlations in bipartite collaboration networks. J. Stat. Mech., IOP Publishing, Bristol, UK P01010, 1–23 (2006)
[30] A.G. Phadke, J.S. Thorp, *Computer Relaying for Power Systems* (Wiley, New York, NY, 1988)
[31] J. Saramaki, K. Kaski, Scale-free networks generated by random walkers. Phys. A **341**, 80 (2004)
[32] A. Vásquez, Growing network with local rules: Preferential attachment, clustering hierarchy, and degree correlations. Phys. Rev. E, American Physical Society, College Park, MD **67**, 056104-1–056104-15 (2003)
[33] V. Zlatić, M. Božičević, H. Štefančić, M. Domazet, Wikipedias: collaborative web-based encyclopedias as complex networks. Phys. Rev. E, American Physical Society, College Park, MD **74**, 016115-1–016115-9 (2006)

On the Routability of the Internet

Pau Erola, Sergio Gómez, and Alex Arenas

The journey of a thousand miles begins with one step.
– *Lao Tzu,* Tao Te Ching

1 Introduction

The Internet is increasingly changing the way we do everyday tasks at work, at home, and how we communicate with one another. In its entrails, the Internet is structured as a network of networks. From a bottom-up perspective, the Internet is made up of networks of routers, each one under the control of a single technical administration. These networks are called autonomous systems (AS). An AS can use an exterior gateway protocol to route packets to other ASes [35] forming one of the largest synthetic complex system ever built. The Internet[1] comprises a decentralized collection of more than 30,000 computer networks from all around the world. Two ASes are connected if and only if they establish a business relationship (customer-provider or peer-to-peer relationships), making the Internet a "living" self-organized system.

The topology of the Internet has been studied at inter-domain level by Faloutsos et al. [18]. The ASes exhibit a power-law degree distribution with an exponent of $\gamma = 2.1$, and the average path length is near 3.2 standing out its small-

[1]This study is focused on the Internet at ASes level. We also refer to this level as inter-domain level.

P. Erola (✉) • S. Gómez • A. Arenas
Departament d'Enginyeria Informàtica i Matemàtiques, Universitat Rovira i Virgili,
43007 Tarragona, Spain
e-mail: pau.erola@urv.cat; sergio.gomez@urv.cat; alexandre.arenas@urv.cat

A. Mukherjee et al. (eds.), *Dynamics On and Of Complex Networks, Volume 2*,
Modeling and Simulation in Science, Engineering and Technology,
DOI 10.1007/978-1-4614-6729-8_3,

Table 1 Properties of the Internet ASes network for June 2009 [1]

Number of nodes	23,752
Number of edges	58,416
Average degree of nodes	4.919
Maximum degree of nodes	2,778
Average clustering coefficient	0.360
Power-law exponent	2.18
Diameter	8
Average path length	3.84

world character. The networks whose degree distribution follows a power-law, at least asymptotically, are called scale-free networks. Table 1 summarizes the main properties of the network.

The main function of the Internet is to forward information from an origin host, traversing switches, routers, and other network nodes, to reach our destination. To do so it uses the Border Gateway Protocol (BGP) [39], the routing protocol of the Internet, which uses a vector with end-to-end paths to guide the routing of information packets. This vector is called the Routing Table (RT), and it represents in fact a distributed global view of the network topology. To maintain this consistent map, routers need to exchange information of reachability through the network. This routing scheme is currently the root of one of the most challenging problems in network architecture: ensure the scalability of the Internet [30].

The Internet is growing abruptly wrapped in nontrivial dynamics. Empirical studies qualified this overall growth as the outcome of a net balance between births and deaths involving large fraction of nodes in the system [33]. This growth is estimated to be exponential, and the number of entries in the RT is currently growing at super-linear rate in the inter-domain level. Other dynamics such as failures to aggregate prefixes, address fragmentation, load balancing, and multi-homing are augmenting the RT demands [11]. This behavior is compromising the BGP scalability due to technological constrains. In addition, the RT updates to maintain the information of the shortest paths involve a huge amount of data exchange and significant convergence times of up to tens of seconds, hindering the communication process.

The poor scaling properties of routing schemes has been studied in depth within the context of compact routing (see [25] for a review). Unfortunately, these studies have concluded that in the presence of topology dynamics, a better scaling on Internet-like topologies is fundamentally impossible: while routing tables can be greatly reduced, the amount of messages per topology change cannot grow slower than linearly. This limitation has raised the need to explore new lines of research. Given the scale-free topology of Internet [33], the work of Boguñá et al. [7, 8] and our recent study [16, 17] rest on the presumption that the complex network theory is the natural framework to analyze and propose solutions. Here we review these works. This chapter is organized as follows. Section 2 briefly reviews some designs in the fields of compact routing and network architecture. In Sect. 3 we present two methodologies based on complex network theory to construct navigable maps of the Internet. Finally, in Sect. 4 we discuss the limitations of these new approaches and future directions to explore.

2 Schemes Based on Compact Routing and Network Architecture

To reduce the cost of communication networks, we should be concerned about the routing of messages through the network, which ideally should follow a shortest path, and the amount of information required, which should be minimal. Simple solutions can guarantee optimal shortest paths at the expenses of keeping in memory big routing tables, but these solutions are too expensive for large systems. In the research field of compact routing, Peleg and Upfal [34] addressed the trade-off between the average stretch (the average ratio of every path length relative to the shortest path) and the space needed to store RTs for general networks.

Earlier, Kleinrock and Kamoun [24] presented an alternative routing strategy to reduce the RT length in large networks. Their method intended to create a nested hierarchical structure of clusters (groups of closer nodes according to some measure), where any node only needs to maintain a small amount of information from distant nodes in other clusters, while it maintains complete information about its neighbors in the cluster. With this approach, the authors achieved a substantial size reduction of the RT (from N to $\ln N$), at the expense of increasing the average message length due to extra labels. However, the stretch analysis is satisfactory only under certain topologies, those in which the shortest path distance between nodes rapidly increases with the network size. Because of this, scale-free networks (under the small-world phenomenon) are not good scenarios for hierarchical routing, suffering severe increments of the path length. Despite the problem described, hierarchical routing is on the basis of the implanted inter-domain address strategy Classless Inter-Domain Routing (CIDR) [20]. CIDR was introduced in 1993 and allows the group (or aggregation) of addresses into blocks using bitwise masks, reducing the number of entries in the RT.

CIDR is an implicit scheme, i.e., the nodes are *a priori* labeled with structural information. The routing process explores this information to choose the neighbor to which a message should be sent [37]. It is also *name-dependent* or *labeled* routing [6], which means that nodes are tagged with topology-dependent information identifiers used to guide the packet forwarding. Among the newer name-dependent solutions, we found the schemes of Thorup and Zwick [40] and Brady and Cowen [9] interesting, which are highlighted here because of their proved efficiency in scale-free networks.

Thorup and Zwick (TZ) presented a scheme with stretch 3 and RT sizes of $O(n^{1/2} \log^{1/2} n)$-bits, being these results for the worst-case graph. In the case of scale-free networks Krioukov et al. estimated the performance of this routing and found on average stretch of 1.1 and very small RT sizes [26]. The basic idea behind TZ is to use a small set of *landmarks* (nodes potentially involved in the process of routing) to guide the navigation. That reveals why we achieve the best possible performance in the TZ scheme: scale-free graphs are optimally structured to exploit high-degree nodes (that will turn out to be landmarks) which are very important for finding shortest paths in such networks [23]. In turn, Brady and Cowen (BC) designed a routing scheme for undirected and unweighted graphs with the basic idea that trees cover scale-free graphs with minor deviations. In scale-free networks BC scheme guarantees an average stretch of 1.1 and logarithmic scaling RT sizes of $O(\log^2 n)$.

Alternatively to name-dependent solutions, we find *name-independent* routing schemes [6]. In this variant, nodes may be labeled arbitrarily making routing generally harder: first we need to know the location of the destination, i.e., we need dictionary tables to translate name-independent labels into locators in a name-dependent map. Abraham et al. [3] presented a nearly optimal name-independent routing for undirected graphs. The scheme has stretch 3, and a size upperbound of the RT of $O(\sqrt{n})$ per node, the same upper limits that in TZ. According to these results, the use of name-independent schemes provides no clear advantage over the use of name-dependent routing [25]. Other name-independent routing solutions have been designed specifically to address the Internet routing problems [38]. Particularly noteworthy are the Locator-Identifier Split (LIS) approaches like LISP [19], ENCAPS [22], and NIMROD [12]. All these LIS proposals separate the identifier and the locator of each node allowing aggressive aggregation techniques. During the communication process, the locators are encapsulated in each packet in a special wrapper, and at inter-domain level, packets are forwarded using only these locators. Krioukov et al. [25] identified the problems that may affect these solutions: they require in general a database to maintain updated locator information, and due to its hierarchical structure, aggressive aggregation is impossible on scale-free topologies.

It must be emphasized that all these proposals are effective for static networks. Under time-varying networks, the proposals assume that each change in the network structure generates a new graph and its routing solution must be recalculated. Krioukov et al. [25] reported the pessimistic scenario that the communication cost lower bound for scale-free graphs is $O(n)$. Nevertheless, adaptive solutions have not been studied in depth yet.

3 Solutions Based on Complex Network Theory

The complex networks theory gives us a new perspective, and new tools, to address the problem of scalability of the Internet routing protocol. The study of the special characteristics of the navigability of complex networks was initiated in 2000 by

Kleinberg [23]. Kleinberg highlighted that in complex networks, without a global view of the network, a message can be routed efficiently between any pair of nodes.

Adamic et al. [4] studied the role that high connectivity nodes (hubs) play in the communication process. Hubs are important actors in the routing process that facilitate search and information distribution, especially in large networks. They also introduce several local search strategies that exploit high-degree nodes which have costs that scale sublinearly with the network size.

Another recent work that studies the routing process in scale-free networks was introduced by Lattanzi et al. [29]. This study is focused on social networks and uses the model of *affiliation networks* that considers the existence of an *interest space* lying underneath. The search is greedily conducted in this space, and their results show that low-degree nodes not connected directly to hubs are hard to find and that large hubs are essential for an efficient routing process.

In this section, we present two heuristic approaches that use the scale-free characteristic of the Internet to propose alternatives to BGP. The first work was presented by Boguñá et al. [8] and builds a navigable map of the Internet. The second is a study of the current authors on the routability of the Internet using local information from its structure [16].

3.1 Hyperbolic Mapping of the Internet

Let us introduce the model proposed by Boguñá et al. with an archetype. There is a classical example in artificial intelligence aimed to find the shortest path between two cities in Romania, from Arad to Bucharest [36]. Let us assume that we know the straight-line distances between all cities of Romania to Bucharest. Having this information, we can use an informed search strategy to find efficiently the shortest path to our destination. In the given example a successful strategy is to choose the neighbor city (with connection by road) whose distance to destination is shorter. Representing the problem as a network, an informed search method chooses a node with lowest cost based on a heuristic function, e.g., minimize the straight-line distance in the above case. Like in this case, if we can build a coordinate system that reveals the network structure in sufficient detail, at each point in our path, we can use this map to determine which direction to choose to take us closer to the destination. That is, a greedy algorithm that mimics the routing process.

According to this philosophy, Boguñá et al. built a map of the Internet on a hyperbolic geometric space [8]. It has been proved that hyperbolic geometry matches strong heterogeneity (in terms of the power-law degree distribution exponent) and clustering properties of complex network topologies [27]. To achieve this goal they used a combination of geometric features, a distance measure between nodes inversely proportional to their probability of being connected, and topological characteristics, in this case the degree of the nodes.

Mapping of the Network

In the first step the nodes are placed in a hyperbolic disk of radius R, uniformly distributed through the angular component $\theta \in [0, 2\pi]$, and with radial coordinates r inversely proportional to the degree of each node. Nodes with higher degree will have smaller r values and will be closer to the center of the disk, and low-degree nodes will be more external. Secondly, Boguñá et al. computed the angular component θ to satisfy the requirement that nearby nodes in the hyperbolic space are connected, i.e., the probability $p(x_{ij})$ that two nodes i and j are connected decreases with the distance x_{ij} between them. This distance can be calculated using the hyperbolic law of cosines

$$\cosh x_{ij} = \cosh r_i \cosh r_j - \sinh r_i \sinh r_j \cos\left(\theta_i - \theta_j\right) . \quad (1)$$

The authors propose the following relationship between probabilities and distances:

$$p\left(x_{ij}\right) = \left(1 + e^{\frac{x_{ij}-R}{2T}}\right)^{-1} , \quad (2)$$

where T is a parameter related with the clustering of the network.

The estimation of these coordinates (θ_i, r_i) is performed by maximization of the likelihood that the Internet topology has been produced by this model. It is given by

$$L = \prod_{i<j} p\left(x_{ij}\right)^{a_{ij}} \left[1 - p\left(x_{ij}\right)\right]^{a_{ij}} , \quad (3)$$

where a_{ij} are the elements of the adjacency matrix of the network.

Navigation on the Hyperbolic Map

Once we have a coordinate pair for each AS, we need to define the routing process over this map. Boguñá et al. took advantage of the characteristics of the underlying hyperbolic space to perform a greedy routing process. Hyperbolic spaces expand exponentially, making the distance between two points approximately the sum of their radial coordinates, with less influence of their angular difference, as can be proved from Eq. (1).

Krioukov et al. [27] established a congruence between the geodesic distances and the shortest paths. Like the trace drawn by the geodesics, shortest paths tend to originate in the outer hyperbolic disk, then getting closer to the center of the embedded space, and finally coming back to the exterior of the disk to reach the destination. Using this similarity, Boguñá et al. designed a greedy packet forwarding that selects at each step of the path the neighbor closest to the destination, following the geodesic. Figure 1 shows an example of a network hyperbolic space and a route path.

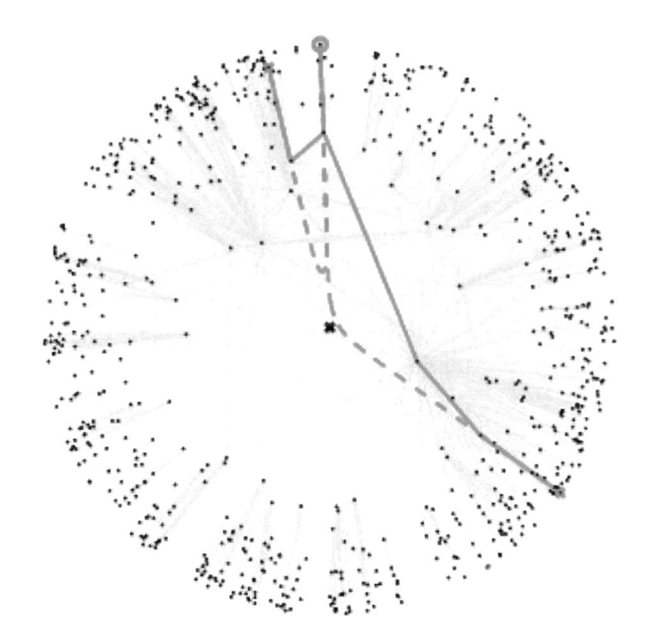

Fig. 1 Hyperbolic mapping of a synthetic network. The *dashed green lines* correspond to two geodesic paths and the *solid lines* to the greedy routing paths (reprinted from [32])

Results

As a result of the high concordance between the network topology and underlying space, the authors achieved near optimal results. The average success ratio is 97%, a very high value considering that the nodes only have knowledge about its neighbors. The greedy paths are very close to the shortest path with an average stretch of 1.1. Furthermore, the local nature of the design minimizes the routing communication needs in front of dynamic topology changes, favoring the network scalability.

More important than the behavior of the scheme in the static network is how it adapts to the dynamic topology. Boguñá et al. have shown that its projection successfully overcomes the failures of links and nodes, with very small losses, and that the addition of new nodes is possible without recalculating the entire map. They have also shown that the projection of the nodes in the metric space is very stable in time eliminating the need to recalculate the coordinates.

3.2 *Routing Using Modular Information*

In 2000, Kleinberg published his work about navigation in small-world networks [23] pointing that the correlation between the local structure of a network and long-range links provides critical cues for finding paths through the network.

We associate these terms of *local structure* and *long-range links* to the modular structure of the network. The modular structure (or community structure) refers to the clustering of nodes in communities, groups of nodes in the network more connected between them than with the rest of the network. Interestingly, most of the real-world networks present modular structure as a typical fingerprint of a self-organized and decentralized evolution, and in particular the Internet [21]. Thus, our initial hypothesis is that the community structure, which proves meaningful insights on the structure and function of complex networks, can be an important actor in the Internet routing properties. To exploit this, we propose to analyze the contribution of each node to the modules using the projection technique presented by Arenas et al. [5]. We will use this information to guide the forwarding of information packets through the network using a simple greedy routing algorithm that aims at finding the neighborhood of the target destination and after to find the node destination within it.

Detecting the Modular Structure

The first step in our study is to detect communities in the ASes network. Whatever strategy is applied to detect these communities (groups), all of them are *blind* to content and only *aware* of the topological structure.

A widespread method to quantify how good is a given partition of a network into communities was proposed in [31]. This measure is known as *modularity*, and it rests on the intuitive idea that random networks do not exhibit a clear community structure. Modularity is a quality function that reads

$$Q = \sum_{\alpha} \left(e_{\alpha\alpha} - a_{\alpha}^2 \right) . \tag{4}$$

where e is a matrix where the elements $e_{\alpha\beta}$ represent the fraction of total links starting at a node in community α and ending at a node in community β and $e_{\alpha\alpha}$ is the fraction of links starting and ending in community α. The vector a_{α} is the sum of any row of e, i.e., the fraction of links connected to community α, and a_{α}^2 is the expected number of intra-community links.

Algorithms that optimize this function yield good community structure compared to a (*equivalent*) random model network. The problem is that the partition space of any graph is huge (the search for the optimal modularity value is a NP-hard problem [10]), and one needs a guide to explore this space and find local maximum values. For more information on the best successful heuristics, Arenas et al. performed a comparison of different methods in [13]. See also this toolset [2]. We have applied a technique to optimize modularity based on the extremal optimization process [14].

Projecting the Network Structure

The use of the modular structure inherent to real complex networks provides a useful information whose exploitation is competitive with global shortest path strategies. To use the information of the modular structure of networks we analyze the contribution of each node to maintain the structure of communities. The object of this analysis is defined as the contribution matrix C of N nodes to M communities.

$$C_{i\alpha} = \sum_{j=1}^{N} W_{ij} S_{j\alpha} \tag{5}$$

where W is the graph matrix, whose elements W_{ij} are the weights of the connections from any node i to any node j, and $S_{j\alpha}$ is the partition matrix, where if node j belongs to community α, then $S_{j\alpha} = 1$, otherwise $S_{j\alpha} = 0$.

Unfortunately, the study of this matrix involves a computationally prohibitive handling of a huge amount of data, especially to be used as a basis for a feasible routing system. To reduce this problem, we propose to analyze the contribution of each node of a network to communities using the projection technique introduced by Arenas et al. [5]. This projection is based on a rank 2 truncated singular value decomposition (TSVD) and allows building a map $\mathcal{U}_2$ where each node n has a coordinate pair or contribution projection vector $\tilde{v}_n$. This two-dimensional plane reveals the structure of communities and their boundaries, and we will use it as the navigable coordinate system of the complex network.

For each coordinate pair we calculate the polar coordinates (R_n, θ_n) where R_n is the length of the contribution projection vector $\tilde{v}_n$ and θ_n is the angle between $\tilde{v}_n$ and the horizontal axis. To interpret correctly this outcome, we need to know also the intramodular projection $\tilde{e}_\alpha$ of each community, the distinguished direction line of the projection of its internal nodes (those that have links exclusively inside the community).

With these values, we can compute a new pair (R_n, ϕ_n), where

$$\phi_n = |\theta_n - \theta_{\tilde{e}_\alpha}|, \tag{6}$$

and the new values

$$R_{\text{int}_n} = R_n \sin \phi_n, \tag{7}$$

and

$$R_{\text{ext}_n} = R_n \cos \phi_n. \tag{8}$$

Here, R_{int} informs about the internal contribution of nodes to their corresponding communities, and R_{ext} reflects the boundary structure of communities. Both values, R_{int} and R_{ext}, are the basis of our routing framework.

Greedy Routing

Now we have a geometrical projection of the modular structure that presumably will be helpful to design a local routing strategy. How to make use of this map for navigation purposes is the aim of this section.

Algorithm 1 Heuristic routing algorithm, where $|\Delta\theta_{k\to j}|$ is the angular distance between the neighbor node k and the destination node j. Since $|\Delta\theta_{k\to j}|$ is small for nodes within the same community, we add a shift value λ to reduce its fast annealing

cost[*] $\leftarrow \infty$
if $\exists\, k \in \alpha_j$ **then**
  for $\forall\, k \in \alpha_j$ **do**
    $\text{cost}[k] \leftarrow \dfrac{\lambda + |\Delta\theta_{k\to j}|}{R_{\text{int}_k}}$
  end for
else
  for $\forall\, k$ with $R_{\text{ext}_j} > 0$ **do**
    $\text{cost}[k] \leftarrow \dfrac{|\Delta\theta_{k\to j}|}{R_{\text{ext}_k}}$
  end for
end if
return k with lower cost[k]

In an unstructured network an algorithm based on hubs' transit (i.e., route toward hubs with the hope that hubs will be directly connected to any target destination) will result in a decent routing even improving classical routing techniques [4]. This basic idea is also used in our modular routing strategy; we will look for nodes in our current community that could act as hub connectors within the community and eventually with other communities. To find an efficient path between two nodes in the network, we choose the local neighbor (node) that has high value of R_n. The nodes with high values of R_n are nodes with many connections that will presumably allow us to lead to the destination quickly reducing the path length. We have to differentiate however two scenarios: when we are into the destination community and when we are not.

The algorithm we propose works as follows (see Algorithm 1). Let us assume we want to go from node i to node j, and let N_i refer to the neighbors of node i. Each node $k \in N_i$ is a candidate in the path. In the routing process, for each node $k \in N_i$, we have to compute a cost function and select the candidate that minimizes it in each step. This process is repeated until the destination is reached, the current node i does not find a feasible successor or a time constraint is violated. We do not allow loops.

This heuristic algorithm sets two scenarios: when our neighbor k belongs to the same community α_j than the destination node j and when it does not. In the first case, when $k \in \alpha_j$, we are interested in finding nodes with an important weight in the community. In the other case, if $k \notin \alpha_j$, we seek for nodes near to the boundaries of other communities. Figure 2 shows a typical path in the Internet projection.

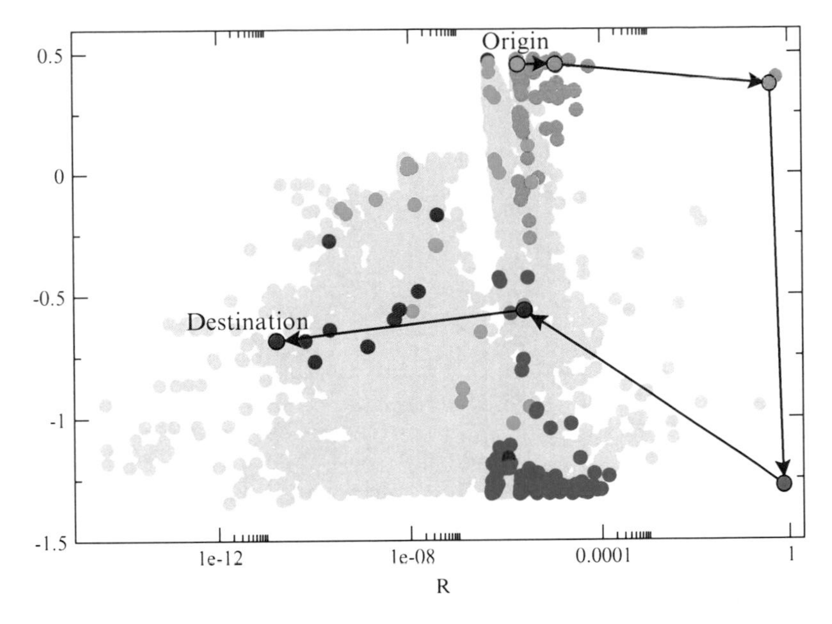

Fig. 2 Example of routing in the projection of the Internet network. The color of the nodes represents their community; some communities have been omitted to simplify the figure. We see in the example a path that begins in a small degree node, which searches the network hubs and gets closer in θ, to enter the destination community and reach the destination node

Results

To evaluate our proposal we used a snapshot of the Internet for June 2009 [1]. In this network, we have found 349 communities using [2] of sizes ranging between 2 and 546 nodes. We have simulated 10^6 paths randomly selected. Figure 3 presents the distribution of path length of our algorithm. We have achieved a success rate of 94% and a median path of 5 steps (shortest path median is 4). Given that we are using only local information, the percentage of success is very good. We attribute the unreachable destinations to nodes that are very far from any hub of the network. Nodes with longer paths are, in most cases, internal nodes connected only with other nodes of small degree. Our projection reflects the boundaries of communities of the network but only provides information about the number of connections a node has in the internal topology of its community. This makes that our search strategy has difficulties to find the pathway to those poorly connected nodes with only intramodular contribution. Because of this, the stretch of our algorithm can get very unfavorable, as is suggested by the long tail of the distribution that shows Fig. 3. The consideration of customer-provider roles can minimize partially this problem by reducing the possible paths.

This study is also concerned about the scalability of the Internet routing protocol. In situations where the data is continuously changing, like in an evolving network, a TSVD projection might become obsolete. It is an interesting question whether the

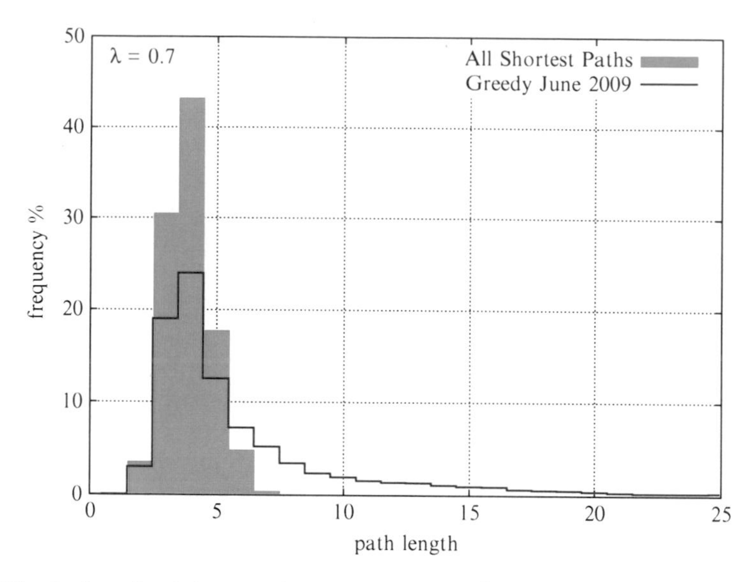

Fig. 3 Distribution of path lengths of the simulation of 10^6 paths using our proposal compared with the shortest path. The value $\lambda = 0.7$ has been determined experimentally

TSVD projection of an initial data set is reliable. In our earlier work [15] we defined two measures to quantify the differences between a sequence of computed TSVD projections of growing Barabási-Albert's scale-free networks. In that situation, we proved that the stability of a TSDV map is very high when considering neighborhood stability (note that R_{int} and R_{ext} are relative modules). Thanks to this, under topology dynamics, addition and removal of nodes in the network can be done without recalculating the projection.

4 Discussion

The theory of complex networks is a powerful tool to address the scalability issues of the Internet routing protocol. Here, we have seen two recent works that are alternative proposals to the classical approaches of compact routing. Boguñá et al. [8] used statistical inference techniques to assign a pair of coordinates to each node in a hyperbolic metric space. In a Poincaré representation, the nodes with high degree go to the center of the space and their angular position is determined by their probability to be connected. Alternatively, in Erola et al. [16], we proposed to exploit the community structure of the network to obtain a linear projection in which nodes of the same community are organized around specific singular directions and hubs have large values of the radial coordinate. At the end, both methods use the same principle: using a projection of the network (map) to define a local routing to the destination using local coordinates on them. The main differences are, while

Boguñá et al. [8] look for the best projection in terms of the statistical distribution of connections, in Erola et al. [16] we look for the best projection in terms of the modular structure inherent in the network. Both approaches prove to be competitive and probably complementary.

Moreover, these proposals have proved to be resilient to changes in the network structure due to failures and growth. Therefore, there is no need to continuously recalculate the corresponding projections, obtaining at the same time high performances of the routing algorithms. Nevertheless, profound changes in the core of large hubs may accelerate their obsolescence. For instance, Labovitz et al. [28] have identified a significant evolution of provider interconnection strategies involving a rapid transition to more densely interconnected and less hierarchical inter-domain topology. This particular time-varying behavior may concern the success of these new routing schemes and should be analyzed in the future.

Acknowledgement This work has been partially supported by the Spanish DGICYT Project FIS2009-13730-C02-02 and the Generalitat de Catalunya 2009-SGR-838. PE acknowledges a URV PhD grant.

References

[1] URL http://www.caida.org/data/active/as-relationships
[2] URL http://deim.urv.cat/~sgomez/radatools.php
[3] I. Abraham, C. Gavoille, D. Malkhi, N. Nisan, M. Thorup, Compact name-independent routing with minimum stretch, in *Proceedings of the Sixteenth Annual ACM Symposium on Parallelism in Algorithms and Architectures* (ACM, 2004), pp. 20–24
[4] L. Adamic, R. Lukose, A. Puniyani, B. Huberman, Search in power-law networks. Phys. Rev. E **64**(4), 046,135 (2001)
[5] A. Arenas, J. Borge-Holthoefer, S. Gómez, G. Zamora-Lopez, Optimal map of the modular structure of complex networks. New J. Phys. **12**, 053,009 (2010)
[6] B. Awerbuch, A. Bar-Noy, N. Linial, D. Peleg, Compact distributed data structures for adaptive routing, in *Proceedings of the Twenty-First Annual ACM Symposium on Theory of Computing* (ACM, 1989), pp. 479–489
[7] M. Boguñá, D. Krioukov, KC. Claffy, Navigability of complex networks. Nat. Phys. **5**(1), 74–80 (2008)
[8] M. Boguñá, F. Papadopoulos, D. Krioukov, Sustaining the internet with hyperbolic mapping. Nat. Comm. **1**, 62 (2010)
[9] A. Brady, L. Cowen, Compact routing on power law graphs with additive stretch, in *Proc. of the 9th Workshop on Algorithm Eng. and Exper*, pp. 119–128, 2006
[10] U. Brandes, D. Delling, M. Gaertler, R. Görke, M. Hoefer, Z. Nikoloski, D. Wagner, On finding graph clusterings with maximum modularity, in *Graph-Theoretic Concepts in Computer Science* (Springer, 2007), pp. 121–132
[11] T. Bu, L. Gao, D. Towsley, On characterizing bgp routing table growth. Comput. Netw. **45**(1), 45–54 (2004)
[12] I. Castineyra, N. Chiappa, M. Steenstrup, The nimrod routing architecture (1996)
[13] L. Danon, A. Díaz-Guilera, J. Duch, A. Arenas, Comparing community structure identification. J. Stat. Mech.-Theory Exp. 2005.09, P09008. (2005)
[14] J. Duch, A. Arenas, Community detection in complex networks using extremal optimization. Phys. Rev. E **72**(2), 027,104 (2005)

[15] P. Erola, J. Borge-Holthoefer, S. Gómez and A. Arenas, Reliability of optimal linear projection of growing scale-free networks, Int. J. Bifurcat. Chaos, **22**(7), (2012)
[16] P. Erola, S. Gómez, A. Arenas, An internet local routing approach based on network structural connectivity, in *Proceedings of IEEE GLOBECOM Workshop on Complex Networks and Pervasive Group Communication*, pp. 7–11 (IEEE, Houston, Texas, USA, 2011)
[17] P. Erola, S. Gómez, A. Arenas, Structural navigability on complex networks. Int. J. Complex Syst. Sci. **1**, 37–41 (2011)
[18] M. Faloutsos, P. Faloutsos, C. Faloutsos, On power-law relationships of the internet topology, in *ACM SIGCOMM Computer Communication Review*, vol. 29–4 (ACM, 1999), pp. 251–262
[19] D. Farinacci, Locator/id separation protocol (lisp). Internet-draft, draft-farinacci-lisp-08 (2008)
[20] V. Fuller, T. Li, Classless inter-domain routing (cidr): the internet address assignment and aggregation plan (2006)
[21] M. Girvan, M. Newman, Community structure in social and biological networks. Proc. Natl. Acad. Sci. **99**(12), 7821 (2002)
[22] R. Hinden, New scheme for internet routing and addressing (encaps) for ipng (1996)
[23] J. Kleinberg, Navigation in a small world. Nature **406**(6798), 845–845 (2000)
[24] L. Kleinrock, F. Kamoun, Hierarchical routing for large networks performance evaluation and optimization. Comput. Network (1976) **1**(3), 155–155 (1977)
[25] D. Krioukov, K. Fall, A. Brady, et al., On compact routing for the internet. ACM SIGCOMM Comput. Comm. Rev. **37**(3), 41–52 (2007)
[26] D. Krioukov, K. Fall, X. Yang, Compact routing on internet-like graphs, in *INFOCOM 2004. Twenty-Third AnnualJoint Conference of the IEEE Computer and Communications Societies*, vol. 1 (IEEE, 2004)
[27] D. Krioukov, F. Papadopoulos, M. Kitsak, A. Vahdat, M. Boguñá, Hyperbolic geometry of complex networks. Phys. Rev. E **82**(3), 036,106 (2010)
[28] C. Labovitz, S. Iekel-Johnson, D. McPherson, J. Oberheide, F. Jahanian, Internet inter-domain traffic, in *ACM SIGCOMM Computer Communication Review*, vol. 40–4 (ACM, 2010), pp. 75–86
[29] S. Lattanzi, A. Panconesi, D. Sivakumar, Milgram-routing in social networks, in *Proceedings of the 20th International Conference on World Wide Web* (ACM, 2011), pp. 725–734
[30] D. Meyer, L. Zhang, K. Fall, et al., Report from the iab workshop on routing and addressing. RFC2439, September (2007)
[31] M. Newman, Fast algorithm for detecting community structure in networks. Phys. Rev. E **69**(6), 066,133 (2004)
[32] F. Papadopoulos, D. Krioukov, M. Boguñá, A. Vahdat, Greedy forwarding in dynamic scale-free networks embedded in hyperbolic metric spaces. arXiv:0805.1266v3 (2010)
[33] R. Pastor-Satorras, A. Vespignani, *Evolution and Structure of the Internet: A Statistical Physics Approach* (Cambridge University Press, Cambridge, 2004)
[34] D. Peleg, E. Upfal, A trade-off between space and efficiency for routing tables. J. ACM **36**(3), 510–530 (1989)
[35] E. Rosen, Exterior gateway protocol (egp) (1982). URL http://tools.ietf.org/html/rfc827
[36] S. Russell, P. Norvig, *Artificial Intelligence: a Modern Approach*. Artificial Intelligence (Prentice-Hall, Englewood Cliffs, 1995)
[37] N. Santoro, R. Khatib, Labelling and implicit routing in networks. Comput. J. **28**(1), 5–8 (1985)
[38] C. Shue, M. Gupta, Packet forwarding: Name-based vs. prefix-based, in *IEEE Global Internet Symposium, 2007* (IEEE, 2007), pp. 73–78
[39] J. Stewart III, *BGP4: Inter-Domain Routing in the Internet* (Addison-Wesley Longman, 1998)
[40] M. Thorup, U. Zwick, Compact routing schemes, in *Proceedings of the Thirteenth Annual ACM Symposium on Parallel Algorithms and Architectures* (ACM, 2001), pp. 1–10

The Evolution of Layered Protocol Stacks Leads to an Hourglass-Shaped Architecture

Saamer Akhshabi and Constantine Dovrolis

1 Introduction

Networking architectures are not designed to work in a static environment. Applications, services, user expectations, communication, and computing technologies, as well as the underlying economics, are all in a constant state of flux. A networking architecture designed now should not just be optimal in terms of what is currently known; instead, it should be able to evolve, and more importantly, maintain, or improve its features while evolving. The Internet has been evolving, without centralized control, over the last 40 years: from the initial studies of packet switching in the 1960s, the development of TCP/IP and Ethernet LANs in the 1970s, the introduction of DNS and TCP congestion control in the 1980s, the adoption of BGP and CIDR in the early 1990s and of NATs in the late 1990s, the introduction of new access technologies in the last mile during the last decade (broadband, wireless LANs, WiMAX, etc.), and of course the persistent expansion of application-layer protocols and services (peer-to-peer, VoIP, video streaming, multiplayer gaming, online social networks, etc.).

Recently, the evolution of the Internet architecture has become a major research issue for two reasons. First, there is a concern that the current Internet architecture has been ossified, meaning that it does not evolve as quickly as it used to, especially at the core protocols (network and transport layers) [14]. Several architectural extensions, such as IP multicast, IPv6, QoS, or new transport protocols such as XCP, have not been adopted in practice, perhaps as a result of this ossification. Second, there is significant interest in designing clean-slate architectures for a future Internet, without aiming to make these architectures backwards compatible with the current Internet. A central objective in these design exercises is evolvability, i.e., to

S. Akhshabi (✉) • C. Dovrolis
College of Computing, Georgia Institute of Technology, Atlanta, GA, USA
e-mail: s.akhshabi@gatech.edu; constantine@gatech.edu

A. Mukherjee et al. (eds.), *Dynamics On and Of Complex Networks, Volume 2*,
Modeling and Simulation in Science, Engineering and Technology,
DOI 10.1007/978-1-4614-6729-8_4, © Springer Science+Business Media New York 2013

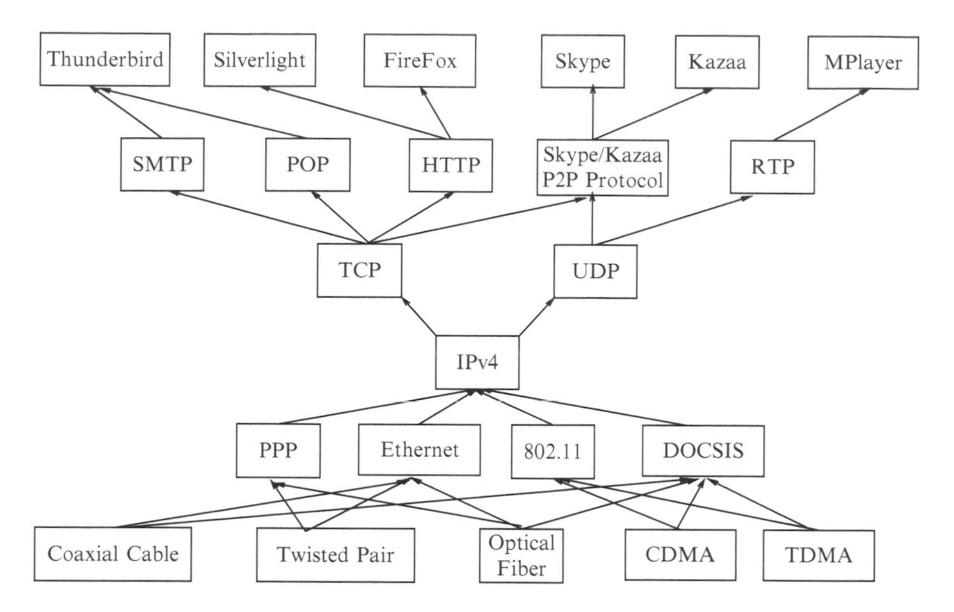

Fig. 1 An (incomplete) illustration of the hourglass Internet architecture

ensure that an architecture will have the ability to evolve under new environments and requirements without getting ossified a few years later [16].

Even though there are several research endeavors that attempt to either accommodate the evolution of the current Internet or to design evolvable future Internet architectures, there is little prior work in the more fundamental issues regarding the evolution of network architectures. We all agree, for instance, that different protocol stacks compete with each other (e.g., TCP/IP vs ATM in the early nineties) and that network architectures change with time as new protocols are created and existing protocols are abandoned [8]. We do not really understand, however, the impact of such competition or birth/death processes on the resulting network architecture.

To state a more concrete but fundamental question: *why does the Internet protocol stack resemble an hourglass?* Is it a coincidence, intentional design, or the result of an evolutionary process in which new protocols compete with existing protocols that offer similar functionality and services? The protocol stack was not always shaped in this way. For instance, until the early nineties, there were several other network-layer protocols competing with IPv4, including Novell's IPX, the X.25 network protocol used in Frame Relay, the ATM network-layer signaling protocol, and several others. It was through a long process that IPv4 eventually prevailed as practically the only surviving protocol at layer 3, creating a very narrow waist at the Internet architecture hourglass (see Fig. 1).

Another important question is the following: *why do we tend to see more frequent innovations at the lower or higher layers of the protocol hourglass, while the protocols at the waist of the hourglass appear to be "ossified" and difficult to replace?* During the last 30–40 years we have seen many new physical and data-link-layer protocols created and surviving. And of course the same can be said about applications and application-layer protocols. On the other hand, the protocols at the waist of the hourglass (mostly IPv4, TCP, and UDP) have been extremely stable, and they have managed to outcompete any protocols that offer the same or similar functionality. *How can a new protocol manage to survive the intense competition with those core protocols at the waist of the Internet hourglass?* In fact, *the ossification of the hourglass waist* has been a major motivation for "clean-slate" efforts to design a novel future Internet architecture [13]. There are two important questions in that context. First, *how can we make it more likely that a new (and potentially better) protocol replaces an existing and widely used incumbent protocol?* And second, how can we make sure that a new architecture we design today will not be ossified 10–20 years later? In other words, *what makes a protocol stack or network architecture evolvable?* The previous questions have generated an interesting debate [9, 10, 17].

In this chapter, our goal is to study protocol stacks (and layered architectures, more generally) as well as their evolution in a rigorous and quantitative manner. Instead of only considering a specific protocol stack, we propose an abstract model in which protocols are represented by nodes, services are represented by directed links, and so a protocol stack becomes a layered directed acyclic graph (or network). Further, the topology of this graph changes with time as new nodes are created at different layers, and existing nodes are removed as a result of competition with other nodes at the same layer.

The proposed evolutionary model, referred to as *EvoArch*, was first proposed in [1] and is based on a few principles about layered network architectures in which an "item" (or service) at layer X is constructed (or composed) using items at layer (X−1). These principles capture the following:

(a) The source of *evolutionary value* for an item
(b) The *generality* of items as we move to higher layers
(c) The condition under which two items compete
(d) The condition under which one item causes the death or removal of a competing item

Perhaps surprisingly, these few principles are sufficient to produce hourglass-shaped layered networks in relatively short evolutionary periods.

As with any other model, *EvoArch* is only an abstraction of reality focusing on specific observed phenomena, in this case the hourglass structure of the Internet protocol stack, and attempting to identify a parsimonious set of principles or mechanisms that are sufficient to reproduce the observed phenomena. As such, *EvoArch* is an *explanatory model* (as opposed to black-box models that aim to only describe statistically some observations). *EvoArch* deliberately ignores many aspects of protocol architectures, such as the functionality of each layer, technological constraints, debates in standardization committees, and others. The fact that these practical aspects are not considered by *EvoArch* does *not* mean that they are insignificant; it means, however, that if the evolution of network architectures follows the principles that *EvoArch* is based on, then those aspects are neither necessary nor sufficient for the emergence of the hourglass structure.

The broader issue of the importance of layering in networking architectures has been discussed in several papers. A "classic" is Clark's 1988 paper [5], which states a largely unknown fact about the evolution of TCP/IP that layering of the architecture was not part of the design from the beginning. A 1983 paper by Cerf and Cain states the following remark about layering [3] that different protocols sit in the same layer because they share the same set of support protocols and not necessarily because their functionality is the same or even similar. Another important step in the study of networking architectures was the 1990 work of Clark and Tennenhouse that advocated vertically integrated layers [6]. The recent book by John Day [8] also includes interesting discussions about past and current network architectures.

At a high level, this work is also related to recent efforts that develop a rigorous theory of network layering and architecture, mostly using mathematical tools from optimization and control systems [4]. We agree with those authors that network architecture can become the subject of more quantitative and rigorous scientific methods. We have a different view, however, on how to get there: instead of thinking about each layer as the solution to an optimization problem, we focus on the evolutionary process that shapes a network architecture over time, and we emphasize the role of robustness and evolvability instead of optimality. Also relevant is the work of Csete and Doyle [7], which has emphasized the role of hierarchical modularity and evolution in both technological and biological systems. Those authors have also identified the significance of the hourglass (or bowtie) structure in the corresponding network architectures.

The rest of the chapter is structured as follows. In Sect. 2, we describe *EvoArch* and explain how the model relates to protocol stacks and evolving network architectures. In Sect. 3, we present basic results to illustrate the behavior of the model and introduce some key metrics. Section 4 is a robustness study showing that the model produces hourglass structures for a wide range of parameter values. The effect of those parameters is studied in Sect. 5 focusing on the location and width of the waist. Section 6 examines the *evolutionary kernels* of the architecture, i.e., those few nodes at the waist that survive much longer than other nodes. Section 7 generalizes *EvoArch* in an important and realistic manner: what if different protocols at the same layer have different qualities (such as performance or extent of deployment)? In Sect. 8, we provide an analytical insight into the emergence of the hourglass structure and illustrate how that analysis could be used to predict the location of the waist in the evolved architecture. We conclude the chapter in Sect. 9.

2 Model Description

In *EvoArch*, a protocol stack is modeled as a directed and acyclic network with L layers (see Fig. 2). Protocols are represented by nodes, and protocol dependencies are represented by directed edges. If a protocol u at layer l uses the service provided

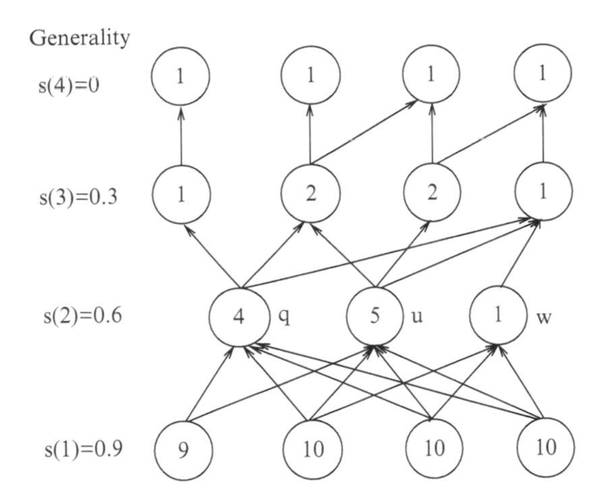

Fig. 2 A toy network with four layers. The value of each node is shown inside the circle

by a protocol w at layer $l-1$, the network includes an "upwards" edge from w to u.[1] The layer of a node u is denoted by $l(u)$. The incoming edges to a node u originate at the *substrates* of u, represented by the set of nodes $S(u)$. Every node has at least one substrate, except the nodes at the bottom layer. The outgoing edges of a node u terminate at the *products* of u, represented by the set of nodes $P(u)$. Every node has at least one product, except the nodes at the top layer.

The substrates of a node are determined probabilistically when that node is created.[2] Specifically, each layer l is associated with a probability $s(l)$: a node u at layer $l+1$ selects independently each node of layer l as substrate with probability $s(l)$. We refer to $s(l)$ as the *generality* of layer l. $s(l)$ *decreases* as we move to higher layers, i.e., $s(i) > s(j)$ for $i < j$. The decreasing generality probabilities capture that *protocols at lower layers are more general in terms of their function or provided service than protocols at higher layers.* For instance, in the case of the Internet protocol stack, a protocol at layer 1 offers a very general bit transfer service between two directly connected points; this is a service or function that almost any higher layer protocol would need. On the other extreme, an application-layer protocol, such as SMTP, offers a very specialized service, and it is only used by applications that are related to email exchanges. Note that if node u does not select any substrate from layer l, we connect it to one randomly chosen substrate from that layer.

[1]In practice, the principle of strict layering is occasionally violated through tunnels or other forms of virtual networks. For the most part, however, layering is the norm in protocol architectures rather than the exception. Considering architectures without strict layering is outside the scope of this paper and an interesting subject for future research.

[2]Of course in practice substrates are never chosen randomly. The use of randomness in the model implies that a realistic mechanism of substrate selection is not necessary for the emergence of the hourglass structure.

Each node u has an *evolutionary value* or simply *value* $v(u)$ that is computed recursively based on the products of u,

$$v(u) = \begin{cases} \sum_{p \in P(u)} v(p) & l(u) < L \\ 1 & l(u) = L \end{cases} \quad (1)$$

The value of the top-layer nodes is assumed to be fixed; in the simplest version of *EvoArch*, it is equal to one. So, the model captures that the value of a protocol u is driven by the values of the protocols that depend on u. For instance, TCP has a high evolutionary value because it is used by many higher-layer protocols and applications, some of them being highly valuable themselves. A brand new protocol, on the other hand, may be great in terms of performance or new features, but its value will be low if it is not used by important or popular higher-layer protocols.

The value of a node largely determines whether it will survive the competition with other nodes at the same layer that offer similar services. Consider a node u at layer l. Let $C(u)$ be the set of *competitors* of u: this is the set of nodes at layer l that share at least a fraction c of node u's products, i.e.,

$$w \in C(u) \text{ if } l(w) = l(u) \text{ and } \frac{|P(u) \cap P(w)|}{|P(u)|} \geq c \quad (2)$$

The fraction c is referred to as the *competition threshold*. In other words, a node w competes with a node u if w shares a significant fraction (at least c) of u's products, meaning that the former offers similar services or functions with the latter. Note that the competition relation is not symmetric: w may provide a generic service, having many products, and thus, competing with several protocols at the same layer; the latter may not be competitors of w if they provide more specialized functions and have only few products.

Given the set of competitors of a node u, we can examine whether u would survive the competition or die. The basic idea is that u dies if its value is significantly less than the value of its *strongest* (i.e., maximum value) competitor. Specifically, let $v_c(u)$ be the maximum value among the competitors of u:

$$v_c(u) = \max_{w \in C(u)} v(w) \quad (3)$$

If u does not have competitors, $v_c(u)$ and the death probability for u are set to zero. Otherwise, we introduce the *death probability ratio* $r = \frac{v(u)}{v_c(u)}$. The death probability $p_d(r)$ is then computed as follows:

$$p_d(r) = \begin{cases} e^{\frac{-zr}{1-r}} & 0 < r < 1 \\ 0 & r \geq 1 \end{cases} \quad (4)$$

The death probability function $p_d(r)$ is shown in Fig. 3 for three different values of the *mortality parameter* z. This parameter captures the intensity of the competition

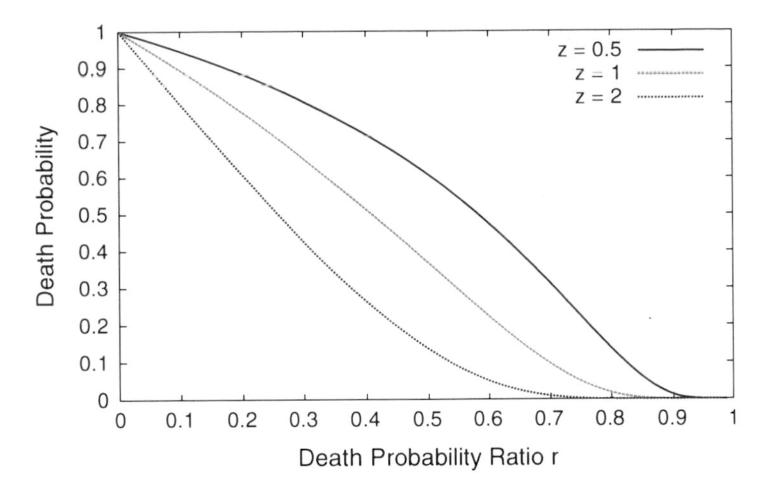

Fig. 3 The death probability for three values of the mortality parameter z

among protocols. As z decreases, the competition becomes more intense and it is more likely that a protocol will die if at least one of its competitors has higher value than itself.

When a node u dies, its products also die *if their only substrate is u*. This can lead to a *cascade effect* where the death of a node leads to the death of several nodes in higher layers.

To illustrate the previous concepts, Fig. 2 shows a toy network with $L = 4$ layers. The generality probability for each layer is shown at the left of the corresponding layer. Note that, on average, the number of products per node decreases as we move to higher layers because the generality probability decreases in that direction. Assuming that $c = 3/5$, nodes u and q are competitors of node w in layer 2. It is likely (depending on the parameter z) that w would soon die because its value is much less than that of its maximum-value competitor, u. u is also a competitor of q, but this competition is much less likely to be lethal for the latter because its value is comparable to that of u.

EvoArch captures the inherent competition between nodes at the same layer and specifically, between nodes that offer about the same service. For instance, FTP and HTTP are two application-layer protocols that can both be used for the transfer of files. The large overlap of the services provided by HTTP with the services provided by FTP (i.e., HTTP is a competitor of FTP) and the fact that HTTP acquired over the years a larger evolutionary value from its own higher-layer products (applications such as web browsers) leads to the extinction of FTP. On the other hand, TCP and UDP are two transport layer protocols that offer largely different services. Their competition, in terms of products (i.e., application-layer protocols), is minimal, and the two protocols have coexisted for more than 30 years.

In the simplest version of *EvoArch*, the creation of new nodes follows *the basic birth process*. Specifically, the number of new nodes at a given time is set to a small fraction (say 1–10%) of the total number of nodes in the network at that time, implying that the larger a protocol stack is, the faster it grows. Each new node is assigned randomly to a layer. In Sect. 6, we also examine a *death-regulated birth process*, in which the frequency of births at a layer depends on the death rate at that layer.

EvoArch is a discrete-time model. By t_k, we denote the *kth round*. In each round, the model execution includes the following steps in the given order:

(a) Birth of new nodes and random assignment to layers.
(b) Examine each layer l, in top-down order, and perform three tasks:
(b.1) Connect any new nodes assigned to that layer, choosing substrates and products for them based on the generality probabilities $s(l-1)$ and $s(l)$, respectively.
(b.2) Update the value of each node at layer l (note that the value of a node in the k'th round can be affected by nodes added in that same round).
(b.3) Examine in order of decreasing value in that layer, whether any node should die (considering the case of cascade deaths).

Initially, we start with a small number of nodes at each layer and form the edges between layers as if all nodes were new births. Unless noted otherwise, the execution of the model stops when the network reaches a given number of nodes.[3] We refer to each execution as an *evolutionary path*.

3 Basic Results

In this section, we illustrate the behavior of the *EvoArch* model focusing on the width of each layer across time. We also introduce the main metrics we consider and the default values of the model parameters.

The default values of *EvoArch*'s parameters are as follows: $L = 10$ layers, $s(l) = 1 - l/L$ (i.e., the generality decreases as 0.9, 0.8, ..., 0.1, 0, as we go up the stack), $c = 0.6$ (i.e., at least 3 out of 5 shared products), and $z = 1$ (see Fig. 3). Each evolutionary path starts with 10 nodes at every layer; the average birth rate at each round is 5% of the current network size, and an evolutionary path ends when the network size reaches 500 nodes (but not sooner than 100 rounds). Unless noted otherwise, we repeat each experiment 1,000 times, while the graphs show the median as well as the 10th, 25th, 75th, and 90th percentiles across all evolutionary paths. We emphasize that the previous default values do *not* correspond, obviously, to the characteristics of the Internet stack. A parameterization of the model for that

[3]We have also experimented with a termination condition based on the number of rounds instead of the number of nodes. There is no significant difference as long as the network can evolve for at least few tens of rounds.

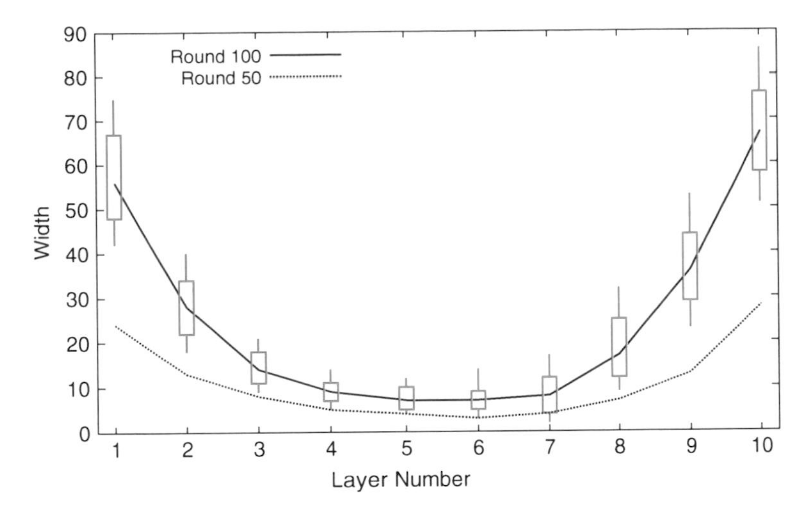

Fig. 4 The median width of each layer at round 50 and round 100 (percentiles are only shown for the latter)

specific architecture is given in Sect. 5.4. *EvoArch* is a general model for layered protocol stacks, and it does not aim to *only* capture the existing Internet architecture.

Figure 4 shows the width of each layer at the 50th and 100th rounds of the evolutionary process (typically, the network reaches 500 nodes in about 100–150 rounds). Note that, at least in terms of the median, the width decreases as we move from the bottom layer to a middle layer, around layer 5, and then it increases again as we move towards the top layer. There is some variability across evolutionary paths, however, and so we further examine if the network structure has the shape of an hourglass in every evolutionary path.

To do so, we introduce a metric that quantifies the resemblance of a layered network structure to an hourglass. Let $w(l)$ be the width of layer l, i.e., the number of nodes in that layer at a given round. Let w_b be the minimum width across all layers, and suppose that this minimum occurs at layer $l = b$; this is the *waist* of the network (ties are broken so that the waist is closer to $\lfloor L/2 \rfloor$). Consider the sequence $X = \{\{w(l)\}, l = 1, \ldots b\}$ and the sequence $Y = \{w(l)\}, l = b, \ldots L\}$. We calculate the normalized univariate *Mann-Kendall statistic for monotonic trend* on the sequences X and Y as coefficients τ_X and τ_Y, respectively [11]. The coefficients vary between -1 (strictly decreasing) and 1 (strictly increasing), while they are approximately zero for random samples. We define $H = (\tau_Y - \tau_X)/2$; H is referred to as the *hourglass resemblance* metric. $H = 1$ if the network is structured as an hourglass, with a strictly decreasing sequence of b layers, followed by a strictly increasing sequence of $L - b$ layers. For example, the sequence of layer widths $\{10,6,8,2,4,7,10,12,9,16\}$ (from bottom to top) has $w_b = 2$, τ_X = -0.67, τ_Y = 0.81, and $H = 0.74$. Note that we do *not* require the hourglass to be symmetric, i.e., the waist may not always be at the middle layer.

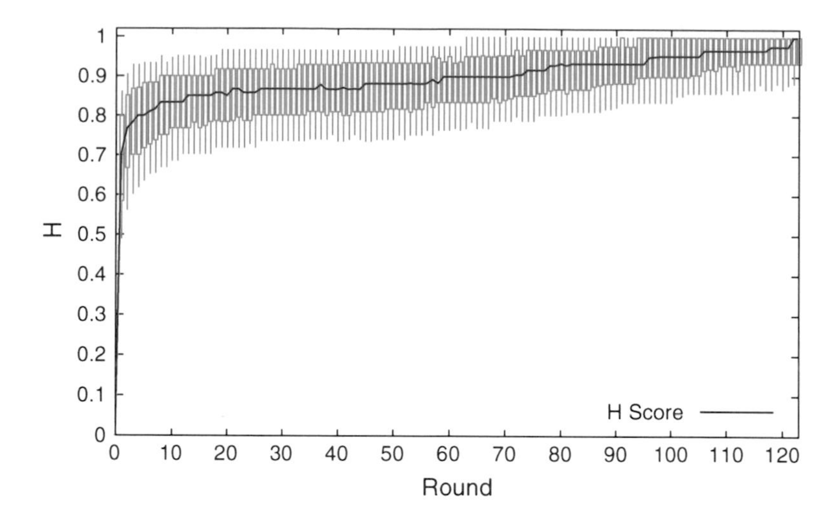

Fig. 5 The hourglass resemblance metric H over time

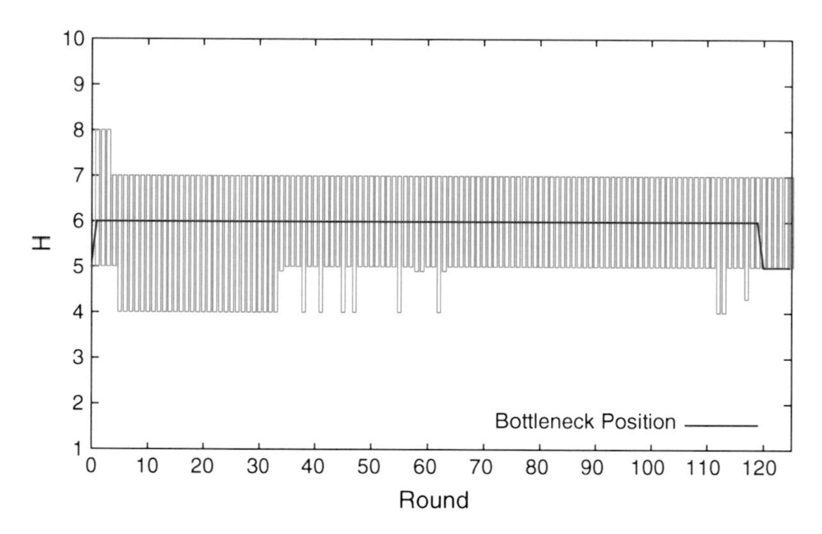

Fig. 6 The location of the waist of the hourglass over time

Figure 5 shows H (median and the previous four percentiles) as function of time. Note that it only takes few rounds, less than 10, for the median H to exceed 80%. By the 100th round, the median H is almost 95% and even the 10th percentile is more than 80%. This illustrates that *EvoArch* generates networks that typically have the shape of an hourglass. Even though the accuracy of the hourglass structure improves with time, the basic hourglass shape (say $H > 0.8$) is formed within only few rounds. Figure 6 shows the location of the waist as function of time and the associated 10th and 90th percentiles across 1,000 evolutionary paths. With the

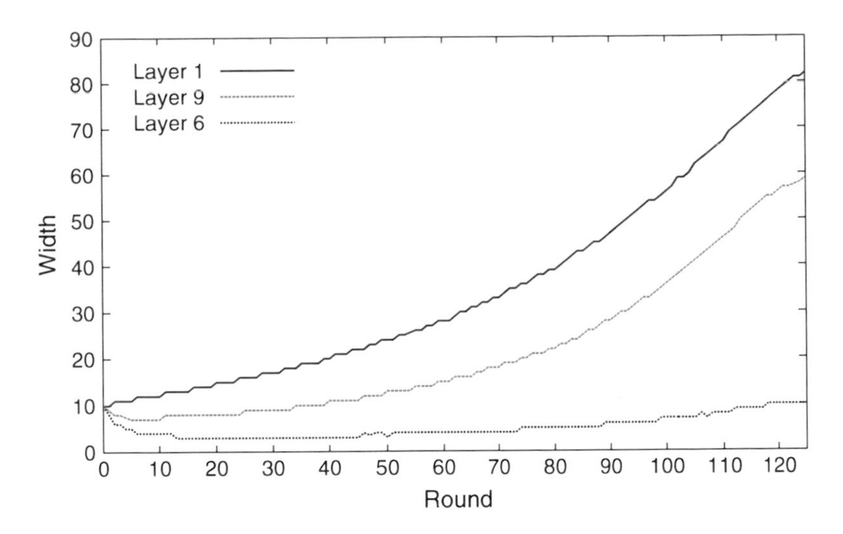

Fig. 7 The median width of three layers over time

default parameter values, the median waist is almost always located at layer 6, while the 10th and 90th percentiles correspond to layers 5 and 7, respectively. So, even though there is some small variability in the exact location of the waist across time and across different evolutionary paths, the narrowest layer of the hourglass does not fluctuate significantly.

Figure 7 shows the median width of the typical waist (layer 6), as well as the median width of layers 1 and 9, as functions of time. Even though all three layers start from the same number of nodes, layers 1 and 9 become significantly wider with time, implying a low death probability. On the other hand, the width of the waist remains relatively low compared to other layers. It typically decreases significantly in the first few rounds, as several of the initial nodes are "unlucky" in terms of products and die soon. It then slowly increases because higher layers become much wider, the birth rate increases with the size of the network, and few additional nodes at the waist can acquire significant value compared to the maximum-value node in that layer.

Obviously, the major question is: *why does EvoArch generate hourglass-shaped networks?* In Sect. 8, we present a mathematical insight on how an hourglass pattern is formed in *EvoArch*, taking into account some assumptions that simplify it significantly. Let us discuss here separately what happens at layers close to the top, close to the bottom, and close to the waist.

Because the generality probability $s(l)$ is quite low at layers close to the top, those nodes typically have a small number of products. This means that they rarely compete with each other, and so the death probability is close to zero. For instance, in the application layer a new protocol can compete and replace an incumbent only if the former provides a very similar service with the latter (e.g., recall the example with FTP and HTTP).

At layers close to the bottom, the generality probability is close to one, and so those nodes have many shared products and thus several competitors. Their value is often similar, however, because those nodes typically share almost the same set of products. Thus, the death probability at layers close to the bottom is also quite low.

At layers close to waist, where the generality probability is close to 50%, the variability in the number of products is maximized-recall that the variance of a Bernoulli random variable $X(p)$ is maximum when $p = 50\%$. So, few nodes in that layer may end up with a much larger number of products than most other nodes in the same layer and so with a much higher value. Those nodes would compete with most others in their layer, often causing the death of their competitors. In other words, the death probability at bottom and top layers is quite low, while the death probability close to the waist is higher. The birth rate, on the other hand, is the same for all layers, and so the network's middle layers tend to become narrower than the bottom or top layers.

The reader should not draw the conclusion from the previous simplified discussion that the waist is always located at the layer with $s(l) = 0.5$. As will be shown in Sect. 5, the competition threshold c also affects the location of the waist. Also, it is not true that the node with the maximum value at the waist never dies. Section 6 focuses on these "extraordinary" nodes, showing that, even though they live much longer than almost all other nodes in their layer, under certain conditions they can also die.

4 Robustness

In this section, we focus on the robustness of the hourglass resemblance metric H with respect to the parameters of the *EvoArch* model. The robustness study has two parts. First, we show that wide deviations from the default value, for a single parameter at a time, do not cause significant changes in H. Second, we show that even if we simultaneously and randomly vary all *EvoArch* parameters, the model still produces hourglass-like structures with high probability.

Let us first focus on the three most important *EvoArch* parameters: the competition threshold c, the generality probability vector s, and the mortality parameter z. We have also examined the robustness of H with respect to the number of layers L, the birth rate, the number of initial nodes at each layer, or the stopping criterion, but those parameters have a much smaller impact on H.

Figure 8a shows the median H score (together with the previous four percentiles) as we vary c between 0 and 1. The value $c = 0$ corresponds to "global" competition, meaning that two nodes of the same layer compete with each other even if they do not share any products. When the competition threshold is so low, the death probability becomes significant even at higher layers, as those nodes start competing even without sharing many products. Thus, the upper part of the network deviates from the hourglass shape.

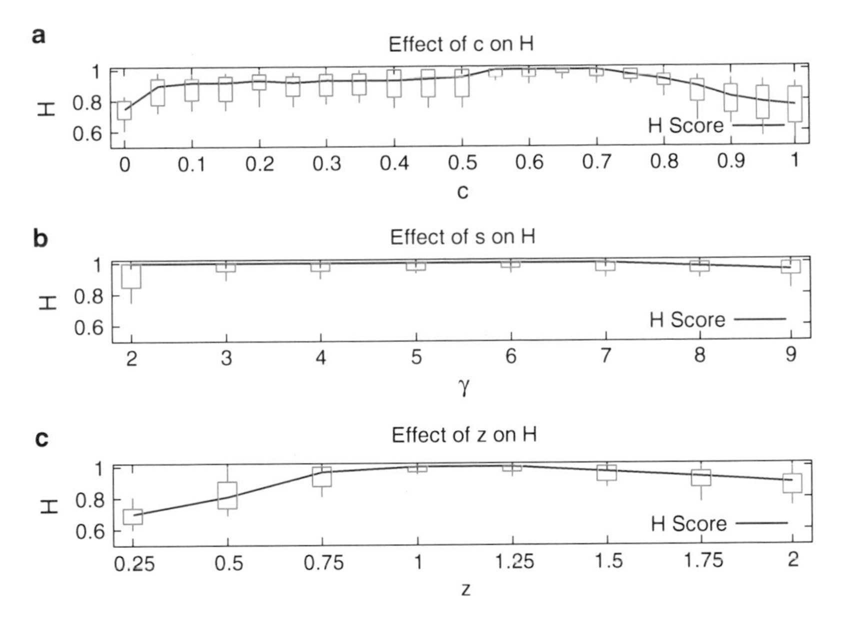

Fig. 8 The hourglass resemblance score H as a function of the competition threshold c, the layer γ at which the generality is 50%, and the mortality parameter z

When $c = 1$, on the other hand, a node u competes with node w only if the former shares *all* products of w. This means that nodes rarely compete, and so most layers grow randomly, without a significant death probability. There is a wide range of c (between 0.1 and 0.9) in which we get reasonably good hourglass structures ($H > 0.8$) in most evolutionary paths. The best results, however, are produced in the range $0.5 < c < 0.8$. *In that range, almost all evolutionary paths produce* $H > 0.9$.

To study the robustness of the model with respect to the generality vector s, we consider a function $s(l)$ comprising of two linear segments that satisfy the constraints: $s(1) = 0.9$, $s(\gamma) = 0.5$, and $s(L) = 0$, where γ is any layer between layer 2 and layer (L−1). This function allows us to place the layer γ at which the generality probability is 50% at any (interior) layer of the architecture. Figure 8b shows H as we vary the layer γ. The model is extremely robust with respect to variations in the generality vector s and layer γ.

Figure 8c shows H as we vary the mortality parameter z. We limit the range of z to less than 2.0 so that the death probability is almost zero only if the value ratio r is close to one; this is not true for higher values of z (see Fig. 3). Note that H is typically higher than 0.9 when $0.75 < z < 1.5$. For lower values of z, the death probability becomes so high that only the most valuable node tends to survive in several layers. When z is higher than 1.5, the death probability becomes too low and several layers grow randomly.

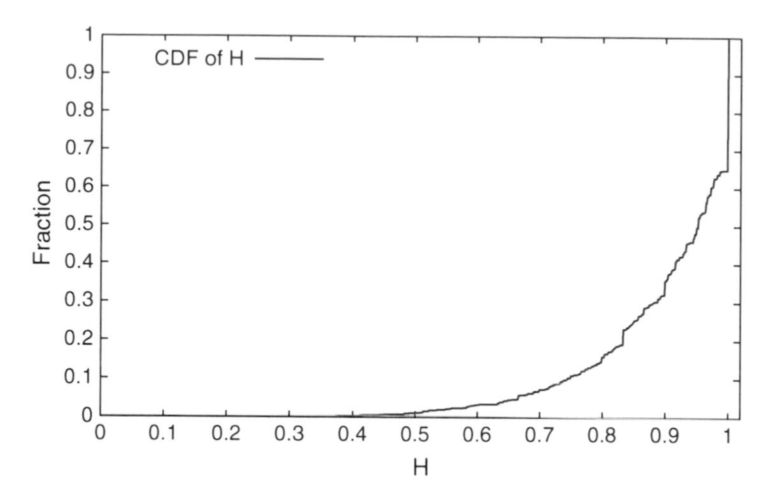

Fig. 9 CDF of the hourglass resemblance score H when all parameters vary randomly in a certain range (see text)

In the previous experiments, we varied one parameter at a time. We now examine the robustness of the model when we randomly sample each parameter value simultaneously from a certain range (a Monte Carlo method). Together with the previous three parameters (c, s, and z), we also consider here variations in the number of layers L, the random number of initial nodes n_0 separately at each layer, the birth rate μ, and the maximum network size N_{max} at the end of an evolutionary path. A subtle point here is that, as L increases, we need to make sure that N_{max} is also increased (with a larger number of layers, the network should be allowed to grow larger). For this reason we set $N_{max} = \eta L$ and vary the factor η instead of N_{max}. We consider the following range for each parameter: $0.25 \leq c \leq 0.75$, $3 \leq \gamma \leq L - 2$, $0.75 \leq z \leq 1.5$, $5 \leq L \leq 15$, $1\% \leq \mu \leq 10\%$, $5 \leq n_0 \leq 20$, and $25 \leq \eta \leq 55$.

We generate 1,000 evolutionary paths, each with a randomly chosen value for all previous parameters. The CDF of the hourglass resemblance scores is shown in Fig. 9. Even when we vary all parameters randomly in the given ranges, the score H is still higher than 0.9 in 68% of the evolutionary paths and higher than 0.75 in 90% of the evolutionary paths. We manually examined some evolutionary paths in which the score H is lower than 0.5. They typically result from "bad" combinations of parameter values. For instance, a large value of c in combination with a large value of z severely suppresses deaths in all layers, allowing the network to grow randomly. Or a small value of c pushes the waist towards higher layers, while a small γ pushes the waist towards lower layers, causing deviations from the basic hourglass shape (e.g., a double hourglass shape with two waists).

5 Location and Width of Waist

In this section, we focus on the effect of the three major *EvoArch* parameters (competition threshold c, generality vector s, and mortality parameter z) on the location and width of the waist.[4] We also estimate the value of these three parameters in the case of the current Internet architecture (TCP/IP stack) and discuss several implications about the evolution of the latter and its early competition with the telephone network. We also discuss how to design a new architecture so that it has higher diversity (i.e., larger width) at its waist compared to the TCP/IP stack. In Sect. 8, we also compare the location of the waist obtained through simulations with those predicted by our mathematical analysis.

5.1 *Effect of Competition Threshold*

Figure 10 shows the location and width of the waist as c increases. Recall from Sect. 4 that the model produces high values of H when c is between 0.1 and 0.9. As c increases in that range, the waist moves lower and its width increases. The competition threshold c quantifies how similar the services or products of two protocols must be before they start competing. As c increases, it becomes less likely that two nodes compete. Especially at higher layers, where the generality is low and nodes have few products, increasing c decreases the frequency of competition and

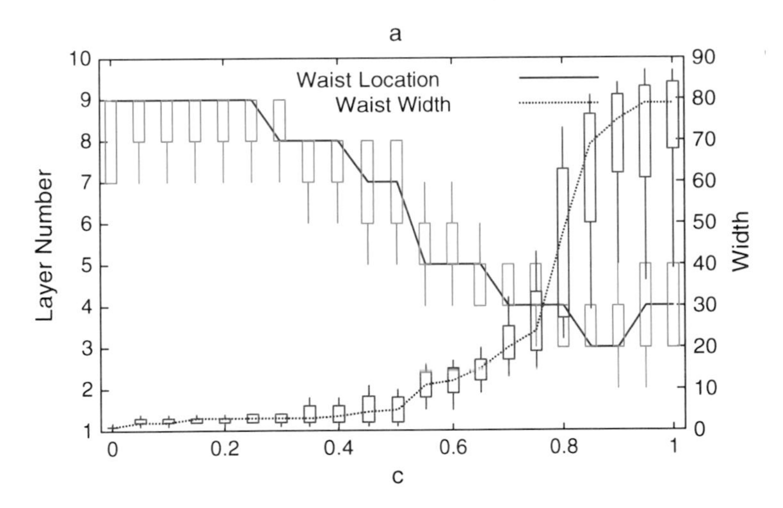

Fig. 10 Location and width of the waist as a function of competition threshold c

[4]Any parameter we do not mention is set to the default value given in Sect. 3.

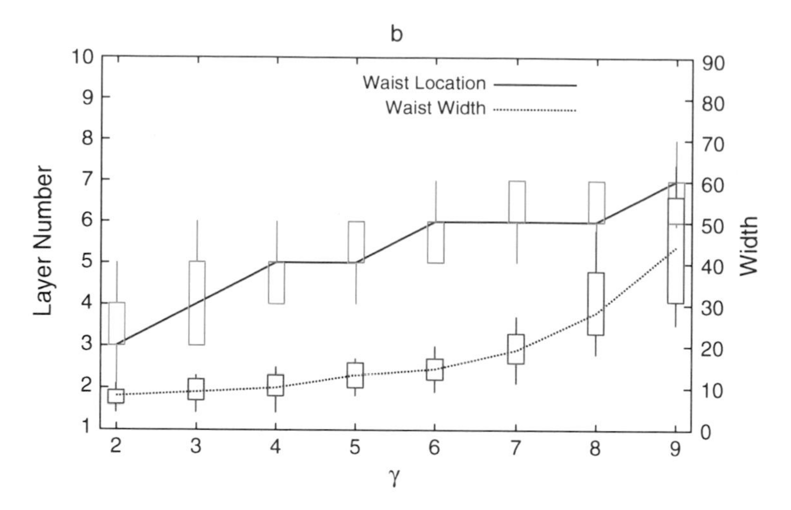

Fig. 11 Location and width of the waist as a function of the layer γ with 50% generality

thus the death probability. This means that those higher layers grow faster. Nodes at lower layers, where $s(l)$ is close to one, have many overlapping products and so they are less affected by c. Thus, *increasing c pushes the waist towards lower layers.* The same reasoning (increasing c decreases the death probability) explains why *the waist becomes wider as c increases.*

5.2 *Effect of Generality Vector*

As in Sect. 4, we focus on a two-segment piecewise linear generality vector: the first segment extends between $s(1) = 0.9$ and $s(\gamma) = 0.5$, and the second extends between $s(\gamma)$ and $s(L) = 0$. This function allows us to control the layer at which the generality is 50% (and the variance of the number of products is maximized) by modifying the parameter γ. Figure 11 shows the location and width of the waist as γ increases from layer 2 to layer 1 (L−1). Recall that *EvoArch* produces high hourglass resemblance scores throughout that range. The general observation is that *as γ increases, the location of the waist increases.* It is important, however, that *the location of the waist is* not *exactly equal to γ*; in other words, the variance in the number of products is not sufficient to predict the layer at which the death probability is highest (and the width is lowest). The competition threshold c also influences the location of the waist, as previously discussed.

As γ increases, the width of the waist also increases. The reason is that the location of the waist moves to layers with larger generality. For instance, Fig. 11 shows that when $\gamma = 5$, the median waist is also at layer 5, while when $\gamma = 8$, the median waist is at layer 6. Thus, when $\gamma = 5$, the generality of the waist is 50%, while when $\gamma = 8$, the generality of the waist is approximately 61%. Higher

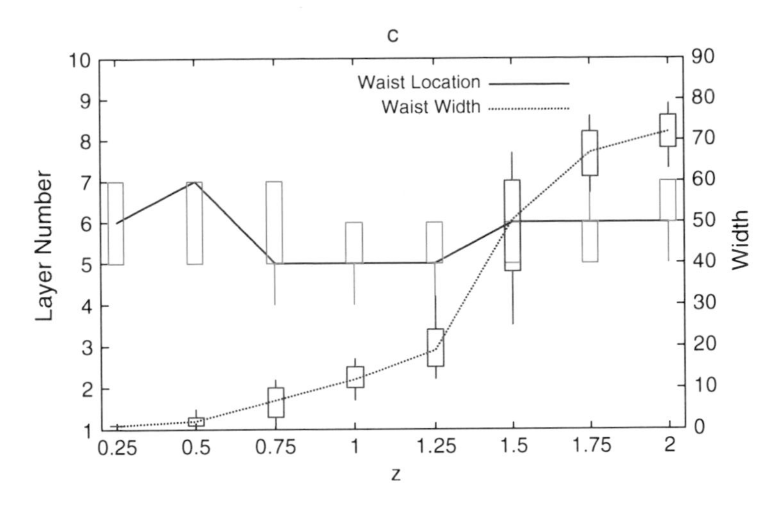

Fig. 12 Location and width of the waist as a function of the mortality parameter z

generality, however, means a larger number of products for new nodes at the waist, a higher evolutionary value relative to the node with the maximum number of products in that layer and thus a higher probability of survival.

5.3 *Effect of Mortality Parameter*

Recall that z controls the shape of the death probability (see Fig. 3), with lower values of z causing more lethal competition. Figure 12 shows the location and width of the waist when z varies between 0.25 and 2.0. As expected, *as z increases, the width of the waist increases*-the reason is that the death probability decreases, allowing more nodes to survive even though they compete with other nodes. On the other hand, *the parameter z does not have a significant effect on the location of the waist.*

5.4 *Implications for the TCP/IP Stack*

In the current Internet architecture, the waist is located at the network layer and so it is practically at the midpoint of the protocol stack (see Fig. 1). Further, the waist is very narrow: just one dominant networking protocol (IPv4) and two major transport protocols (TCP and UDP). We have estimated a good parameterization of the *EvoArch* model for the case of the TCP/IP stack (based on trial and error and also exploiting the trends shown in Figs. 10, 11, and 12). The values are as follows: $L = 6$ (we distinguish between application-layer protocols such as HTTP at layer 5 and individual applications such as Firefox at layer 6), $c \approx 0.7$, $\gamma = 3$, and $z \approx 0.3$. With these parameter values the waist is almost always located at layer 3 and it consists of only few nodes (typically less than three). The median H score is 1, and the 10–90th percentiles are 0.66 and 1, respectively.[5]

What do these parameter values imply about the evolutionary characteristics of the current Internet architecture? In terms of the parameter c, a competition threshold around 70% implies that two protocols can coexist in the TCP/IP stack as long as their relative product overlap (see Eq. (2)) is no more than about 70%; otherwise, at least one of them will compete with the other. A good example of two protocols that coexist at the same layer with little overlap in their services and functionality are TCP and UDP. The reason is that one of them is mostly used by applications that require reliability, while the other is chosen by a largely non-overlapping set of applications that prefer to avoid TCP's retransmissions, congestion control, or byte-stream semantics. It is only few applications (e.g., DNS or Skype) that use both TCP and UDP.

The low value of z (approximately 0.3) implies that competition between protocols at the TCP/IP stack is very intense: *a protocol can survive only if its value is higher than about 90% of the value of its strongest component!* A good survival strategy for a new protocol u would be to *avoid competition* with the highest-value protocol in that layer w. This can be achieved if u has largely non-overlapping products with w; in other words, *the new protocol should try to provide mostly different services or functionality than the incumbent.* The relatively high value of c (70%) means that a significant degree of service overlap would be tolerated, making it easier for the new protocol to also support some of the legacy applications.

The previous point also suggests an intriguing answer to a historical question. *How can we explain the survival of the TCP/IP stack in the early days of the Internet, when the telephone network was much more powerful?* During the 1970s or 1980s, the TCP/IP stack was not trying to compete with the services provided by the telephone network. It was mostly used for FTP, E-mail, and Telnet, and those services were not provided by the incumbent (telephone) networks. So, TCP/IP managed to grow and increase its value without being threatened by the latter. In the last few years, on the other hand, the value of the TCP/IP protocols has exceeded the value of the traditional PSTN and cable-TV networks, and it is now in the process of largely replacing them in the transfer of voice and video.

In terms of the parameter s, the fact that the waist of the TCP/IP stack is located at the network layer implies that the generality of that layer is close to 50%. This means that a new protocol at the network layer would see the highest variability (i.e., maximum uncertainty) in terms of whether it will be selected as substrate from protocols at the next higher layer. So, from an architect's perspective, the network layer of the TCP/IP stack is the layer at which a new protocol would experience the maximum uncertainty in terms of deployment and ultimate success.

---

[5] A corresponding parameterization using the more realistic death-regulated birth process is given in Sect. 6.

5.5 *Future Internet Architectures*

EvoArch also gives some interesting insights about the evolvability of future Internet clean-slate architectures. Suppose that a network architect would like to ensure that there is more diversity (i.e., larger width) in the waist of a new architecture compared to the TCP/IP stack-this goal has been suggested, for instance, by Peterson et al. [14]. *How can the network architect increase the likelihood that the evolution of a new architecture will lead to a wider waist, with several surviving protocols?* Based on the previous results, this will happen if we increase z-it is unlikely, however, that a network architect can control the intensity of competition; that is largely determined by economic and deployment considerations. A second and more pragmatic approach is to design protocols that are largely non-overlapping in terms of services and functionality, as previously discussed, so that they do not compete with each other. This approach was discussed in the previous section.

A third approach is to design the architecture so that its waist is located at a layer with a high generality. As we saw in Fig. 11, as we increase γ, increasing the generality of all layers, the waist moves higher, at a layer with higher generality. This also means that the waist is getting wider, allowing the coexistence of several protocols at that layer. How can this be done in practice? Suppose that we start from a 6-layer architecture X in which the waist is located at layer 3, and we want to redefine the functionality of each layer so that the waist of the new architecture Y is located at a higher layer. We should increase the generality of each layer (but still maintaining that $s(l)$ decreases as l increases) so that the corresponding protocols provide more general services in Y than in X. For instance, instead of defining HTTP as an application-specific protocol that is only used by web browsers, HTTP can be redefined and used as a very general content-centric delivery protocol. This specific example, actually, has been recently proposed as a rather simple way to provide the benefits of clean-slate content-centric architectural proposals using an existing protocol [15].

6 Evolutionary Kernels

It is often said that the core protocols of the Internet architecture (mostly IPv4, TCP, and UDP) are "ossified," meaning that they are hard to modify or replace, creating an obstacle for network innovations [14]. At the same time however, they can be viewed more positively as the protocols that form the core of the architecture, creating a common interface between a large number of nodes at lower layers and a large number of nodes at higher layers. This is why we refer to them as *evolutionary kernels*, based on a similar concept about certain genes and gene regulatory networks in biology [10]. What can we learn from *EvoArch* about such ossification effects and evolutionary kernels? *Does the model predict the emergence*

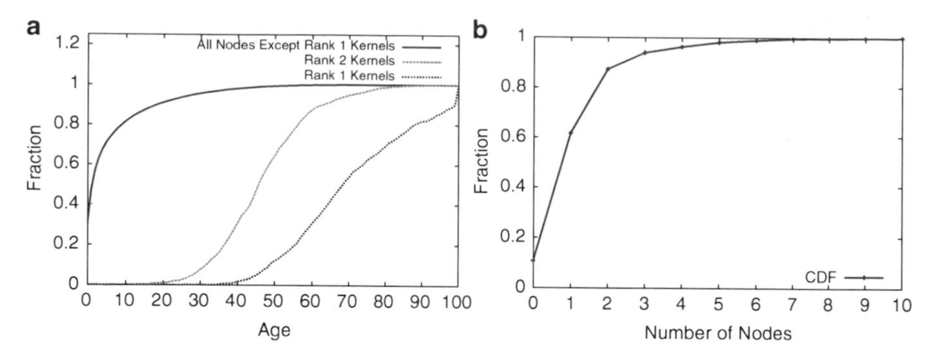

Fig. 13 (**a**) CDF of the age of various node subsets at layer 6 and (**b**) CDF of the number of nodes at layer 6 surviving more than half of the evolutionary path

of long-surviving nodes at the hourglass waist? What is the reason that those nodes manage to survive much longer than their competitors? Do they ever get replaced by other nodes, and if so, under what conditions?

Let us focus on the waist-under the default parameters ($L = 10$), the waist is typically at layer 6. As previously discussed, the waist has the highest death probability, and so one may expect that it is unlikely to find any long-living nodes at that layer. We generate 1,000 evolutionary paths, each lasting 100 rounds. At the end of the evolutionary path, we calculate the *maximum age* among all nodes that were ever born at layer 6. Figure 13a shows the CDF of the maximum age for various subsets of nodes: (a) *the node with the maximum age* (we refer to such nodes as *rank 1 kernels* or simply *kernels*), (b) the second older node (we refer to them as *rank 2 kernels*), and (c) all nodes, excluding only rank 1 kernels.

Note that almost all (rank 1) kernels survive for at least 50–60% of the entire evolutionary path. Actually, about 40% of the kernels are still alive at the end of the evolutionary path, meaning that their age is only determined by their birth time. On the other hand, the remaining nodes have much shorter life span. About 40% of them do not survive for more than a round or two, and 90% of them survive for less than 20 rounds. The rank 2 kernels have much larger age than most nodes, but there is still a significant gap in the age of rank 1 and rank 2 kernels. So, our first observation is that *EvoArch predicts the emergence of very stable nodes at the waist that tend to survive for most of the evolutionary path.*

Figure 13a shows that only a small fraction of nodes survive for more than 50 rounds in an evolutionary path. So, let us identify in each evolutionary path those few nodes that survive for at least that long-we refer to them as "higher-rank kernels." Figure 13b shows the CDF of the number n of higher-rank kernels in each evolutionary path. In about half of the evolutionary paths only the rank 1 kernel exists. In almost all cases, $n \leq 4$. So, the number of nodes that can coexist with the rank 1 kernel for more than 50 rounds is typically at most three. This confirms that it is difficult to survive for long at the waist in the presence of a rank 1 kernel. The nodes that manage to do so either have almost the same set of products with

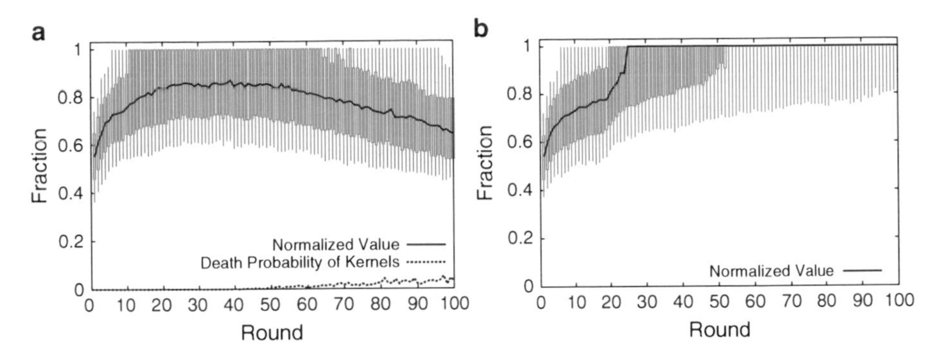

Fig. 14 Normalized value for kernels at layer 6 (**a**) with basic birth process and (**b**) with death-regulated birth process

the rank 1 kernel (and thus almost the same value) or they have mostly different products than the rank 1 kernel, not competing with it. Examining the birth times of those nodes, we can observe that in about 70% of the evolutionary paths, rank 1 kernels are born earlier than higher-rank kernels.

How large is the value of a kernel, and how can a kernel die? We define the *normalized value* $\hat{v}(u)$ of a node u at a given round as its value $v(u)$ divided by the value that u would have *if it was connected to all products at the next higher layer* in that round. So, $\hat{v}(u) \leq 1$. Note that because the death probability is almost zero when $r > 0.90$ (for the default value of z), if the normalized value of a node is higher than 90%, that node cannot die even if it had a higher-value competitor.

Figure 14a shows the normalized value of all (rank 1) kernels, based on 1,000 evolutionary paths. In the first 10–20 rounds, the normalized value increases as the upper layer grows larger and the kernels acquire new products. Then, during the next 30–40 rounds, their normalized value varies around 80–90%, which means that those kernels are unlikely to die, even if they face competition. During the last 30–40 rounds, however, the normalized value of many kernels gradually drops. To understand this trend and to explain how a kernel can be replaced, we have to examine the birth process. As time progresses, the upper layers of the network grow larger (recall that the death probability is low at higher layers due to low competition). In the basic birth model, however, the birth rate is proportional to the size of the network, and new nodes are distributed uniformly across all layers. Thus, the birth rate at the waist also increases with time.

The previous increase has two consequences. First, as the layer above the waist increases in size, new potential products appear for both the kernel and its competitors. Each of these new nodes will select the kernel as substrate with probability $s(l_b)$, where l_b is the waist. Second, as the birth rate at the waist increases, it becomes more likely that a new node at that layer will acquire enough new products so that its value becomes comparable, or even higher, than the value of the kernel. In other words, the death of kernels is largely due to the birth of several new nodes at the next higher layer: *if the kernel fails to quickly acquire most of those new potential products at the next higher layer, it will experience a decrease in its normalized value, becoming more vulnerable to new or existing competitors at its own layer.*

The previous discussion raises the question: *what if the birth rate is not the same at all layers?* Specifically, what if the birth rate at a layer is negatively correlated with the death probability at that layer? This modification of the model, which we refer to as *death-regulated birth process*, captures situations in which the implementation or deployment of new protocols at a certain layer is discouraged by the intense competition that one or more incumbent protocols create at that layer. Arguably, this is a more realistic model than the basic birth model we considered earlier.

In the death-regulated birth process, we maintain an estimate of the death probability $\tilde{d}(l)$ at layer l since the start of the evolutionary path. As in the basic birth process, the overall birth rate is proportional to the network size, and the allocation of births to layers is random. However, a birth at layer l is successful with probability $1 - \tilde{d}(l)$; otherwise, the birth fails and it is counted as a death in that layer.

The death-regulated birth process creates a positive feedback loop through which the emergence of one or more kernels at the waist reinforces their position by decreasing the rate at which new nodes (and potential competitors) are created. Figure 14b shows the normalized value of the rank 1 kernel at layer 6, when we switch from the basic birth model to the death-regulated birth model after round 20. Note that the median normalized value (as well as the 25th percentile) of the kernel becomes 100% within just few rounds. In other words, *with a death-regulated birth process, it becomes practically impossible to replace a kernel at the waist of the hourglass.* Even when a new node u is somehow successfully born at the waist, the number of new nodes at the next higher layer is limited (because the birth process is also death-regulated at that layer) and so u's products will be most likely shared by the kernel. This means that u will face the kernel's competition, and so u will most likely die.

We can estimate a good parameterization of the *EvoArch* model for the case of the TCP/IP stack using the death-regulated birth process (again based on trial and error and exploiting the trends shown in Figs. 10, 11, and 12): $L = 6$, $c \approx 0.7$, $\gamma = 3$, and $z \approx 0.5$. With these values the waist is located at layer 3, its median width is one node, the median width of layer 4 is four nodes, while the width of the remaining layers increases with time. The median H score is 1 and the 10–90th percentiles are 0.66 and 1, respectively.

6.1 Kernels in the Internet Architecture

There are several interesting connections between what *EvoArch* predicts about kernels and what happens in the Internet architecture. There is no doubt that IPv4, as well as TCP and UDP, are the kernels of the evolving Internet architecture.

They provide a stable framework through which an always expanding set of physical and data-link-layer protocols, as well as new applications and services at the higher layers, can interoperate and grow. At the same time, those three kernel protocols have been difficult to replace or even modify significantly. Further, the fact that new network or transport layer protocols are rarely designed today implies that the birth process at those layers is closer to what we call "death-regulated," i.e., limited by the intense competition that the kernel protocols create.

EvoArch suggests an additional reason that IPv4 has been so stable over the last three decades. Recall that a large birth rate at the layer *above the waist* can cause a lethal drop in the normalized value of the kernel, if the latter is not chosen as substrate by the new nodes. In the current Internet architecture, the waist is the network layer, but the next higher layer (transport) is also very narrow and stable. So, the transport layer acts as an *evolutionary shield* for IPv4 because new protocols at the transport layer are unlikely to survive the competition with TCP and UDP. On the other hand, a large number of births at the layer above TCP or UDP (application protocols or specific applications) are unlikely to significantly affect the value of those two transport protocols because they already have many products. In summary, the stability of the two transport protocols adds to the stability of IPv4 by eliminating any potential new transport protocols that could select a new network-layer protocol instead of IPv4.

In terms of future Internet architectures, *EvoArch* predicts that even if these architectures do not have the shape of an hourglass initially, they will probably do so as they evolve. When that happens, the emergence of new ossified protocols (kernels) will be a natural consequence. If the architects of such clean-slate designs want to proactively avoid the ossification effects that we now experience with TCP/IP, they should try to design the functionality of each layer so that the waist is wider, consisting of several protocols that offer largely distinct but general services, as discussed in Sect. 5.5.

7 Quality Differentiation

So far we have assumed that the value of a protocol is only determined by the value of its products. It would be more realistic, however, to consider that the evolutionary value of a protocol also depends on other factors, which we refer to as *quality*. The "quality factor" should be interpreted broadly; it can capture properties such as performance, extent of deployment, reliability or security, clarity of the corresponding specification, or other features. The quality factor also allows *EvoArch* to capture the effect of incremental improvements in existing protocols: such improvements do not create a new node in the model, but they increase the quality parameter of an existing node. In the following, we assume that the quality factor of a node is constant-an interesting extension of the model will be to consider time-varying quality factors.

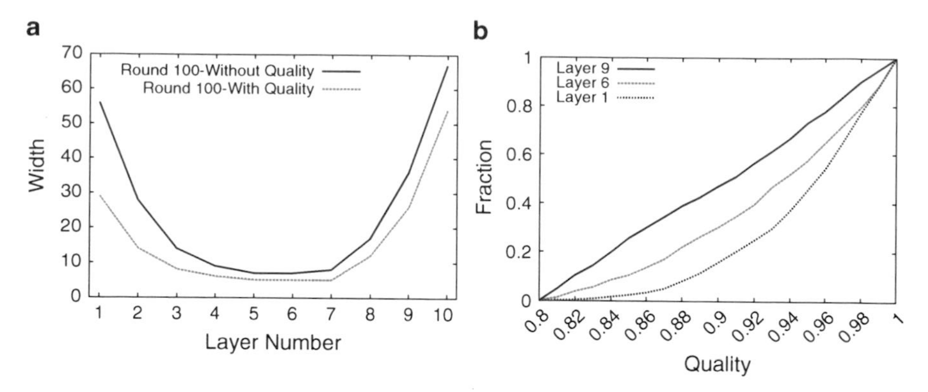

Fig. 15 (**a**) The median width of each layer when nodes have different quality factors (compared to the case that all nodes have the same quality). (**b**) The CDF of the quality factor $q(u)$ of the kernel node at the waist (layer 6). The corresponding CDFs for the oldest node at layer 1 and layer 9 are also shown

In this section, we conduct a simple extension to the *EvoArch* model so that each protocol u has a certain quality factor $q(u)$. We are mostly interested in two questions. First, *how does this quality differentiation affect the shape of the resulting architecture?* And second, *focusing on the kernel nodes at the waist, do they tend to be nodes with the highest quality factor?* The quality of a node u is represented by a multiplicative factor $q(u)$, uniformly distributed in $[q_{min}, 1]$ (with $0 < q_{min} < 1$). The value of node u is then the product of the quality factor $q(u)$ with the quality-independent value of u:

$$v(u) = \begin{cases} q(u) \sum_{p \in P(u)} v(p) & l(u) < L \\ q(u) & l(u) = L \end{cases} \tag{5}$$

7.1 *Effect of Quality Differentiation*

We have repeated the analysis of Sect. 3 with the previous model extension, with $q_{min} = 0.8$. Figure 15a shows the median width of each layer with the default parameter values, based on 1,000 evolutionary paths. *The resulting network continues to have the hourglass shape* (decreasing width up to a certain layer and then increasing width), even when nodes have different quality factors; the median H score is 1.0, and the 10th and 90th percentiles are 0.9 and 1.0, respectively. The location of the waist does not change as a result of the quality factor heterogeneity.

There are two interesting differences, however. First, *the network grows at a slower pace.* This is a result of the increased death probability because of nodes with low-quality factor. Second, *the lower part of the hourglass, below the waist, is now smaller in size than the upper part of the hourglass.* The reason is that nodes at lower layers have more competitors (due to higher generality at those layers) than nodes at higher layers. When all nodes have the same quality, nodes at lower layers compete widely, but they usually survive the competition because their product sets are similar, and thus, their values are similar. With heterogeneous qualities, on the other hand, the value of nodes at lower layers can be significantly different, increasing the death probability compared to the case of homogeneous qualities. This increased death probability makes the size of the lower part of the hourglass smaller than the upper part.

It is interesting that the TCP/IP stack has a similar asymmetry, with the bottom part of the hourglass being smaller in terms of size than the upper part. *EvoArch* offers a plausible explanation for this effect: the heterogeneity among different protocols at the same layer, in terms of performance, security, extent of deployment, etc., increases the death probability at lower layers more than at higher layers.

7.2 *Quality of Kernel Nodes*

Section 6 focused on the evolutionary kernels at the waist of the hourglass. In the case of heterogeneous qualities, *an important question is whether the kernels tend to also have the highest quality factor.* In other words, *can we expect that the "best" nodes (i.e., highest quality factor) are also the strongest nodes in terms of value?*

Figure 15b shows the CDF of the quality factor of the kernel node at the waist (layer 6), based on 1,000 evolutionary paths. As a reference point, we also show the CDF of the quality factor of the oldest node at layer 1 and at layer 9.

If the quality of a node has no effect on its death probability, we would expect that the CDF of the quality factor of the oldest node would be a straight line between 0.8 and 1 (recall that the quality factor varies uniformly in that range). This is the case at layer 9. In that layer, which is almost at the top of the stack, nodes rarely compete with each other because of their low generality and small number of products. So, their quality does not influence their age, and the quality factor of the oldest node varies randomly between 0.8 and 1.

At the bottom layer (layer 1), on the other hand, we observe a strong bias towards higher quality values. As discussed earlier in this section, the nodes at the bottom layer have high generality, many competitors, and so their quality factor strongly affects their value and death probability. *It is mostly high-quality nodes that survive at the bottom layers.*

At the waist, where only few kernel nodes survive for most of the evolutionary path, the bias towards higher quality values is much weaker. In about 30% of the evolutionary paths the quality factor of the kernel is less than 0.9, which means that *those kernels are in the bottom half of all nodes in terms of quality.* Similarly, the probability that a node with high-quality factor (say $q(u) > 0.95$) becomes the kernel of the waist is only 40%.

7.3 Implications for IPv4 and IPv6

In the Internet architecture, it would be hard to argue that IPv4 has dominated at the network layer because it was an excellent protocol in terms of design. During the last 30–40 years, several other network-layer protocols have been proposed, and probably some of them were better than IPv4 in several aspects. *EvoArch* gives us another way to understand the success of IPv4 and to also think about future protocols that could potentially replace it. It will help if such future protocols are better than IPv4; that is not a sufficient condition to replace IPv4, however. If the potential replacements attempt to directly compete with IPv4, having a large overlap with it in terms of applications and services but without offering a major advantage in terms of the previous "quality factor," it will remain difficult to replace IPv4.

What does EvoArch suggest about IPv6 and the difficulty that the latter faces in replacing the aging IPv4? We should first note that IPv6 does not offer new services compared to IPv4; it mostly offers many more addresses.[6] This means that IPv6 has, *at most*, the same products with IPv4, and so the latter is its competitor. Further, because IPv6 is not widely deployed, it is reasonable to assume that its quality factor is much lower than that of IPv4. So, even if the two protocols had the same set of products, IPv4 has much higher value and it wins the competition with IPv6. The situation would be better for IPv6 under two conditions. First, if IPv6 could offer some popular new services that IPv4 *cannot* offer-that would provide the former with additional products (and value) that the latter does not have. Second, IPv6 should avoid competition with IPv4, at least until it has been widely deployed. That would be the case if IPv6 was presented, not as a replacement to IPv4, but as "the second network-layer protocol" that is required to support the previous new services.

8 Emergence of Hourglass Structures: An Analytical Insight

In this section we mathematically analyze a simpler version of *EvoArch* and derive an expression for the death probability ratio $r(l)$ for a node with the average number of products at layer l. Numerically, that expression suggests that the ratio $r(l)$ has a unique minimum at a certain layer $\hat{l}$ that only depends on the generality probabilities and the competition threshold. Because the death probability decreases monotonically with $r(l)$ (see Fig. 3), the previous observation means that the death probability has a unique maximum at layer $\hat{l}$, and it decreases monotonically at

[6]The original proposals for IPv6 included several novel services, such as mobility, improved auto-configuration, and IP-layer security, but eventually, IPv6 became mostly an IPv4-like protocol with many more addresses.

layers above and below $\hat{l}$. It is this death probability pattern that pushes, over several evolutionary rounds, the network to take the shape of an (generally asymmetric) hourglass with a waist at layer $\hat{l}$. Using the expression for $r(l)$, we can also predict the location of the waist of the resulting hourglass structure and compare it with the simulation results in Sect. 5.

With a simplified version of the model, we can derive the death probability ratio $r(l)$ for an "average node" at layer l, i.e., for a node with the mean number of products at layer l.

We can do so under several assumptions that simplify the *EvoArch* model significantly:

1. The network is static (i.e., non-evolving).
2. Each layer has the same number of nodes n.
3. $n\,s(L-1) \gg 1$ (recall that $s(L-1)$ is the lowest nonzero generality across all L layers).
4. A node can only have one competitor: the node with the largest number of products at that layer.

The following analysis focuses on two nodes at each layer l: a node u with the mean number of products at layer l and a node u_m with the expected value of the *maximum* number of products at layer l. Based on assumption (4), the death probability of node u depends on the ratio $r(l) = v(u)/v(u_m)$ between the value of node u and the value of node u_m. Let $\rho(l)$ be the number of products of node u and $\rho_m(l)$ the number of products of node u_m. We need a fifth assumption regarding nodes u and u_m:

$$\frac{v(u)}{v(u_m)} = \frac{\rho(l)}{\rho_m(l)} \tag{6}$$

Based on the previous five assumptions, we derive next the following mathematical expression for the death probability ratio $r(l)$:

$$r(l) = 1 - \left(1 - \frac{\lfloor n\,s(l)\rceil}{\lfloor \rho_m(l)\rceil}\right) \sum_{i=\lceil c\times \lfloor n\,s(l)\rceil\rceil}^{\lfloor n\,s(l)\rceil} \frac{\binom{\lfloor \rho_m(l)\rceil}{i}\binom{n-\lfloor \rho_m(l)\rceil}{\lfloor n\,s(l)\rceil - i}}{\binom{n}{\lfloor n\,s(l)\rceil}} \tag{7}$$

where $\lfloor x\rceil$ represents the "round to closest integer" operator, while

$$\rho_m(l) = \sum_{i=0}^{n} i\left(F(i)^n - F(i-1)^n\right) \tag{8}$$

where

$$F(x) = \sum_{i=0}^{x} \binom{n}{i} s(l)^i \left(1 - s(l)\right)^{n-i}. \tag{9}$$

Unfortunately the previous expression for $r(l)$ is not tractable, and it does not allow us to examine whether $r(l)$ has a unique minimum. We observed numerically,

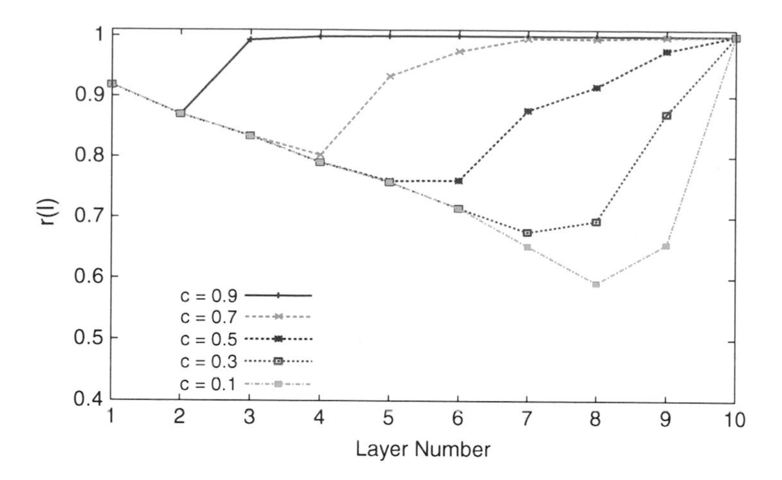

Fig. 16 The death probability ratio $r(l)$ as a function of the layer l for different values of the competition threshold c. The parameters are $L = 10, n = 50$, and $s(l) = 1 - l/L$

however, that the ratio $r(l)$ has a unique minimum at a certain layer $\hat{l}$ that only depends on the generality probabilities and the competition threshold (see Fig. 16).

In the following, we derive the expression for $r(l)$ given in (7).

Proof

Based on the earlier definition of nodes u and u_m at layer l, the death probability ratio for node u is

$$r(l) = \begin{cases} \frac{v(u)}{v(u_m)} & \text{if } u_m \text{ is a competitor of } u \\ 1 & \text{otherwise} \end{cases} \tag{10}$$

while based on assumption (1.6), the previous value ratio is equal to the corresponding number of products ratio

$$r(l) = \begin{cases} \frac{\rho(l)}{\rho_m(l)} & \text{if } u_m \text{ is a competitor of } u \\ 1 & \text{otherwise.} \end{cases} \tag{11}$$

We can rewrite the ratio $r(l)$ as

$$r(l) = P[u_m \text{ is competitor of } u] \times \frac{\rho(l)}{\rho_m(l)} + \Big(1 - P[u_m \text{ is competitor of } u]\Big) \times 1. \tag{12}$$

So, we need to derive $\rho(l)$, $\rho_m(l)$, and the probability $P[u_m$ is competitor of $u]$.

The number of products for a node at layer l is given by the random variable

$$X = \max\{B(n, s(l)), 1\}. \tag{13}$$

where $B(n, s(l))$ is a random variable that follows the binomial distribution with parameters n and $s(l)$. Based on assumption (3), the expected value $n \times s(l)$ of this binomial distribution is much larger than one at all layers, and so the average number of products at layer l can be approximated as

$$\rho(l) \approx n \times s(l). \tag{14}$$

We derive the expected value $\rho_m(l)$ of the maximum number of products at layer l using ordered statistics.

Specifically, if $X_1, X_2, \ldots, X_n$ are n *I.I.D.* random variables with cdf and pmf $F(x)$ and $f(x)$, respectively, the pmf of the k'th-ordered statistic $X_{(k)}$ (in increasing order) is given by

$$P(X_{(k)} = x) = \sum_{j=0}^{n-k} \binom{n}{j} \Big((1 - F(x))^j \left(F(x)\right)^{n-j} - (1 - f(x))^j \left(F(x) - f(x)\right)^{n-j} \Big). \tag{15}$$

The n'th order statistic represents the maximum of $X_1, X_2, \ldots, X_n$, and its pmf $f_m(x)$ is

$$f_m(x) = P(X_{(n)} = x) = F(x)^n - \left(F(x) - f(x)\right)^n. \tag{16}$$

Suppose now that the random variable X_i represents the number of products of a node i at layer l ($i = 1 \ldots n$), as defined in (13). If $n \times s(L-1) >> 1$, X can be approximated as a binomial random variable with parameters n and $s(l)$, and so

$$f(x) = P(X = x) = \binom{n}{x} s(l)^x \left(1 - s(l)\right)^{n-x} \tag{17}$$

and

$$F(x) = P(X \leq x) = \sum_{i=0}^{\lfloor x \rfloor} \binom{n}{i} s(l)^i \left(1 - s(l)\right)^{n-i}. \tag{18}$$

Replacing $f(x)$ and $F(x)$ in $f_m(x)$ and assuming that x is integer, we get that

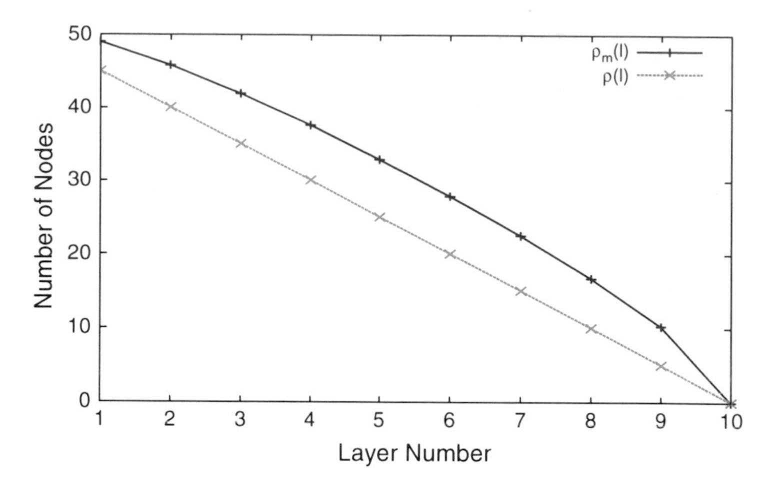

Fig. 17 $\rho(l)$ and $\rho_m(l)$ as a function of layer l. The parameters are $L = 10, n = 50$, and $s(l) = 1 - l/L$

$$
\begin{aligned}
f_m(x) = & \left[\sum_{i=0}^{x} \binom{n}{i} s(l)^i \left(1 - s(l)\right)^{n-i} \right]^n \\
& - \left[\sum_{i=0}^{x} \binom{n}{i} s(l)^i \left(1 - s(l)\right)^{n-i} \right. \\
& \left. - \binom{n}{x} s(l)^x \left(1 - s(l)\right)^{n-x} \right]^n \\
= & F(x)^n - F(x-1)^n.
\end{aligned}
\tag{19}
$$

$\rho_m(l)$ is then given by

$$
\rho_m(l) = \sum_{i=0}^{n} i f_m(i). \tag{20}
$$

Replacing (19) in (20), we get (8).

Figure 17 shows $\rho(l)$ and $\rho_m(l)$ as a function of layer l.

The final step is to derive the probability that u_m is a competitor of u. This probability, denoted by $c(l)$, refers to the event that u shares at least a fraction c of its products with u_m at layer l, where c is the competition threshold.

We define the function

$$
\begin{aligned}
h_l(x, y, z) = P[& \textit{A node with x products at layer l shares exactly} \\
& \textit{y products with a node that has z products}]
\end{aligned}
$$

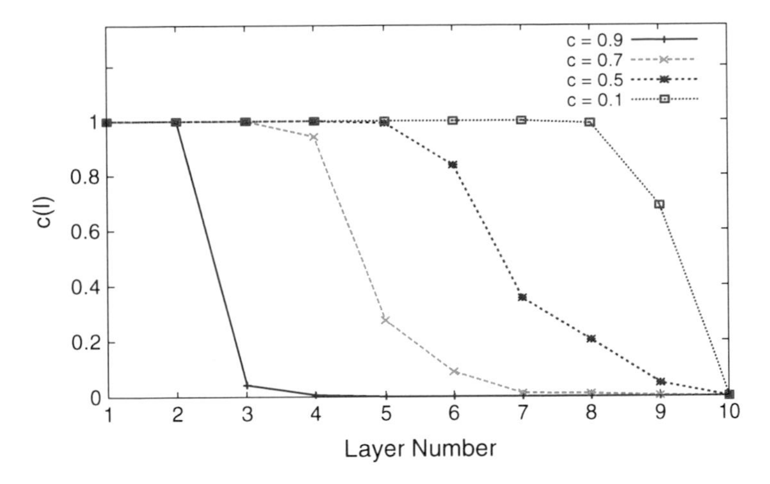

Fig. 18 The probability $c(l)$ as a function of layer l for different values of the competition threshold c. The parameters are $L = 10$, $n = 50$, and $s(l) = 1 - l/L$

which follows a hypergeometric distribution and is equal to

$$h_l(x, y, z) = \frac{\binom{z}{y}\binom{n-z}{x-y}}{\binom{n}{x}}. \tag{21}$$

$c(l)$ can then be expressed as

$$\begin{aligned} c(l) &= P[u_m \textit{ is a competitor of } u] \\ &= P[\textit{a node with } \rho(l) \textit{ products at layer } l \textit{ shares at least} \\ &\quad c\,\rho(l) \textit{ products with a node that has } \rho_m(l) \textit{ products}] \\ &= \sum_{i=\lceil c\times\lfloor\rho(l)\rceil\rceil}^{\lfloor\rho(l)\rceil} h_l(\lfloor\rho(l)\rceil, i, \lfloor\rho_m(l)\rceil) \\ &- \sum_{i=\lceil c\times\lfloor\rho(l)\rceil\rceil}^{\lfloor\rho(l)\rceil} \frac{\binom{\lfloor\rho_m(l)\rceil}{i}\binom{n-\lfloor\rho_m(l)\rceil}{\lfloor\rho(l)\rceil-i}}{\binom{n}{\lfloor\rho(l)\rceil}}. \end{aligned} \tag{22}$$

Figure 18 shows the probability $c(l)$ as a function of layer l.

Finally, combining the expressions (14), (8), (22), and (12), we get the final result in (7).

The expression (7) for $r(l)$ can also be used to predict the location of the waist as a function of the two important parameters of the model, the competition threshold c and the generality vector s. The predicted waist is the layer with the smallest death

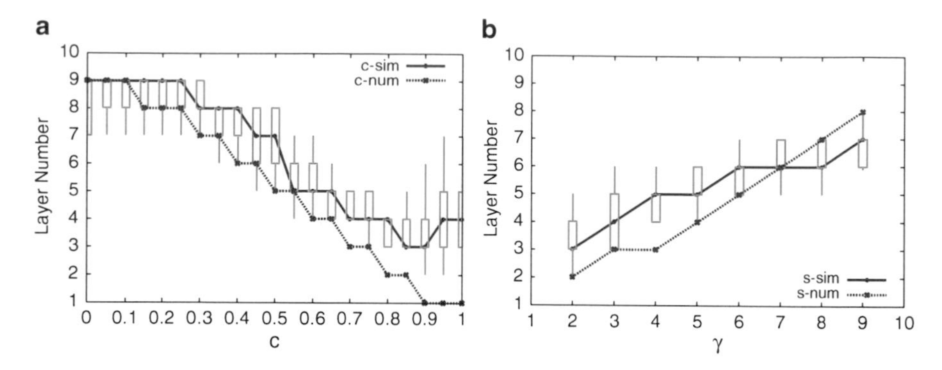

Fig. 19 The predicted location of the waist through numerical evaluation of equation (7) for $r(l)$ compared to its location through discrete-time simulation of *EvoArch* (**a**) as a function of the competition threshold c and (**b**) as a function of γ

probability ratio $r(l)$. To study the effect of s, we consider a two-segment piecewise linear generality vector as in Sect. 5. Figure 19b show the location of the waist as a function of c and γ, respectively, obtained from both discrete-time simulations of *EvoArch* and numerically from the expression (7) for $r(l)$. The trend in the location of the waist as we vary the parameters can be well predicted by the numerical results. Furthermore for most of the combinations of parameters, the numerical waist falls within one layer of the *median* simulation waist. Note, however, that for most of those parameters, the numerical waist falls below the simulation waist. This is a direct result of our simplifying assumption to consider competition only with the node with maximum number of products, u_m. When generality is high, i.e., in lower layers, this assumption is good because nodes have many products, and it is very likely that they compete with u_m. However when generality is low, i.e., in higher layers, it becomes less likely that an average node competes with u_m. In *EvoArch*, however, that node might compete with some other nodes that have fewer products than u_m. This results in an overestimation of $r(l)$ and consequently underestimation of the death probability for layers with low generality. Due to this underestimation in higher layers, the predicted waist through this mathematical analysis falls below the waist obtained from discrete-time simulations of *EvoArch*.

9 Conclusions

The main thesis behind this work is that we can study network architectures in a quantitative and scientific manner in the same way that we study, for instance, the performance of transport protocols or the stability of routing protocols. In this spirit, we proposed a model for the evolution of layered protocol stacks. *EvoArch* is based on few principles about the generality of protocols at different layers, the competition between protocols at the same layer, and how new protocols are created. Even though *EvoArch* does not capture many practical aspects and protocol-specific or layer-specific details, it predicts the emergence of an hourglass architecture and the appearance of few stable nodes (not always of the highest quality) at the waist. Further, *EvoArch* offers some intriguing insights about the evolution of the TCP/IP stack, the competition between IPv4 and IPv6, and the evolvability of future Internet architectures. Possible extensions of *EvoArch* include a dynamic notion of quality (to capture, for instance, how the deployment of a protocol can change with time depending on the protocol's value), a growing number of layers as the complexity of the provided services increases with time, and architectures without strict layering.

Finally, we note that the presence of hourglass (or bowtie-like) architectures has been also observed in metabolic and gene regulatory networks [7, 18], in the organization of the innate immune system [2], as well as in gene expression during development [12]. Even though it sounds far-fetched, it is possible that there are similarities between the evolution of protocol stacks and the evolution of the previous biological systems. We explore these cross-disciplinary connections in ongoing work.

Acknowledgement This research was supported by the NSF award 0831848 ("Towards a Theory of Network Evolution").

References

[1] S. Akhshabi, C. Dovrolis, The evolution of layered protocol stacks leads to an hourglass-shaped architecture, in *ACM SIGCOMM*, 2011
[2] B. Beutler, Inferences, questions and possibilities in toll-like receptor signalling. Nature **430**(6996), 257–263 (2004)
[3] V.G. Cerf, E. Cain, The DoD internet architecture model. Comput. Network **7**(5), 307–318 (1983)
[4] M. Chiang, S.H. Low, A.R. Calderbank, J.C. Doyle, Layering as optimization decomposition: a mathematical theory of network architectures. Proc. IEEE **95**(1), 255–312 (2007)
[5] D.D. Clark, The design philoshopy of the DARPA internet protocols, in *Proceedings of ACM SIGCOMM*, pp. 106–114, 1988
[6] D.D. Clark, D.L. Tennenhouse, Architectural considerations for a new generation of protocols, in *ACM SIGCOMM Computer Communication Review*, vol. 20, pp. 200–208, 1990
[7] M. Csete, J. Doyle, Bow ties, metabolism and disease. Trends Biotechnol. **22**(9), 446–450 (2004)
[8] J. Day, *Patterns in Network Architecture: A Return to Fundamentals*, ISBN-10: 0132252422, ISBN-13: 9780132252423, Prentice Hall, 2008
[9] C. Dovrolis, What would Darwin think about clean-slate architectures? ACM SIGCOMM Comput. Comm. Rev. **38**(1), 29–34 (2008)
[10] C. Dovrolis, T. Streelman. Evolvable network architectures: what can we learn from biology? in ACM SIGCOMM Computer Communication Review, vol. 40, pp 72–77, 2010
[11] M. Hollander, D.A. Wolfe, *Nonparametric Statistical Methods*. 2nd Edition, ISBN-10: 0471190454, ISBN-13: 9780471190455, John Wiley & Sons, 1999
[12] A.T. Kalinka, K.M. Varga, D.T. Gerrard, S. Preibisch, D.L. Corcoran, J. Jarrells, U. Ohler, C.M. Bergman, P. Tomancak, Gene expression divergence recapitulates the developmental hourglass model. Nature **468**(7325), 811–814 (2010)

[13] NSF-10528, *Future Internet Architecture (FIA)*, National Science Foundation, http://www.nsf.gov/pubs/2010/nsf10528/nsf10528.htm, 2010
[14] L. Peterson, S. Shenker, J. Turner, Overcoming the internet impasse through virtualization, in *ACM SIGCOMM HotNets*, 2004
[15] L. Popa, A. Ghodsi, I. Stoica, HTTP as the narrow waist of the future internet, in *ACM SIGCOMM HotNets*, 2010
[16] S. Ratnasamy, S. Shenker, S. McCanne, Towards an evolvable internet architecture, in *Proceedings of ACM SIGCOMM*, August 2005
[17] J. Rexford, C. Dovrolis, Future internet architecture: clean-slate versus evolutionary research. Comm. ACM **53**, 36–40 (2010)
[18] J. Zhao, H. Yu, J.H. Luo, Z.W. Cao, Y.X. Li, Hierarchical modularity of nested bow-ties in metabolic networks. BMC Bioinformatics **7**(1), 386 (2006)

Opinion Dynamics on Coevolving Networks

Federico Vazquez

1 Introduction

Statistical physics provides the ideal framework to study systems composed of many interacting units, no matter whether these units represent particles, organisms, animals, or people. That is why it has been successfully applied not only in the field of physics but also in biology, ecology, economy, and, more recently, in sociology. Within this context, the application of Statistical Physics to social phenomena has rapidly gained importance during the last decade [1] and is nowadays an established field of active research. These studies are based on the assumption that certain regularities observed at large scales, such as the consensus of a population on a given issue, are a consequence of the interaction among the individuals that form the system. A simplistic but yet instructive way of representing individuals and their social relationships is by using models of agents that interact on a complex network, where nodes and links symbolize agents and their interactions, respectively. Each agent is endowed with a state (opinion, culture, language, etc.), represented by a number or a vector in the most complex case, that changes due to the interactions with its neighbors in the network. Thus, these models constitute a virtual laboratory used to study various social dynamics, such as opinion formation, the propagation of culture, language competition, and the spreading of rumors or fads.

Early works consider that the topology of the interaction network remains unchanged during the spreading dynamics, implicitly assuming that contacts between individuals are either fixed over time or they change so slowly that, for practical purposes, network evolution can be neglected. However, recent experiments on face-to-face contacts [2, 3], carried out in a conference venue, showed that the network of human contacts (conversations) changes very rapidly with time and that the

F. Vazquez (✉)
Max Planck Institute for the Physics of Complex Systems, Nöthnitzer straße 38, D-01187 Dresden, Germany
e-mail: federico@pks.mpg.de

A. Mukherjee et al. (eds.), *Dynamics On and Of Complex Networks, Volume 2*, Modeling and Simulation in Science, Engineering and Technology, DOI 10.1007/978-1-4614-6729-8_5,

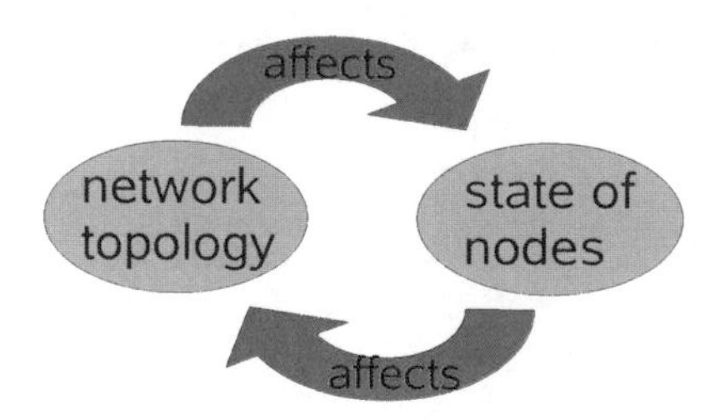

Fig. 1 In coevolutionary social dynamics, there is a feedback loop between the states of the nodes and the topology of interactions, which gives rise to a coupled evolution of the spreading phenomena and the network of contacts

frequency and the duration of such interactions largely vary over the attendees. Even though, to our knowledge, there are no quantitative measurements on how time scales associated to spreading phenomena and network evolution in real human populations relate to each other, the work by Cattuto et al. [2] is a clear evidence of the highly dynamic nature of social networks and stresses the importance of including this ingredient on more realistic models.

In the simplest case, the dynamics of the network might be independent on the spreading phenomena, but in reality both dynamics are coupled. On the one hand, individuals' decisions are influenced by their neighbors through the mechanism of *social pressure* [4] and, on the other hand, individuals have a tendency to continuously change their connections, forming new ties with similar others (*homophily*) [5, 6] or with diverse groups (*heterophily*). This creates a feedback loop between the states of the nodes and the topology of interactions and gives rise to a very rich and complex dynamics (see Fig. 1). Networks that exhibit such property are called *adaptive* or *coevolving* networks [1,7–9]. Experimental evidence of this coevolutionary dynamics can be found in the work by Lazer [10], in which he analyzes the data on the attitudinal behavior of individuals of a government agency, the Office of Information and Regulatory Affairs. The analysis shows that the internal network was quite rigid and that individuals are strongly modeled by this network. Also, individuals differed in their plasticity, that is, in the way they were influenced by others.

Besides social science, adaptive networks appear in many other areas of research, such as biology, ecology, epidemiology, game theory, genetics, neuronal and immune systems, and communication [7, 8]. As pointed out by Gross and Blasius in their review [7], all these systems share some common features at the macroscopic level, such as self-organization towards critical behavior, formation of complex topologies, complex macroscopic behavior (more complex than in static networks), and the emergence of special nodes from a homogeneous population. We shall see that the coevolving social models discussed in this chapter show traces of these four hallmarks.

The field of adaptive networks has been very active since the appearance of the first articles on this topic around 11 years ago. In particular, the study of spreading phenomena on coevolving social networks has become very intense during last

6 years; thus, we believe it deserves some attention. In this chapter, we give a brief overview of some coevolving models for opinion formation that have been extensively studied in the literature. We group them into two main classes: voter models and threshold models. The intrinsic dynamics of these models on static or very slowly varying topologies leads to coarsening and eventually to global order, if no interaction constraints are present. As we describe in the next section, the addition of some kind of link dynamics, in the form of removal or rewiring of connections, induces the appearance of new macroscopic patterns, such as stable or metastable coexistence of opinions, or the fragmentation of the network in communities.

The rest of the chapter is organized as follows. In Sect. 2, we discuss voter and threshold models, with different adaptation rules, and stress possible similitudes with respect to their macroscopic behavior. In Sect. 3, we briefly report on different mean-field approaches used to describe the dynamics of these systems and discuss about their performance. In the last section, we summarize and point out some possible lines of research to be developed in the field of coevolving networks.

2 Coevolution of Opinions and Social Contacts

In real life, people are frequently faced to situations in which they have to make a decision about a certain issue. In many cases, this involves choosing one of a given set of possible options: choose one among several TV cable companies, decide to vote for a given political candidate, buy a product or not, etc. In principle, one assumes that there is no preference a priori for a particular option; thus, they are considered to be equivalent. We usually make decisions based on recommendations or by influence of other people, typically our friends or people we know; thus, social interactions play an important role in the decision process. One of the most studied problems by statistical physicists working on social dynamics [1] is whether a population may reach global agreement (consensus) on a given issue or not (diversity) as a consequence of the interaction among individuals. In this context, *voter models* [11] and *threshold models* [12] are two of the simplest opinion dynamics models that allow to explore how consensus can be reached in a population of interacting agents. These dynamics mimic a simple mechanism for the competition between two (or more) equivalent opinions within a population. Due to an imitation strategy, individuals tend to become equal as they interact, leading to a coarsening dynamics as in the Ising model, with the ultimate dominance of one opinion.

In the next two sections, we describe a series of extensions to voter and threshold models, respectively, which incorporate some type of network adaptation into the dynamics, allowing the coevolution of contacts and opinions. We shall see that coevolution gives rise to new interesting phenomena, besides the global consensus observed in fixed topologies, which depend on the relative speed between network adaptation and opinion spreading. When the network changes very rapidly, compared to opinion updates, two new scenarios may appear. In the case that individuals form new connections with other partners who share their same opinion, a static state of opinion coexistence is ultimately reached, in which the system is polarized into disconnected groups, each composed by same-opinion individuals. Allowing individuals to connect to opposite-opinion partners leads to a metastable active state, where coexistence is characterized by weakly interconnected groups. Given that new neighbors are selected at random, the network always evolves towards a stationary configuration similar to an Erdös–Renyi graph, which is independent of the initial topology. However, real social networks have some properties, like assortativity and scale-free distributions [13], which are not observed in Erdös–Renyi graphs. Therefore, even though the coevolutionary mechanism studied in the models of this chapter leads to the formation of communities (a phenomenon observed in real societies), it does not seem to be enough to explain why real social networks look the way they look.

We also mention that some diffusion processes similar to voter and threshold models have been simulated on real time-varying networks, leading to interesting results. This is the case of the susceptible-infected-susceptible model studied by Stehlé et al. [14] and the naming game recently explored by Maity et al. [15], both on face-to-face contact networks [2]. However, these are not truly coevolving networks, given that the evolution of the network is independent on the state of the nodes.

We would like to point out that the following sets of words are interchangeable in the rest of the chapter: node/individual/person, state/spin/opinion, and link/connection/relation.

2.1 Voter Models

The original voter model was first introduced by Clifford in 1973 [16] as a model for species competition, in which each species try to invade the territory occupied by the other species. But the name *voter model* was given after the work by Holley and Liggett in 1975 [17], two probability theorists who independently introduced the same model as Clifford, as an attempt to study the "simplest" nontrivial model of interacting particle systems that could be exactly solvable. They set a simple rule for the dynamics, in which a system composed of two types of particles evolves by allowing particles to copy the state of randomly chosen nearest-neighbors in a lattice. They made the analogy with a population that has two possible positions on a political issue, where individuals blindly adopt the position hold by a random neighbor. They referred to people as "voters" in their work, which led the name of one of the most popular opinion dynamics models nowadays, even though voters never really vote in the voter model.

In the mathematical implementation of the original voter dynamics [16, 17], each site of a lattice is occupied by a voter that can take, at a given time, one of two possible states, either $+$ or $-$. These states represent the opinion or decision of the voter and, initially, are assigned with the same probability $1/2$. In a single step, a randomly chosen voter updates its state by taking the opinion of a nearest-neighbor, also chosen at random. In finite lattices, these steps are repeated until the system reaches full order (all voters are in the same state) by a large fluctuation. Once the system falls into this consensus situation it can no longer evolve (absorbing state), given that the copy dynamics makes no more changes in the state of voters. Due to the conservation of the magnetization (difference between the fractions of $+$ and $-$ voters) under this particular dynamics, the probability to converge to a given opinion equals the initial fraction of voters of that opinion. In infinite large lattices, the system reaches full order only in dimension $d \leq 2$, while it remains in a disordered active state for $d > 2$ [18, 19].

The properties of the voter model - and also of all models presented in this chapter - can be analyzed by running numerical simulations of the dynamics and plotting the time evolution of different global magnitudes, such as the magnetization or the density of interfaces between $+$ and $-$ states. Also, some theoretical approaches such as master equations, Fokker-Planck, and Langevin equations [20, 21] are sometimes used to gain an analytical insight about the macroscopic behavior of the system. They consist of writing differential equations for the time evolution of the fraction of voters in each state and for the density of different types of connected pairs (links) in the system. These equations are very useful because they can be numerically solved very quickly, and they provide the same results obtained by simulating the microscopic dynamics, which can take a very long time. Besides, in some cases these equations can be analytically solved, giving a good description of the system in terms of closed formulae.

The voter model has also been extensively studied on complex networks [22–29], where voters are placed at the nodes of a network and are allowed to interact only with their neighbors. The number of neighbors $k \geq 0$ of a given node may vary over the population, which is characterized by the degree distribution P_k. It turns out that the dynamical properties of the system strongly depend on the shape of P_k, given by its moments $\langle k^n \rangle = \sum_k k^n P_k$. For instance, the mean consensus time depends on the first two moments $\langle k \rangle$ and $\langle k^2 \rangle$ [26, 29], and it is also proportional to the number of nodes N, while the consensus probabilities are given by the initial weighted magnetization $\sum_{i=1}^{N} S_i k_i$, with $S_i = \pm 1$ [25]. In infinitely large networks, the system never fully orders but reaches a partially ordered stationary state characterized by the average connectivity $\langle k \rangle$ [29].

One of the first coevolutionary versions of the voter model is the one proposed by Zanette and Gil [30, 31]. In this model, pair of voters try to agree on one opinion as they interact but, if they fail, an attempt to drop their social tie is made by breaking the link between them. The initial state consists of a fully connected network (all-to-all interactions) of N nodes, which can take either $+$ or $-$ states with equal probability. The system evolves according to the following rules. In a single time step, a pair of connected nodes is chosen at random. If they have the same state, nothing happens. Otherwise,

1. One node chosen at random takes the state of its neighbor with probability p_1, or
2. The link between the nodes is deleted with probability $(1 - p_1)p_2$, or
3. Nothing happens with probability $(1 - p_1)(1 - p_2)$

Here p_2 controls the rate at which links are broken when no opinion agreement between neighbors is achieved. The system evolves under these rules until no more changes are possible, that is, when every connected pair of nodes in the network share the same state. The final structure of the network typically consists of a set of disconnected communities formed by same-state nodes, with specific configurations that depend on the ratio $q = p_1/[p_1 + (1 - p_1)p_2]$ which measures relative rates between state updates and link deletion. For large values of q, links disappear very slowly; thus, the system remains connected forming a single component until it reaches full order. For intermediate values of q, the deleting mechanism leads to the fragmentation of the network into two large communities of similar sizes and opposite states, while for very small q the system splits into a community larger than half of the population and a set of very small communities. The previous model considers that pair of individuals decide to simply terminate a social relation in case they do not reach opinion agreement and never interact again. One alternative to this way of modifying social contacts is to replace opposite-opinion partners by other individuals with the same opinion. The dynamics of this particular homophilic mechanism was studied by Holme and Newman in a multiple-state voter model [32], in which the number of possible opinions G is proportional to the population size N. The initial configuration is built by attaching M links to pair of nodes chosen at random, forming an Erdös–Renyi network of mean degree $\mu = 2M/N$, and assigning opinions with a uniform distribution over the population. In a time step, a node i with state S_i is picked at random. Then,

1. With probability ϕ, i disconnects the link to a random neighbor and reconnects it to a node j chosen at random among those with the same state as i (i.e., $S_j = S_i$).
2. With the complementary probability $1 - \phi$, i copies the state of a randomly chosen neighbor.

The rewiring probability ϕ controls the rate at which the network evolves relative to the opinion-spreading rate, proportional to $1-\phi$. This competition between rewiring and copy dynamics leads, in the long run, to a stationary state formed by isolated same-sate network components, as in the model by Zanette, but with a number of different opinions that vary between 1 and G. Given that this stationary state is active (the rewiring never ends), the number of components equals the number of different opinions, unlike in Zanette's model, where many isolated components might have the same state.

As ϕ decreases from 1, a phase transition is observed between a phase characterized by many small communities for large ϕ and a phase composed of a giant community of the order of the system size, plus a set of small communities (see Fig. 2). By plotting the size of the largest component S vs ϕ, one can clearly identify the transition point ϕ_c, at which S drops to zero. The transition was proved to be continuous by doing a finite-size scaling of S, with critical exponents that suggest that the transition is not in the universality class of the typical random graph percolation transition.

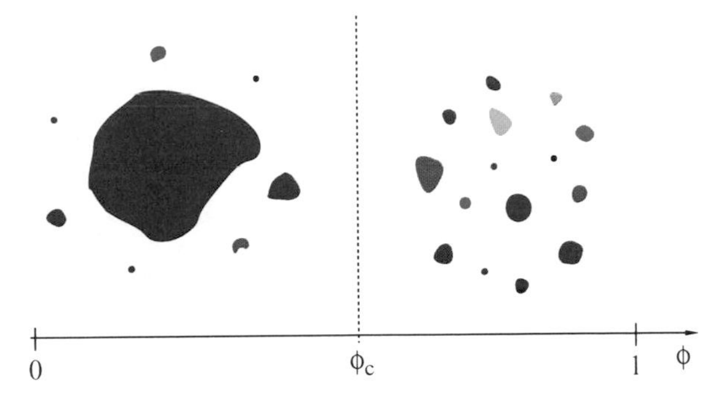

Fig. 2 Schematic representation of the fragmentation transition in the model by Holme and Newman [32]. For small values of the rewiring probability ϕ, a giant network component of same-opinion individuals emerge, whereas for rewiring probabilities above the transition point ϕ_c, the network breaks in multiple disconnected components

As we described in the two previous models, the coevolution of network and states leads to a topological phase transition where the network breaks into multiple components, the so-called *fragmentation transition*. This is a quite generic mechanism of many coevolutionary models, and it seems to be quite robust, independent of the details of the model. Indeed, this transition was observed in models with two or more states [30–36], with interaction constraints [37–42], with continuous opinion states [37, 38], with conservation in the number of links (rewiring) or not (deleting), and with several other rewiring rules.

Motivated by these observations and with the aim of gaining a deeper insight about the origin of the fragmentation transition, Vazquez et al. [33] studied in detail a voter model with homophilic rewiring. This model is particularly simple because it has only two states and the number of links is conserved. The system starts from a degree-regular random graph where each node is randomly connected to μ neighbors, and a uniform distribution of $+$ and $-$ states. Similar to the previous models, at each iteration step, a node i and a neighbor j are chosen at random. If they have the same state, nothing happens. Otherwise, with probability p, node i rewires the link from j to another randomly chosen node with the same state as i or with the complementary probability $1 - p$, node i copies j's state. The competition between the rates associated to the evolution of the network and the spreading of opinion leads to the fragmentation of the network. Roughly speaking, when the ratio $p/(1-p)$ between the rates of these two processes overcomes a threshold, so that the network evolution is faster than the node state propagation, the network evolves and ultimately breaks into two large components of similar size. This phenomenon is depicted in Fig. 3 for the dynamics of a multiple-state voter model [40]. Initially, many states are uniformly distributed on a square lattice (Fig. 3a), and, as the system evolves by copying and random rewiring, the topology quickly becomes similar to a random network (Fig. 3b). Then, some interconnected communities composed by same-state nodes start to appear (Fig. 3c), and after a long time, these communities finally get disconnected (Fig. 3d), and the dynamics

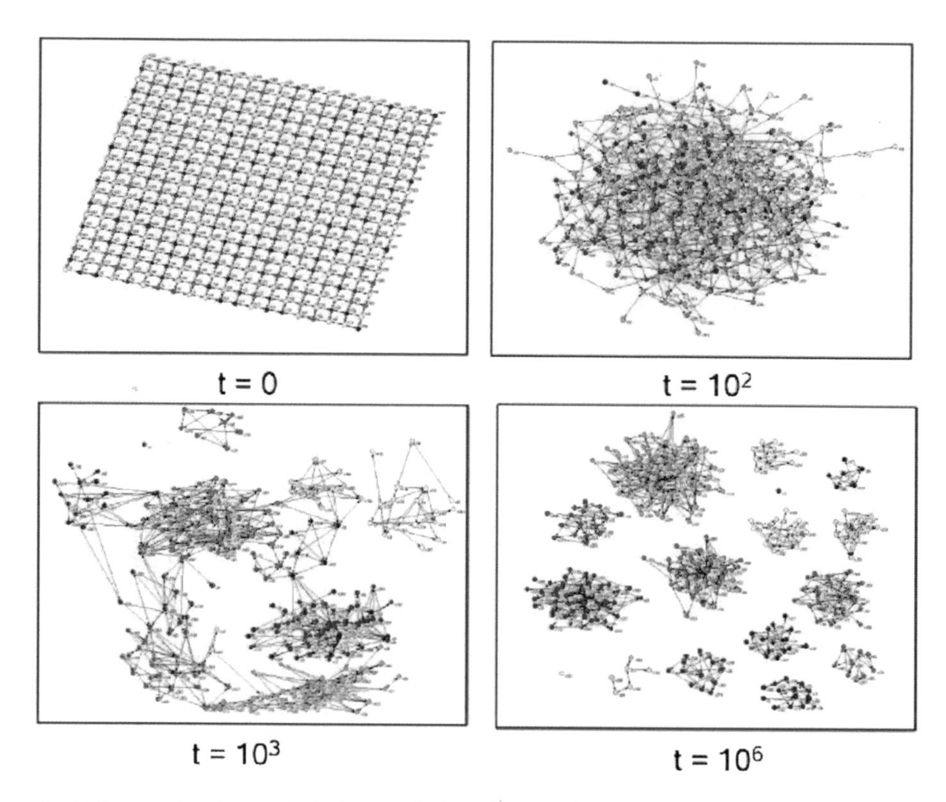

Fig. 3 Snapshots of a typical time evolution of network and individual's opinions under a coevolutionary dynamics. When network evolution is much faster than opinion spreading, the network splits into many components, each composed of individuals that share the same opinion

stops as no more updates are possible. In the opposite scenario, the ratio $p/(1-p)$ is small, and the network remains connected until it freezes in a single component. Assuming that this threshold is proportional to the density of links μ, one can write the fragmentation condition as $p/(1-p) > \mu + a$. Then, given that an Erdös–Renyi network without rewiring gets fragmented below the percolation threshold $\mu = 2$, the critical rewiring should be $p_c = 0$ when $\mu = 2$, from which $a = -2$. Therefore, a rough estimation of the critical rewiring is $p_c = (\mu - 2)/(\mu - 1)$, which is in good agreement with simulations only for large values of μ.

An analytical insight about the fragmentation transition can be obtained by tracing the magnetization m and the density of active links ρ, that is, links between nodes with opposite states. A mean-field equation for the time evolution of ρ in infinite large systems leads to the approximate expression for the transition point p_c described above. In finite networks, the transition point corresponds to a sudden bifurcation of the stochastic trajectory of the system in the (m, ρ) plane.

The models explained above assume homophilic relations among individuals in which connections between opposite-opinion voters are either deleted or replaced by same-opinions connections. However, heterophilic relations are also possible in human populations. With the aim of exploring how heterophily affects the dynamics of opinion formation, Kimura and Hayakawa [34] studied a voter model with both homophilic and heterophilic reconnections. In each step, either a rewiring to a same-state node happens with probability ϕ (homophilic rewiring), or a rewiring to opposite-state happens with probability ψ (heterophilic rewiring), or a copy state dynamics happens with probability $1 - \phi - \psi$. Without heterophily ($\psi = 0$), the network splits into isolated communities (fragmentation), but when a minimum amount of heterophily ($\psi > 0$) is introduced, these communities get connected by links, whose number increases with ψ. Therefore, heterophily leads to an active state with no global consensus, where the number of voters of each class fluctuates but opinion domains seem to be stable. Also, this particular structure of stable communities is characterized by a small-world property.

In the type of rewiring discussed above, the reconnection rate to a node with the same (opposite) state is controlled by a fixed probability ϕ (ψ). An alternative to this picture was studied by Nardini et al. [35], in a 2-state voter model. Reconnections are to randomly chosen nodes, independent of their state; thus, both homophilic and heterophilic rewiring rates are not fixed but vary over time, depending on the fraction of voters in each class. They explored two types of node dynamics: the direct voter model (d VM) and the reverse voter model (r VM). In the d VM, a selected node *adopts* the state of a random neighbor (like in the models described above), while in the r VM, a chosen node *gives* the state to a random neighbor. This simple difference in the update rule has striking consequences on the mean consensus time τ. In the d VM, consensus is achieved very quickly, growing as $\tau \sim \ln N$ with the number of nodes in the network N, whereas in the r VM, consensus is reached very slowly, with a scaling given by $\tau \sim e^{\alpha N}$. This difference is due to the appearance of a positive (negative) feedback in the dynamics of the d VM (r VM). In both cases, when there are more $+$ than $-$ nodes ($N_+ > N_-$), a randomly rewired link is more likely to get connected to a $+$ than a $-$ node, increasing the degree of $+$ nodes and decreasing the degree of $-$ nodes. Then, this increases (decreases) the probabilities of choosing a $+$ ($-$) neighbor in an update event. Thus, in the d VM, a node will more likely copy the state of a $+$ node, increasing even more N_+ and thus inducing a positive feedback that leads to a monotonic increase in N_+ until full consensus ($N_+ = N$) is reached. Due to the positive feedback, there is a bias that makes N_+ increase exponentially fast, leading to a logarithmic in N convergence times. In the r VM, a node will more likely give the state to a $+$ node, which may change to $-$ state, decreasing N_+ and leading, on an average, to a negative feedback that drives N_+ to the mean value $N/2$. Consensus is reached by an exponentially rare fluctuation that happens in a time that grows exponentially fast with the system size.

2.2 Threshold Models

In 1978, Granovetter published a seminal work about the effects human behavioral thresholds on the collective dynamics of populations [12]. He investigates a simple model in which individuals have to choose between two distinct and mutually exclusive behavioral alternatives, like to riot or not, vote for a particular candidate or not, and migrate to another city or not. The model assumes that the decision to make one or the other choice usually depends on how many people already made which choice, a special mechanism for social influence. Thus, a given person makes a particular choice only when the total number of individuals that accepted that choice crosses the intrinsic threshold of the person.

A popular type of threshold models applied to opinion formation is the *majority rule*, where individuals adopt the opinion of the majority in their immediate neighborhood on a square lattice [43, 44]. The model consists of a population of N agents located on square lattice, where each agent can take one of two possible opinions (spin 1 or -1). In a single step, a group of G contiguous spins is chosen at random (with G odd), and all spins adopt the state of the local majority. This event is repeated until the system ultimately reaches consensus. In dimension $d > 1$ and mean-field (all-to-all interactions), consensus is always in the state of the initial majority, while in dimension $d = 1$, the initial minority has a chance to win [44]. The mean approach to the final state is very slow, because a fraction of the realizations fall into configurations that contain same-spin bands, which slowly diffuse and coalesce until one band takes over the system [45].

Another prototypical example of opinion threshold dynamics is the zero-temperature Ising model with Glauber dynamics, originally proposed by Glauber in 1963 [46], as a stochastic Ising model for ferromagnetism, that evolves towards thermal equilibrium. At each step of the dynamics, a randomly chosen spin adopts the state of the majority in its neighborhood on a square lattice ($2d$ neighbors in d dimensions), and, in case of a tie, it adopts one of the two states with equal probability $1/2$. Unlike the majority rule dynamics where all spins in the neighborhood update their states, here only the chosen spin is updated. In one dimension, the dynamics of the zero-temperature Glauber model is equivalent to that of the voter model; thus, the system reaches consensus in a time that scales as L^2 with the system size L [46]. In dimension $d > 1$, there is a probability that the system falls into an absorbing disordered configuration composed of frozen same-spin bands; thus, the system does not always reaches full consensus [47, 48].

The zero-temperature Glauber dynamics was also studied on small-world and complex networks (see [28, 49, 50] for a reference on the first investigations). It was found that in small-world and Erdös–Renyi networks of infinitely large size, the system does not reach full ordering. In finite homogenous and heterogeneous networks, the system usually gets trapped into partially ordered metastable states. However, sometimes the dynamics leads to complete order, but with a probability that vanishes as the system size increases; thus, full consensus is never reached in the thermodynamic limit.

One of the first models that explores the coevolution of the majority rule with the interaction network was studied by Benczik et al. [51, 52]. In this model, each spin can take either value 1 or -1 and interacts with a set of neighbors that is updated at every iteration. In a single step, a spin i adopts the majority state of its neighborhood; that is, if most of its neighbors are positive, then σ_i is set to $\sigma_i = 1$ or, otherwise, is set to $\sigma_i = -1$. In the case of a tie, σ_i remains unchanged. Besides, links are updated in the following way: spins that share the same state are connected with probability p, while spins in different states are connected with probability $1 - p$.

The analysis of this model in infinite large systems reveals four different phases: two consensus phases, one active disordered phase, and one static disordered phase. The consensus phases happen for intermediate values of p, where the system reaches global order in the state of the initial majority. The active disordered phase is observed for small values of p. Here, spins are usually connected to more neighbors with opposite state than with equal state; thus, they are always in the minority state among their neighbors, and they have to flip to the majority state at every time step. This creates a high flipping rate in the system, leading to a disordered active state with similar fractions of $-$ and $+$ spins. Finally, the static disordered phase is seen for large values of p where, contrary to the previous case, most spins are usually in the majority state; thus, they rarely flip. The dynamics of spins quickly reaches a static configuration similar to the initial one, where the fraction of spins in each class remains unchanged. In finite systems, due to fluctuations, the system ultimately reaches one of the two consensus states, which are the only stable (absorbing) states. For intermediate p, the system falls into metastable states before reaching the consensus state, which takes exponentially long times.

The results of this model suggest that the final configuration of a system subject to a threshold dynamics might depend on the initial conditions. In order to stress the importance of initial conditions on coevolutionary dynamics, Lambiotte et al. [53] studied a threshold model on a directed network. They found that the fragmentation transition induced by coevolution is driven by the initial configuration of the system. In this binary-state model, each node has two incoming links and a Poissonian distribution of outgoing links. The system starts with a fraction $n_+(0)$ and $n_-(0) = 1 - n_+(0)$ of $+$ and $-$ nodes, respectively, uniformly distributed over an Erdös–Renyi graph. Only if both incoming neighbors disagree with the state of a focal node i, then i takes their state. Otherwise, i rewires the link connected to the disagreeing neighbor to another node chosen at random, keeping its in-degree constant. The dynamics runs until the network splits into a $+$ and a $-$ component, whose size depends on the initial fraction $n_+(0)$ of $+$ nodes. When $n_+(0)$ is approximately below 0.25, the system reaches $-$ consensus ($n_+(\infty) = 0$), as the fraction of $+$ nodes is not enough to survive and form an isolated cluster. When $n_+(0)$ is above 0.25, the size of the final $+$ cluster increases monotonically with $n_+(0)$ and reaches $n_+(\infty) = 1$ for $n_+(0) \simeq 0.75$. Finally, above 0.75, the system reaches the $+$ consensus ($n_+(\infty) = 1$). This model illustrates how the initial configuration may affect the final topology of the network and points out that memory effects seem to be very strong in some coevolving models.

Up till now, we have presented models where links are rewired to nodes chosen at random in the network, creating long-range connections and leading to topologies that resemble random graphs, with Poissonian degree distributions. An alternative to this sort of rewiring was proposed by Fu and Wang [54], who studied a majority rule model with short-range rewiring, that is, to neighbors of neighbors, or "second nearest-neighbors" using lattice terminology. They found that the system reaches a stationary state that consists of an evolving network formed by interconnected communities of same-opinion individuals. They also show that clustering speeds up global consensus. As in the model by Holme and Newman, initially there are many opinions uniformly distributed on an Erdös–Renyi graph. In a single step, a chosen node i takes the state of the majority of its neighbors with probability p. With complementary probability $1 - p$, i disconnects the link to an opposite-state random neighbor. Then, with probability ϕ, this link is reconnected to a random neighbor of its neighbors who hold the same state, or with probability $1 - \phi$, the link is reconnected to a node chosen at random in the network. For small values of p, the fast evolution of the network leads to a state characterized by the coexistence of many opinions and the formation of interconnected communities. Within each community, nodes share the same state and are densely connected, while there are a few links that connect to other communities. For large p, global consensus in one opinion is achieved. Also, by increasing ϕ, the network develops into a structure with high clustering. This effect seems to facilitate the propagation of states, reducing the number of coexistent opinions and, therefore, fostering global agreement.

As mentioned before, the majority rule is a particular case of a model with threshold $s = 1/2$. A more general case in which the threshold can take an arbitrary value $0 \leq s \leq 1$ was studied by Mandrà et al. in [55], where they found a rich variety of stationary states, depending on s and the rewiring vs copy rate. They consider a binary-state population on an initially Erdös–Renyi graph that evolves according to the following rules. In a single step, a node i with state σ_i is picked at random. If the number of neighbors of i with state σ_i is larger than or equal to a threshold s, then i is considered to be stable, and nothing happens. Otherwise, with probability ϕ, i detaches the link from a neighbor j with opposite state $\sigma_j = -\sigma_i$ and reattaches it to a random node with state σ_i (rewiring), and with probability $1 - \phi$, i adopts j's state (state change). Note that the threshold s plays the role of a parameter that controls the rate at which the dynamics evolves, either by updating node states or by rewiring links.

There are five different phases in the (s, ϕ) space: An *active phase* for large s, where a stationary dynamic state is observed, with ongoing spin flip and link rewiring and with coexistence of stable domains. This is the only stable active phase, all the rest are frozen phases. A *disordered phase* for small s, in which the system quickly freezes because most nodes are initially stable, so it remains in a disordered frozen configuration with many small domains, overall magnetization $m \simeq 0$, and a fraction of active links $\rho \simeq 1/2$. A *consensus phase* for intermediate s and small ϕ, where the system evolves and freezes in a single large component. A *community phase* for intermediate s and large ϕ, where there are various large components interconnected by links. And a *fragmented phase* for large s and ϕ, where the system splits into two large components. The consensus and fragmented phases are analogous to the ones found in the voter models of Sect. 2.1, while the active phase also appears in the model by Benczik, and a phase similar to the community phase is seen in the model by Fu and Wang. However, the disordered frozen phase is new to this particular model, possibly due to the fact that the majority rule dynamics allows for the existence of trapped disordered configurations.

3 Analytical Descriptions of Spreading Phenomena on Evolving Networks

In this section, we briefly discuss some analytical techniques specially developed to deal with simple dynamical processes on complex networks and argue that some better analytical approaches are required to properly describe dynamics on coevolving networks.

During the last years, a series of mean-field approximations have been proposed in the literature in order to describe the dynamics of simple models for spreading phenomena, such as contact processes and voter models, on static complex networks [26, 27, 29, 56–63]. To account for the effects of the heterogeneity of contacts on the system's evolution, various heterogeneous approximations have been developed. In the *heterogeneous node approximation* (HNA) [56–58], also known as *compartmental approach*, nodes with the same state and degree are considered to be equivalent; thus, they are grouped in the same compartment. Starting from the microscopic dynamics of each node, equations for the time evolution of the density of nodes in each compartment or class are derived by doing a proper coarse graining. These are sets of coupled ordinary differential equations that can be numerically integrated, providing information at the mesoscopic level. Using this approach, Sood and Redner [26, 27] showed that the consensus time in the voter model on static complex networks depends, as a first approximation, on the first two moments of the degree distribution, and it increases linearly or sublinearly with the system size, depending on whether the network is homogeneous or heterogeneous, respectively. In some cases, effective equations for macroscopic variables, like global densities of nodes in a given state, are obtained by adding densities of all degree classes associated to that state. These equations turn to be very useful because they can be analytically solved, and the solutions provide, in many cases, a quite accurate description of the system's evolution and its stationary states.

The HNA, like any other mean-field approach, matches numerical results from Monte Carlo simulations very well, as long as the system is "well mixed," that is, when different classes of nodes are uniformly distributed over the network. In these homogeneous systems, nearest-neighbor state correlations are negligible, as the HNA assumes. State correlations are weak probably due to the random nature of the network that lacks structural correlations (loops, clustering, degree-degree correlations, etc.) or may be due to the small diameter of complex networks that leaves no space to build long-range correlations.

However, the HNA does not always work well in all range of parameters, because long- or intermediate-range correlations become important in some models, specially close to transition points. In this respect, pair approximations analogous to the ones used in lattice models [64, 65] have been developed to take into account correlations between nearest-neighbors, by taking pairs of connected nodes (links) as elemental units of the mathematical description of the system. This is the case of the *homogeneous pair approximation* (HomoPA), in which links are grouped by the state of the nodes at their ends. This approximation has a significantly better performance with respect to the HNA in the voter model on static networks [29], specially at low mean degrees, where nearest-neighbor correlations seem to be strong. Also, the HomoPA has proven to be a useful description of the dynamics of voter models [29, 34, 35] and threshold models [55] on coevolving networks, providing a rather simple framework to obtain approximate analytical expressions for the system's evolution and the transition points. More accurate analytical descriptions can be obtained with the *heterogeneous pair approximation* (HetePA) [59], in which links formed by nodes with the same state and degree are placed in the same group. Within this approach, Pugliese and Castellano [59] found expressions for consensus times and link densities in the voter model on static networks, which match numerical solutions with an incredibly high accuracy (larger than 99%), for any value of mean degree. Finally, the *improved compartmental formalism* [60–62], specially introduced to study coevolving networks, appears to be the most refined mean-field approach for spreading phenomena on complex networks. In this approach, nodes are grouped according to their states, degrees, and number of neighbors in a given state. Unlike the pair approximation, the information about the neighborhood of a given node is not approximated, because it is already contained in the definition of the class of the node. This approach was found to improve the analytical description of the stationary and final states of the voter model on evolving networks [63], with respect to the simple HomoPA.

Remarkably, all mean-field approaches mentioned above do not properly describe the dynamics of the coevolving voter model over all range of parameters. Specifically, the stationary density of active links predicted by the homogeneous and heterogeneous pair approximations, as well as the improved compartmental formalism, deviates from the one obtained from Monte Carlo simulations. These deviations become very large as the system approaches to the fragmentation transition point and, as a consequence, the critical point is completely incorrectly estimated. We speculate that the bad performance of mean-field approaches are due to the appearance of dynamical correlations induced by the changing topology, giving rise to the formation of communities, analogous to the same-state domains that appear in extended systems. Indeed, it is known that spatial lattice models with coarsening dynamics, like the Ising model, are not well described by mean-field theories [66, 67]. This is because coarsening induces long-range order (distant spins tend to be aligned or correlated), which is directly related to fluctuations in the density of spins along the lattice, totally neglected by the mean-field approximation.

It is also known that correlations become increasingly large and eventually diverge as one approaches the transition point of spin systems, which might explain why the mean-field approximation becomes worst close to the fragmentation point of the coevolving voter model. Other techniques, based on the spreading of a seed (active link) from the absorbing fragmented phase, are able to predict with very good accuracy the critical point of the fragmentation transition [68], but they do not give a good estimate of the density of active links away from this point.

In a general context, it was shown by Gleeson [69] that existing mean-field approaches fail to properly describe contact processes and voter models on real correlated networks. Therefore, we believe that new approaches should be developed to deal with this scenario, given that most static networks in nature present some degree of correlation. In this respect, predicting the correct temporal evolution and transition points in simple models on coevolving networks remains a challenge.

4 Summary and Conclusions

In this chapter, we have described some of the most representative opinion dynamics models on evolving networks, grouped in two categories based on the node update dynamics: voter models and threshold models. These models represent only a small fraction of all coevolutionary models for social phenomena that appeared in the literature during the last 5 years, but we cannot cover all of them in this short review. Many of these models are extensions to the models presented in this chapter, while some others describe totally different dynamics, such as the spread of ideologies in a society [70], represented by three non-equivalent states like in the susceptible-infected-recovered epidemic model. Also, there have been studies about the effects of coevolution on models with interaction constraints, such as the Axelrod model for the propagation of culture [39–42] and the bounded-confidence model for opinion formation [37, 38].

The state and link dynamics of the models described in this chapter are very different in nature: copy vs threshold dynamics for node updates, deleting vs rewiring dynamics for link updates. However, we have shown that many models share the same macroscopic behavior, such as global consensus, and static or dynamic coexistence of opinions. *Consensus* seem to appear when the opinion-spreading process dominates over network evolution and is characterized by a static state of single-opinion dominance, represented by a giant network community (and possibly some small communities) composed of like-minded individuals. A situation where two or more *static communities* of similar size emerge (opinion coexistence) is usually obtained when network evolution becomes relevant, and cannot be neglected. Finally, some *dynamic metastable communities* appear when the system is highly disordered due to special types of rewiring or node updates, with opinions and connections continuously changing.

We have also discussed about the performance of different mean-field techniques used to describe the time evolution of simple models for spreading phenomena on coevolving networks. These approaches work well for dynamics like the contact process on adaptive networks, but they fail in some others like in the coevolution voter model, mainly close to the fragmentation transition point, where state correlations seem to play an important role. We argue that some new analytical approaches need to be developed in order to cope with dynamics that induce long-range correlations.

These mean-field approaches are meant to work well in infinite large networks, because the description of the system is in terms of ordinary differential equations, which neglect finite-size fluctuations. However, fluctuations may be relevant in real finite networks, given that they are responsible for driving the system to the final absorbing state. Therefore, a more accurate description of coevolving finite networks should be based on master equations or stochastic differential equations, which are also very useful to calculate ordering or convergence times.

Finally, we understand that most coevolving social models proposed in the literature assume, for simplicity, that people behave similarly, either by the way they change their opinions or by the way they interact with their neighbors. However, we know that in reality people are very different from each other and that human interactions are very heterogeneous, as the face-to-face experiments mentioned in the introduction have confirmed. Therefore, more refined models should incorporate this heterogeneity in the connections between agents. This can be done by adding some type of disorder in the system, like for instance, considering rewiring rates that depend on the pair of agents or taking interaction rates that vary over the network.

References

[1] C. Castellano, S. Fortunato, V. Loreto, Statistical physics of social dynamics. Rev. Mod. Phys. **81**, 591–646 (2009)

[2] C. Cattuto, W. Van den Broeck, A. Barrat, V. Colizza, J.-F. Pinton, A. Vespignani, Dynamics of person-to-person interactions from distributed RFID sensor networks. PLoS ONE **5**, e11596 (2010)

[3] L. Isella, J. Stehlé, A. Barrat, C. Cattuto, J.-F. Pinton, W. Van den Broeck, What's in a crowd? Analysis of face-to-face behavioral networks. J. Theor. Biol. **271**, 166 (2011)

[4] B. Latane, Pressure to uniformity and the evolution of cultural norms: Modeling dynamics of social impact, in *Computational Modeling of Behavior in Organizations*, ed. by C.L. Hulin, D.R. Illgen (American Psychological Association, Washington, DC, 2000), pp. 189–215

[5] J.M. McPherson, L. Smith-Lovin, J. Cook, Birds of a feather: Homophily in social networks. Ann. Rev. Sociol. **27**, 415–44 (2001)

[6] D. Centola, An experimental study of homophily in the adoption of health behavior. Science **334**, 1269 (2011)

[7] T. Gross, B. Blasius, Adaptive coevolutionary networks: a review. J. R. Soc. Interface **5**(20), 259–271 (2007)

[8] T. Gross, H. Sayama (eds.), *Adptive Networks: Theory, Models and Applications* (Springer, New York, 2009)

[9] S. Lozano, Dynamics of social complex networks: Some insights in recent research, in *Dynamics On and Of Complex Networks: Aplications to Biology, Computer Science and the Social Sciences*. Modeling and Simulation in Science, Engineering and Technology (Springer-Birkhauser), pp. 133–143 (2009)

[10] D. Lazer, The co-evolution of individual and network. J. Math. Sociol. **25**, 69–108 (2001)
[11] T.M. Liggett, *Interacting Particle Systems* (Springer, New York, 1985)
[12] M. Granovetter, Threshold models of collective behavior. Am. J. Sociol. **83**(6), 1420 (1978)
[13] M.E.J. Newman, Assortative mixing in networks. Phys. Rev. Lett. **89**, 208701 (2002)
[14] J. Stehlé, N. Voirin, A. Barrat, C. Cattuto, V. Colizza, L. Isella, C. Régis, J.-F. Pinton, N. Khanafer, W. Van den Broeck, P. Vanhems, Simulation of a SEIR infectious disease model on the dynamic contact network of conference attendees. BMC Med. **9**, 87 (2011)
[15] S.K. Maity, T.V. Manoj, A. Mukherjee, Opinion formation in time-varying social networks: The case of Naming Game, Phy. Rev. E 86, 036110 (2012)
[16] P. Clifford, A. Sudbury, A model for spatial conflict. Biometrika **60**(3), 581–588 (1973)
[17] R.A. Holley, T.M. Liggett, Ergodic theorems for weakly interacting infinite systems and the voter model. Ann. Probab. **3**, 643 (1975)
[18] P.L. Krapivsky, Kinetics of monomer-monomer surface catalytic reactions. Phys. Rev. A **45**, 1067 (1992)
[19] L. Frachebourg, P.L. Krapivsky, Exact results for kinetics of catalytic reactions. Phys. Rev. E **53**, R3009 (1996)
[20] G.W. Gardiner, *Handbook of Stochastic Methods* (Springer-Verlang, Berlin), (1997)
[21] F. Vazquez, C. Lopez, Systems with two symmetric absorbing states: relating the microscopic dynamics with the macroscopic behavior. Phys. Rev. E **78**, 061127 (2008)
[22] C. Castellano, D. Vilone, A. Vespignani, Incomplete ordering of the voter model on small-world networks. Europhys. Lett. **63**, 153 (2003)
[23] D. Vilone, C. Castellano, Solution of voter model dynamics on annealed small-world networks. Phys. Rev. E **69**, 016109 (2004)
[24] K. Suchecki, V.M. Eguíluz, M. San Miguel, Voter model dynamics in complex networks: Role of dimensionality, disorder, and degree distribution. Phys. Rev. E **72**, 036132 (2005)
[25] K. Suchecki, V.M. Eguíluz, M. San Miguel, Conservation laws for the voter model in complex networks. Europhys. Lett. **69**, 228 (2005)
[26] V. Sood, S. Redner, Voter model on heterogeneous graphs. Phys. Rev. Lett. **94**, 178701 (2005)
[27] V. Sood, T. Antal, S. Redner, Voter models on heterogeneous networks. Phys. Rev. E **77**, 041121 (2008)
[28] C. Castellano, V. Loreto, A. Barrat, F. Cecconi, D. Parisi, Comparison of voter and Glauber ordering dynamics on networks. Phys. Rev. E **71**, 066107 (2005)
[29] F. Vazquez, V.M. Eguíluz, Analytical solution of the voter model on uncorrelated networks. New J. Phys. **10**, 063011 (2008)
[30] D.H. Zanette, S. Gil, Opinion spreading and agent segregation on evolving networks. Phys. D **224**, 156 (2006)
[31] S. Gil, D.H. Zanette, Coevolution of agents and networks: Opinion spreading and community disconnection. Phys. Lett. A **356**, 89 (2006)
[32] P. Holme, M.E.J. Newman, Nonequilibrium phase transition in the coevolution of networks and opinions. Phys. Rev. E **74**, 056108 (2006)
[33] F. Vazquez, V.M. Eguíluz, M. San Miguel, Generic absorbing transition in coevolution dynamics. Phys. Rev. Lett. **100**, 108702 (2008)
[34] D. Kimura, Y. Hayakawa, Coevolutionary networks with homophily and heterophily. Phys. Rev. E **78**, 016103 (2008)
[35] C. Nardini, B. Kozma, A. Barrat, Who's talking first? Consensus or lack thereof in coevolving opinion formation models. Phys. Rev. Lett. **100**, 158701 (2008)
[36] G. Demirel, R. Prizak, P.N. Reddy, T. Gross, Opinion formation and cyclic dominance in adaptive networks. Eur. Phys. J. B **84**, 541–548 (2011)
[37] B. Kozma, A. Barrat, Consensus formation on adaptive networks. Phys. Rev. E **77**, 016102 (2008)
[38] B. Kozma, A. Barrat, Consensus formation on coevolving networks: groups' formation and structure. J. Phys. A Math. Theor. **41**, 224020 (2008)
[39] D. Centola, J.C. Gonzalez-Avella, V.M. Eguiluz, M. San Miguel, Homophily, cultural drift, and the co-evolution of cultural groups. J. Conflict Resolut. **51**, 905–929 (2007)

[40] F. Vazquez, J.C. González-Avella, V.M. Eguíluz, M. San Miguel, Time-scale competition leading to fragmentation and recombination transitions in the coevolution of network and states. Phys. Rev. E **76**, 46120 (2007)
[41] B. Wang, Y. Han, L. Chen, K. Aihara, Limited ability driven phase transitions in the coevolution process in Axelrod's model. Phys. Lett. A **373**, 1519 (2009)
[42] C. Gracia-Lázaro, F. Quijandría, L. Hernández, L.M. Floría, Y. Moreno, Coevolutionary network approach to cultural dynamics controlled by intolerance. Phys. Rev. E **84**, 067101 (2011)
[43] S. Galam, Minority opinion spreading in random geometry. Eur. Phys. J. B **25**, 403–406 (2002)
[44] P.L. Krapivsky, S. Redner, Dynamics of majority rule in an interacting two-state spin system. Phys. Rev. Lett. **90**, 238701 (2003)
[45] P. Chen, S. Redner, Majority rule dynamics in finite dimensions. Phys. Rev. E **71**, 036101 (2005)
[46] R.J. Glauber, Time-dependent statistics of the ising model. J. Math. Phys. **4**, 294 (1963)
[47] V. Spirin, P.L. Krapivsky, S. Redner, Fate of zero-temperature ising ferromagnets. Phys. Rev. E **63**, 036118 (2001)
[48] V. Spirin, P.L. Krapivsky, S. Redner, Freezing in ising ferromagnets. Phys. Rev. E **65**, 016119 (2001)
[49] D. Boyer, O. Miramontes, Interface motion and pinning in small-world networks. Phys. Rev. E **67**, 035102 (2003)
[50] C. Castellano, R. Pastor-Satorras, Zero temperature Glauber dynamics on complex networks. J. Stat. Mech. P05001 (2006)
[51] I.J. Benczik, S.Z. Benczik, B. Schmittmann, R.K.P. Zia, Lack of consensus in social systems. EPL **82**, 48006 (2008)
[52] I.J. Benczik, S.Z. Benczik, B. Schmittmann, R.K.P. Zia, Opinion dynamics on an adaptive random network. Phys. Rev. E **79**, 046104 (2009)
[53] R. Lambiotte, J.C. González-Avella, On co-evolution and the importance of initial conditions. Phys. A **390**, 392–397 (2011)
[54] F. Fu, L. Wang, Coevolutionary dynamics of opinions and networks: From diversity to uniformity. Phys. Rev. E **78**, 016104 (2008)
[55] S. Mandrà, S. Fortunato, C. Castellano, Coevolution of Glauber-like Ising dynamics and topology. Phys. Rev. E **80**, 056105 (2009)
[56] R. Pastor-Satorras, A. Vespignani, Epidemic spreading in scale-free networks. Phys. Rev. Lett. **86**, 3200 (2001)
[57] R. Pastor-Satorras, A. Vespignani, Epidemic dynamics and endemic states in complex networks. Phys. Rev. E **63**, 066117 (2001)
[58] Y. Moreno, R. Pastor-Satorras, A. Vespignani, Epidemic outbreaks in complex heterogeneous networks. Eur. Phys. J. B **26**, 521 (2002)
[59] E. Pugliese, C. Castellano, Heterogeneous pair approximation for voter models on networks. EPL **88**, 58004 (2009)
[60] P.-A. Noël, B. Davoudi, R.C. Brunham, L.J. Dubé, B. Pourbohloul, Time evolution of epidemic disease on finite and infinite networks. Phys. Rev. E **79**, 026101 (2009)
[61] V. Marceau, P.-A. Noël, L. Hébert-Dufresne, A. Allard, L.J. Dubé, Adaptive networks: Coevolution of disease and topology. Phys. Rev. E **82**, 036116 (2010)
[62] J.P. Gleeson, High-accuracy approximation of binary-state dynamics on networks. Phys. Rev. Lett. **107**, 068701 (2011)
[63] R. Durrett, J.P. Gleeson, A.L. Lloyd, P.J. Mucha, F. Shi, D. Sivakoff, J.E.S. Socolar, C. Varghese, Graph fission in an evolving voter model. Proc. Natl. Acad. Sci. USA **109**, 3682–3687 (2012)
[64] H. Matsuda, N. Ogita, A. Sasaki, K. Sato, Stochastical mechanics of population: The lattice Lotka-Volterra model. Prog. Theor. Phys. **88**, 1035 (1992)
[65] M.J. Keeling, The effects of local spatial structure on epidemiological invasions. Proc. R. Soc. Lond. B **266**, 859 (1999)

[66] R.K. Pathria, *Statistical Mechanics* (Butterworth-Heinemann), (1996)
[67] H.E. Stanley, *Introduction to Phase Transitions and Critical Phenomena* (Oxford University Press, Oxford, 1971)
[68] G.A. Böhme, T. Gross, Analytical calculation of fragmentation transitions in adaptive networks. Phys. Rev. E **83**, 035101(R) (2011)
[69] J.P. Gleeson, S. Melnik, J.A. Ward, M.A. Porter, P.J. Mucha, Accuracy of mean-field theory for dynamics on real-world networks. Phys. Rev. E **85**, 026106 (2012)
[70] M.S. Shkarayev, I.B. Schwartz, L.B. Shaw, Recruitment dynamics in adaptive social network, arXiv:1111.0964

Part II
Community Analysis

A Template for Parallelizing the Louvain Method for Modularity Maximization

Sanjukta Bhowmick and Sriram Srinivasan

1 Introduction

Large-scale networks are used to model systems of interacting entities, such as those arising in biology, sociology, and software engineering. The vertices in the network represent the entities of the system and the edges and the interactions between them. Analyzing the properties of these networks can provide us with important insights about the underlying systems. An important operation in analyzing networks is detecting naturally occurring communities in the system. A popular method for finding communities in a network is by maximizing modularity. Modularity measures how better the vertices in a community are connected as opposed to a random connection [8]. Finding the maximum modularity is an NP-complete problem [3]; consequently, there exist several efficient heuristics for finding a near-optimal solution [2, 4, 9, 10].

As networks increase in size, for example, Facebook currently has 845 million vertices, it is essential to develop parallel implementations for the modularity maximization algorithms. However, even though parallel algorithms for graphs is a well-researched subject, to date there exist few parallel algorithms for modularity maximization [12–14]. We posit that this is because most agglomerative methods for obtaining high modularity require frequent synchronization, which reduces the scope of parallelization. Additionally, it has been observed that the results of modularity maximization, both the value of modularity and the community allocation, are affected by the ordering of the vertices. Therefore, it is difficult to evaluate the correctness of the parallel algorithm.

S. Bhowmick (✉) • S. Srinivasan
Department of Computer Science, University of Nebraska at Omaha,
6001 Dodge Street, Omaha, NE 68182-0500, USA
e-mail: sbhowmick@unomaha.edu; ssrinivasan@unomaha.edu

A. Mukherjee et al. (eds.), *Dynamics On and Of Complex Networks, Volume 2*,
Modeling and Simulation in Science, Engineering and Technology,
DOI 10.1007/978-1-4614-6729-8_6,

In this paper, we introduce a shared-memory parallel algorithm for the Louvain method. A resolution-limit-free variant of this method was proposed by [14]. However, to the best of our knowledge, this is the first tightly coupled parallel implementation of the original Louvain method. This method includes several local updates which are amenable to parallel algorithms; however, the algorithm also requires frequent synchronization of the communities across levels that render parallelism difficult. We use this algorithm to demonstrate a template for parallelizing similar agglomerative modularity maximization methods and discuss how parallelization affects the variation in modularity values.

The rest of this paper is organized as follows. In Sect. 2, we provide a brief introduction to graph theory terminology with emphasis on modularity and community detection. We also discuss some of the existing parallel algorithms for modularity maximization. In Sect. 3, we describe a simple shared-memory algorithm for parallelizing the Louvain method. In Sect. 4, we present results that demonstrate the scalability and correctness of our results. In Sect. 5, we discuss how our method can be extended to other agglomerative techniques and time-varying networks. We conclude in Sect. 6 with a discussion on future research.

2 Background

In this section we define some terms associated with network analysis. A network $G = (V, E)$ is defined as a set of vertices V and a set of edges E. An edge $e \in E$ is defined by two vertices u, v which are its *endpoints*. A vertex u is a *neighbor* of v if they are joined by an edge. The *degree* of a vertex u is the number of its neighbors. In this paper we will only consider an unweighted undirected network.

Finding densely connected groups of vertices or communities is an important network analytic operation. Many community detection methods are based on modularity maximization. The metric of modularity was proposed by Newman and Girvan [8] and is based on the concept that random networks do not form strong communities. Given a partition of a network into groups of vertices, let C_{ij} represent the fraction of total edges across in group i and group j and $a_i = \sum_j C_{ij}$ be the fraction of edges with one endpoint in group i. Therefore, the probability of edges that have one endpoint at a node in i (j) is a_i (a_j). Thus, the expected number of intra-community edges with a group i is a_i^2. The total fraction of links in group i is C_{ii}. So, a comparison of the actual and expected values, summed over all groups of the partition, gives us the modularity, i.e., the deviation of the partition from random connections: $Q = \sum(C_{ii} - a_i^2)$.

Generally, high modularity indicates a good division to communities. However, computing the optimal modularity is an NP-hard problem [3]. Heuristics for maximizing modularity include spectral partitioning, divisive, and agglomerative methods [9]. Detecting communities using modularity maximization can be hampered by the resolution limit, that is, the algorithms are unable to detect communities smaller than a certain size [5]. The Louvain method [2] addresses this problem by creating a hierarchy of communities, with the smaller ones discovered in the initial iterations and the larger super-communities in the subsequent iterations. This somewhat ameliorates the resolution-limit problem, compared to the CNM algorithm, as the communities in each level of hierarchy can point to finer resolutions of the community.

As networks increase in size, the use of parallel algorithms to handle large data becomes essential. However, we know of only two approaches to parallelize maximum modularity algorithms. The first implementation is based on the label propagation algorithm by Raghavan et al. [10]. This is a linear-time algorithm for community detection. Initially all vertices are assigned a unique label, and with subsequent iterations, the vertices adopt labels of their neighbors to denote the community in which they belong. Label propagation is based on local updates which render it relatively easier to parallelize, and a highly scalable implementation of this algorithm has been implemented for GPGPUs by Soman et al. [13].

The second implementation is based on the method proposed by Clauset et al. [4]. In this method, each vertex is initially assigned to a separate community. The edges in the network are weighted to represent the increase in modularity if the endpoints are combined to one community. At each subsequent iteration, the pair of vertices with highest edge weight is combined. This process of finding the heaviest weight is similar to the maximum matching problem. In the multithreaded parallel algorithm (on CRAY XMT and OpenMP) by Reidy et al. [12], the combination process is implemented as a parallel maximal matching algorithm.

3 A Shared-Memory Algorithm for Parallelizing the Louvain Method

In this section we describe our parallel implementation of the Louvain method [2] for modularity maximization. Agglomeration methods for modularity maximization initialize each of the n vertices into a separate community. At each iteration a pair of vertices whose merging produces the maximum increase in modularity is combined. The Louvain method improves the agglomerative process of modularity maximization due to three main contributions.

The *first contribution is to increase the speed*—instead of considering *all* vertex pairs, the Louvain method considers the maximum increase in modularity amongst every vertex and its neighbors. Thus, in one pass through the network, instead of identifying only one new community and reducing the number of communities to $n - 1$, this method can identify multiple communities and potentially reduce the number of communities to $n/2$.

The *second contribution is to improve flexibility*—a greedy heuristic does not always produce optimal results. The Louvain method attempts to improve on modularity maximization by removing vertices from their assigned communities and checking if modularity can be improved by reassigning the vertex to any of the other neighboring communities. This process is repeated over several iterations.

The *third contribution focuses on the hierarchical nature* of social networks. Once the initial allocation of communities is determined, in the second phase, the Louvain method joins the vertices within a community into supervertices. The community detection algorithm is then repeated over these supervertices. Thus, if required we can identify different levels of connection between the vertices.

These three contributions are the key features of the Louvain method and any parallel implementation of this algorithm should preserve these features. Algorithm 2 provides the pseudocode for the Louvain method. The primary technique is the same as described in the paper by Blondel et al. [2]; however, the statements have been rearranged to highlight regions most amenable to parallelization.

3.1 Parallel Implementation of the Louvain Method

Regions with loops (*for* and *while*) are the most natural parts of the code to exhibit parallelism. Many parts of the initialization process can be trivially parallelized. These include *(i)* assignment of values to $Degree$ (`Line 3`), *(ii)* computing value of Q by reduction (`Line 4`), and *(iii)* assignment of vertices to communities in parallel (`Lines 6--8`). However, the main body of the algorithm is not as easy to parallelize. We will now consider areas of iteration in this section of the code (`Lines 10--36`) and evaluate how amenable they are to parallelization.

We first consider the *while* loop at `Line 13` which signals the beginning of the first phase. At a first glance, this loop seems easy to parallelize, since the number of iterations depend on a constant. However, this loop deals with the flexibility of the Louvain method. Each iteration evaluates the community assignment obtained from the previous iteration and attempts to find a better one if possible. If each of the iterations was executed in parallel, then the results from earlier iterations cannot be incorporated.

The next loop we encounter is at `Line 14`, which represents going through each supervertex (or community) to identify the best assignment. If the latest mapping of vertices to communities is known, then this process can be executed in parallel. However, to ensure that no process reads an older value of community, the code in `Lines 28--33` has to be in a critical section.

Within the body of the loop at `Line 13`, we encounter two other regions of potential parallelization. The first is at `Line 21`, where we find the set of neighboring communities N_c. This operation requires us to first find the neighbors of the vertices within community c and then add the communities of the neighbors to N_c. This process can be implemented in parallel for each vertex.

The code for finding the best community amongst the members of N_c (`Lines 23--27`) also lends itself to be easily parallelized. For each neighboring community n, the change in modularity, dQ_n, due to adding c can be computed in parallel. Moreover, if we store the dQ_n of each community in an array, then finding the maximum increase in modularity becomes a reduction operation. Finally, based on

Algorithm 2 Louvain Method for Modularity Maximization

Input: A graph $G = (V, E)$. Vector A to store fraction of edges of each community
Output: A vector VID for mapping vertices to communities
Q to store value of modularity.

```
 1: procedure Initialization
 2: IT_int=0
 3: Degree = A                                        /* Store the values of A in Degree */
 4: Q = -Σ_{v=1}^{|V|} A[v]^2
 5: Old_Q = Q - 1                                     /* Initialize Modularity Value */
 6: for all v ∈ V do                                  /* Assign individual communities to each vertex */
 7:     Set VID[v].node = v
 8:     Set VID[v].comm = v
 9: end for
10: Set Total_Comms to |V|
11: procedure Louvain Method
12: while Old_Q < Q do
13:     Old_Q = Q                                     /* Beginning Phase 2 */
14:     while IT_int < 4 do
15:         for all c < Total_Comms do                /* Going Through Each Community */
16:             Set Cur_Comm to c                     /* Initialize current community of c */
                                                      /* Remove c from Cur_Comm */
17:             A[Cur_Comm] = A[Cur_Comm] - Degree[Cur_Comm]
18:             Set dQ to increase in modularity by adding c to Cur_Comm
19:             Q = Q - dQ
                                                      /* Find best community for c */
20:             Find set of neighboring communities N_c of c
21:             Max_dQ = dQ
22:             for all n ∈ N_c do
23:                 Compute dQ_n, change in modularity by adding c to n
24:                 if dQ_n > Max_dQ then
25:                     Max_dQ = dQ_n
26:                     Set New_Comm to n
27:                 end if                            /* Move c to New_Comm */
28:                 A[Cur_Comm] = A[Cur_Comm] - Degree[Cur_Comm]
29:                 A[New_Comm] = A[New_Comm] + Degree[Cur_Comm]
30:                 for all v ∈ V do
31:                     if VID[v] = Cur_Comm then
32:                         VID[v] = New_Comm
33:                     end if
34:                     Update Q = Q + Max_dQ
35:                 end for
36:             end for
37:         end for                                   /* End of Phase 1 */
38:     end while
39:     Combine communities to supervertices
40:     Total_Comms = max(VID.comm)
41:     Reduce size of A to only contain valid communities
42: end while
```

the identity of the most suitable community to join, we can update the community assignments from `Lines 28--33` in a critical section as discussed earlier.

Based on this analysis we see that the primary bottleneck in phase I (`Lines 13--33`) is the update of (i) the fraction of edges associated with each community, A vector (`Line 17 and Lines 28--29`); (ii) the community assignment, VID vector (`Lines 30--32`); and (iii) the modularity Q (`Line 19, Line 33`). These updates reduce the parallel potential of the code since they have to be computed sequentially. We can, however, reduce some of the operations. The value of Q need not be computed during the iterations in phase 1 but can be computed later from the community assignment stored in VID. We can thus update the value of Q in phase 2, where it will be a perfectly parallel operation. Since the value of dQ with respect to the current community Cur_Comm is already computed, we do not have to recompute it in (`Lines 23--27`). We therefore do not have to update the value of A at (`Line 17`). Moreover the updates in (`Lines 28--32`) need be implemented only when the vertex moves from its earlier community, i.e., New_Comm is different from Cur_Comm. These updates are implemented as an atomic operation on A as is the reduction operation in obtaining the maximum change in modularity. This implicit ordering ensures that the communities are combined to increase modularity only and therefore guarantees the convergence.

The second phase has fewer opportunities for parallelization, and these depend on the implementation choices of the operation. For example, to easily identify the vertices belonging to the same community, we sorted the vector VID based on increasing order of the communities such that vertices within the same community are arranged consecutively. This sorting was done in parallel using a parallel mergesort algorithm. The parallel implementation of the Louvain code is given in Algorithm 3.

4 Empirical Results

In this section we present our experimental results that demonstrate that the algorithm is highly scalable and the modularity values are not significantly impacted by parallelization. We also observe that if a network has a well-defined community structure, then the algorithm is faster and the deviations amongst the timings and the values are less than for networks with more amorphous communities.

While results for most parallel network algorithms are presented on massive networks, we believe that executing the algorithms on relatively smaller datasets is a more rigorous test for scalability. Algorithms such as agglomerative-based modularity maximization require frequent synchronization amongst neighbors within distance k, and the smaller the network, the more likely are the potential conflicts. Moreover, our goal is to present a parallel template for agglomeration-based methods that can be easily executed on any shared-memory machine, and in that spirit we implemented our algorithm on an Opteron quad-core system with only 8

Algorithm 3 Parallel Implementation Louvain Method

Input: A graph $G = (V, E)$. Vector A to store fraction of edges of each community
Output: A vector VID for mapping vertices to communities
Q to store value of modularity.

1: **procedure** Initialization
2: IT_int=0
3: $Total\_its$=4 **The number of outer iterations**
4: $Degree = A$ **Values assigned in parallel**
5: $Q = -\Sigma_{v=1}^{|V|} A[v]^2$ **Obtained by parallel reduction**
6: $Old\_Q = Q - 1$
7: **for all** $v \in V$ **do**
8:     **in parallel**
9:     Set $VID[v].node = v$
10:     Set $VID[v].comm = v$
11: **end for**
12: Set $Total\_Comms$ to $|V|$
13: **procedure** Louvain Method
14: **while** $Old\_Q < Q$ **do**
15:     $Old\_Q = Q$
16:     **while** $IT\_int < Total\_its$ **do**   /* Beginning Phase 1 */
17:         **for all** $c < Total\_Comms$ **do**
18:             **in parallel**
19:             Set Cur_Comm to c
20:             Set dQ to increase in modularity by adding c to Cur_Comm
21:             Find set of neighboring communities N_c of c **in parallel**
22:             $Max\_dQ = dQ$
23:             Set New_Comm to Cur_Comm
24:             **for all** $n \in N_c$ **do**
25:                 **in parallel**
26:                 Compute dQ_n, change in modularity by adding c to n
27:                 **if** $dQ_n > Max\_dQ$ **then**
28:                     **use parallel reduction**
29:                     $Max\_dQ = dQ_n$
30:                     Set New_Comm to n
31:                 **end if**
32:                 **if** $Cur\_Comm != New\_Comm$ **then**
33:                     **use atomic operations to update** A
34:                     $A[Cur\_Comm] = A[Cur\_Comm] - Degree[Cur\_Comm]$
35:                     $A[New\_Comm] = A[New\_Comm] + Degree[Cur\_Comm]$
36:                     **for all** $v \in V$ **do**
37:                         **if** $VID[v] = Cur\_Comm$ **then**
38:                             $VID[v] = New\_Comm$
39:                         **end if**
40:                     **end for**
41:                 **end if**
42:             **end for**
43:         **end for**   /* End of Phase 1 */
44:     **end while**
45:     Combine communities to supervertices **parallel mergesort**
46:     Compute Q **in parallel**
47:     $Total\_Comms = max(VID.comm)$
48:     Reduce size of A to only contain valid communities
49: **end while**

GB RAM. Such a machine is easily accessible and highlights the portability of our algorithm. As the results show the algorithm is scalable and can therefore perform equally well on larger systems and networks.

Our experimental setup is as follows: we created two sets of LFR benchmarks [7] of 8,000 and 10,000 vertices, respectively, with mixing parameters, μ being .1, .3, .5, and .7. Lower mixing parameters indicate a more distinctive community distribution. The average degree was set to 15 and the maximum degree was set to 50. The power-law exponent for degree distribution was 2 and the exponent for community size distribution was 1. The community size range was from 7 to 50.

4.1 Scalability Results

A parallel algorithm is scalable if the execution time decreases as the number of processing units are increased. We present results of strong scalability which measures how the execution time changes as the number of processing units are varied for a fixed problem size. We experiment by changing number of threads from 2, 4, 8, 16, and 32. However the inner parallel loops at `Line 17` and `Line 20` are not dependent on the size of the network but rather on the average size of the community. Note that this value (7–50) is much smaller compared to the number of vertices, and using 16 or 32 threads on this region of the code will be inefficient. Therefore, for these two lines, the number of threads was set to constant at 4 for all the executions.

Figure 1 shows that the execution time progressively decreases as the number of processing units are increased. The plots for networks with 10K vertices are smoother than the one for networks with 8K vertices. This is because the 8K networks are smaller, and as the communities in the networks become less distinct (for higher μ values), there are more dependencies, hence more conflicts at the critical section, amongst neighboring vertices. We also notice that even though the plots of the 10K networks are smoother, there is a slight increase at 32 threads. This is because as the number of threads increases, there is more contention for the critical section, and the benefits of parallelism outweigh the sequential costs. Thus, the results show that scalability is affected by both the network properties and the size.

We now compare how the execution time changes over the iterations in phase 2. Recall that in phase 1, which is over 4 iterations, the algorithm removes and joins vertices (or supervertices) from communities to find the highest modularity at that level. At the end of phase 1, the vertices within a community are combined to a supervertex, which effectively reduces the size of the network. Therefore, it is not surprising that at each outer (phase 2) iteration, the execution time significantly decreases. This result indicates that the focus for improvement due to parallelism should be on the earlier iterations, notably iteration 1. We also observe that as the mixing value increases, the item for the inner (phase 1) iteration also increases. This again conforms to our theory that the more ambiguous the community, the more time

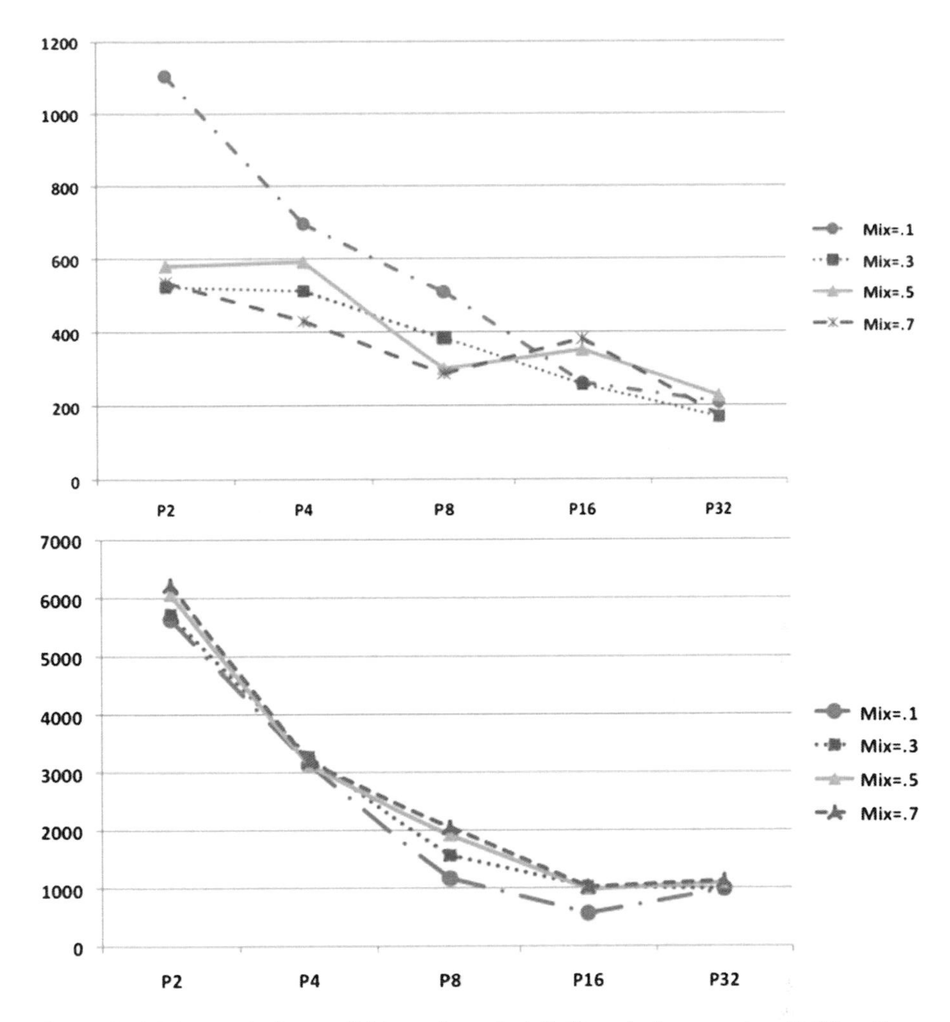

Fig. 1 *Scalability results for parallel Louvain method:* (*left*) results for networks with 8K vertices; (*right*) results for networks with 10K vertices. Each point represents the total execution time of one network for a given mixing parameter and a processor

is required at the critical section. The individual times within each phase are about the same except for a few outliers. We have more outliers for the 8K networks. We believe that the drastic change in results is reflective of a major shift in community assignments across iterations. Also, the time taken for the 10K networks is slightly larger than the time for the 8K networks, which is natural due to the difference in size. Figure 2 shows the execution time for the 8K networks with 16 threads, and Fig. 3 shows the execution time for the 10K networks with 8 threads. We have also conducted preliminary experiments on a larger AMD Magny-Cours machine that gave about 4 times speedup.

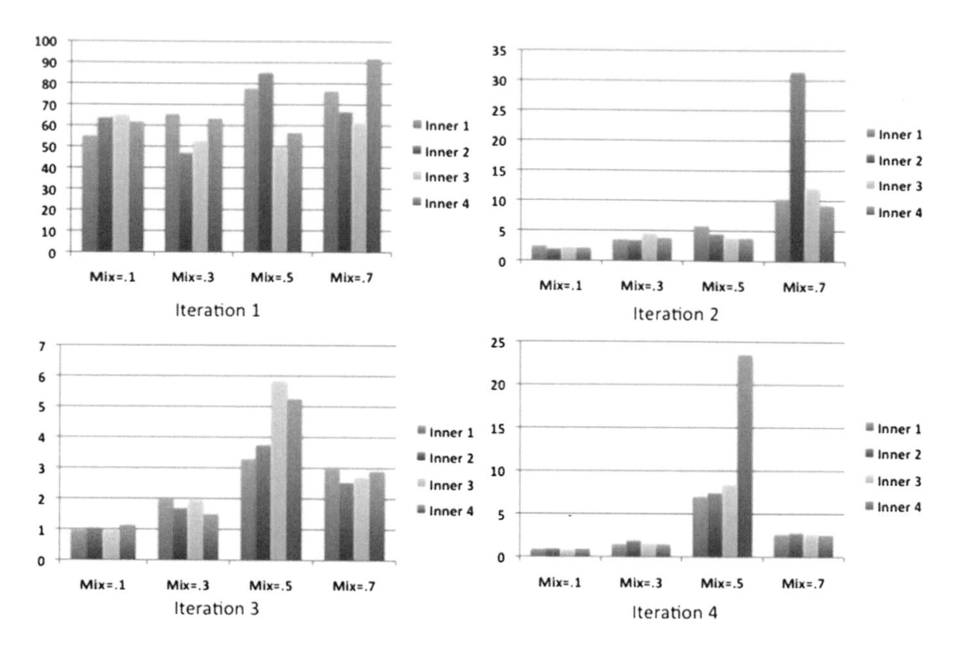

Fig. 2 *Time taken per inner iteration for 8K networks:* the bars in each plot show the parallelized inner iterations, and the individual plots represent the outer loop. The results are for executions on 16 threads

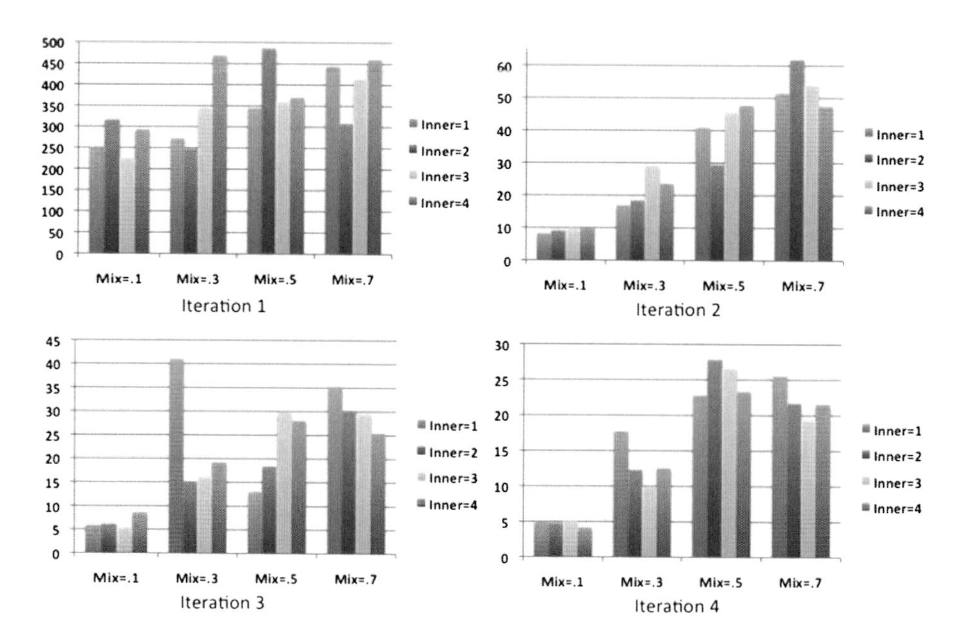

Fig. 3 *Time taken per inner iteration for 10K networks:* the bars in each plot show the parallelized inner iterations, and the individual plots represent the outer loop. The results are for executions on 8 threads

4.2 *Evaluation of Correctness*

An empirical method for evaluating the correctness of parallel programs is by comparing the results with those obtained by the sequential counterpart. However, as discussed in [2], results of the Louvain method, like all other combinatorial optimization techniques, are dependent on the order in which vertices are processed. In other words, there is no one single result to compare. This effect is aggravated in the parallel case, because the sequence in which processors execute the code can change for each execution cycle. Indeed, even in sequential cases comparison between the results of different community detection methods is an area of active research [11]. We compared the communities using normalized mutual information (NMI). The values of NMI range from 1 to 0, the higher the number, the better the similarity between two communities. We observed that for lower mixing parameters, the NMI values across the processors were around .90; however, as the mixing value increased, the difference between communities obtained using different processors was also higher. For mixing parameter of .7, the difference was as much as .76. These results clearly show that as communities become more distinct, the ordering of the vertices (which is affected by parallelization) plays a key role in the community distribution.

However, the Louvain method is ultimately designed to increase modularity. Therefore, a more accurate evaluation of our algorithm is to compare the standard deviation of the modularity value across the different processors. Figure 4 gives the values of modularity and the standard deviation across networks and across processors. All the standard deviation values are quite low, though the modularity values are more consistent when $\mu = .1$. We also observe that lower mixing parameters, i.e., distinct communities, produce higher modularity. Once again the variations are more common in the 8K network, because of increased number of contentions. This supports our claim that experiments on smaller networks provide stronger testing examples for parallel algorithms.

5 Application to Time-Varying Networks

Parallel algorithms for networks that vary over time are more challenging. This is because, as the network evolves, the distribution of the data to processing units should also change. Although there is no explicit distribution of data in shared-memory architectures, each addition and deletion of edges and vertices can trigger change in the allocation of data. Let us examine what data structures are the most suitable for dynamic networks. A natural choice for representing networks is by adjacency lists, since it is easy to add or remove elements. However, link lists are not stored contiguously in memory, and therefore, the cache is not properly utilized. Adjacency arrays on the other hand store data in contiguous location taking full advantage of the cache. However, adding or deleting elements from an array requires several operations to readjust the storage locations. We propose using data structures

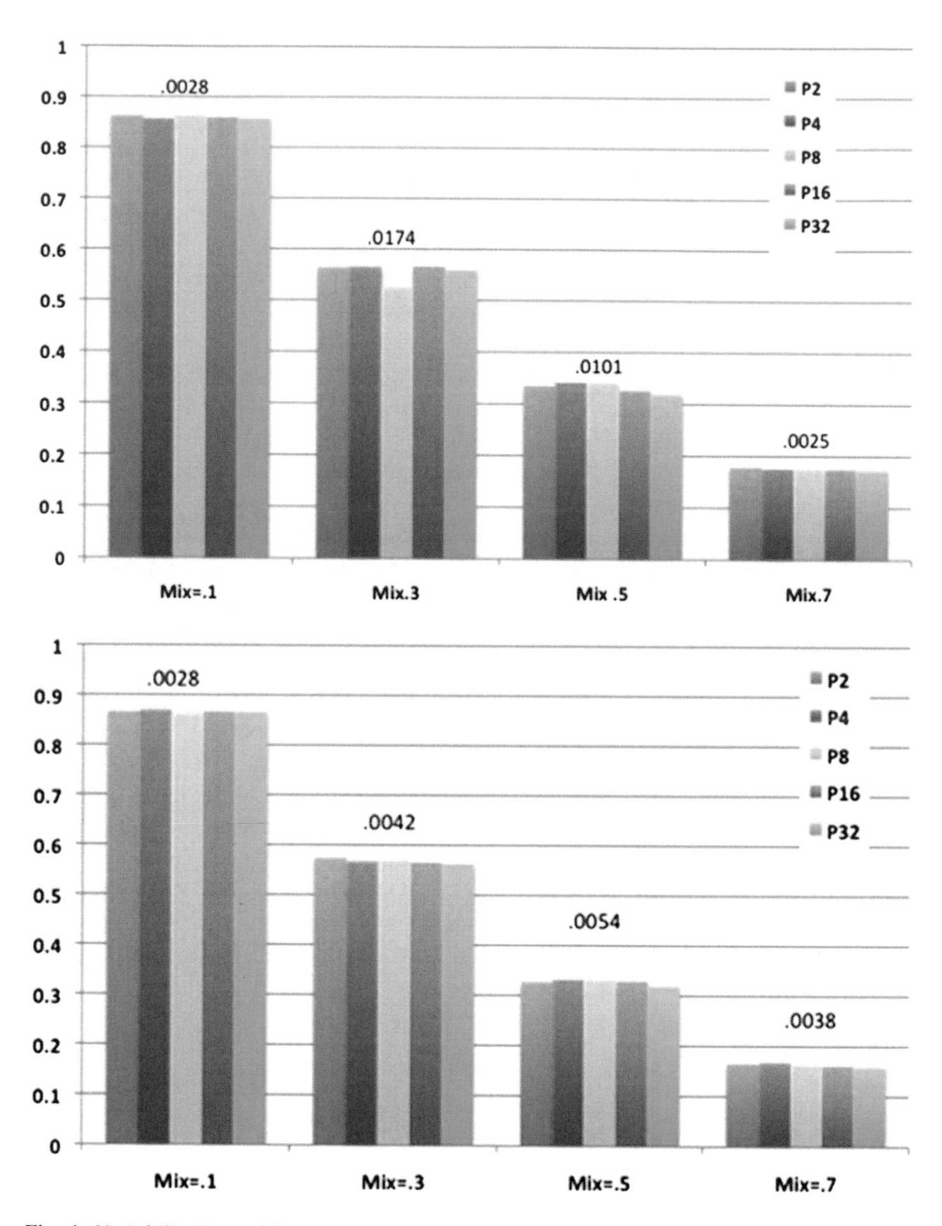

Fig. 4 *Variability in modularity across processors:* (*left*) results for networks with 8K vertices; (*right*) results for networks with 10K vertices

that combine the flexibility of link lists with the locality of arrays. Each vertex is associated with an array of its neighbors and some extra buffer spaces, marked by -1. The buffers are replaced by the values of the corresponding neighbors as new edges are added. Conversely, as edges are deleted, the values of the associated neighbors change to -1. Thus, by including some extra storage, we can effectively get the advantages of both arrays and link lists. Such combination of link lists, arrays, and buffers is a natural choice for updating dynamic networks, and a version using one link list per vertex is used in multithreaded graph algorithms [6]. Note that this data structure requires only local updates, and only the processing units associated with the changed edges and vertices need be involved.

Before discussing the potential for parallel modularity maximization on dynamic networks, we show how the underlying principles of parallelizing the Louvain method can be applied to other agglomerative modularity maximization methods as well. The initialization process is almost identical for most of the methods and can be trivially parallelized. The primary parallel operation deals with combining the vertex pairs at each iteration. If the updates are based on local information (such as the label propagation method or the Louvain method), then there can potentially be multiple combinations per iteration. The challenge is to identify the regions of conflict in writing to data structures and use a critical section (or atomic operations) for those regions. Alternatively the update can be based on global information such as the CNM method (although the parallel version described in [12] uses weighted matching which is based on local updates), and then, reduction can be used to find the optimal pair of vertices to be merged. The final part of the algorithm is the update of the data structures to the new value. Updates usually have to be implemented in a critical section, so there is not much scope for parallelization. However, the algorithm can be modified to eliminate some data structures whose values can be implicitly obtained (e.g., computing the value of modularity).

These ideas also translate in the case of dynamic networks. One of the primary objectives in designing algorithms for dynamic networks is to avoid recomputation. As discussed, the updates to the edges and vertices and consequently the updates at the initialization stage are local. Therefore, they can be easily parallelized. In the iteration stage, we only need to recompute the portions of the network affected by the change. To identify these regions we store the list of vertex pairs that were combined, from the first time step. For each subsequent atomic change, we designate endpoints of the modified edge as perturbed vertices. We replicate the operations from the list if the associated vertices are not marked as perturbed. When we find a perturbed vertex, we invoke the original parallel algorithm for the remainder of the iterations. The new list of operations is used for the next time step. This method of dynamic updates has been shown to be effective for the CNM algorithm, although for a sequential implementation [1].

6 Conclusions and Future Work

In this paper, we have presented a shared-memory parallel algorithm for the Louvain method for modularity maximization and discussed how the template can be extended to other agglomeration methods and also to finding modularity in dynamic networks. Our results show that while our implementation is scalable and produces modularity values commensurate with those expected from a sequential case, the performance of our algorithm and variability of the results depend on the properties of the network as well as its size.

We plan to apply this initial template to other modularity maximization techniques as discussed as well as on dynamic networks. We also plan to experiment with different vertex orderings, such as reverse Cuthill Mckee, in order to bring neighboring vertices closer in memory. This we anticipate will result in fewer contentions at the critical sections. Finally, since networks with stronger community structures produce more stable results, we also plan to extend our studies to classify networks based on their community structure.

Acknowledgement This work was completed with the support of the College of Information Science and Technology, University of Nebraska at Omaha (UNO), and the FIRE grant from the UNO Office of Research and Creative Activity.

References

[1] S. Bansal, S. Bhowmick, P. Paymal, Fast community detection for dynamic complex networks, communications in computer and information science, vol. 116, in *Proceedings of the Second Workshop on Complex Networks*, 2010
[2] V.D. Blondel, J.-L. Guillaume, R. Lambiotte, E. Lefebvre, Fast unfolding of communities in large networks. J. Stat. Mech. Theory Exp. **2008**(10), P10008 (25 Jul 2008). doi:10.1088/1742-5468/2008/10/p10008
[3] U. Brandes, D. Delling, M. Gaertler, R. Gorke, M. Hoefer, Z. Nikoloski, D. Wagner, On modularity clustering. IEEE Trans. Knowl. Data Eng. **20**(2), 172–188 (2008)
[4] A. Clauset, M.E.J. Newman, C. Moore, Finding community structure in very large networks. Phys. Rev. E **70**(6), 66111 (2004)
[5] B.H. Good, Y.-A. de Montjoye, A. Clauset, The performance of modularity maximization in practical contexts. Phys. Rev. E **81**, 046106 (2010)
[6] D. Ediger, J. Riedy, H. Meyerhenke, D.A. Bader, Tracking structure of streaming social networks, in *5th Workshop on Multithreaded Architectures and Applications* (MTAAP) (2011)
[7] A. Lancichinetti, S. Fortunato, Benchmarks for testing community detection algorithms on directed and weighted graphs with overlapping communities. Phys. Rev. E **80**, 016118 (2009)
[8] M.E.J. Newman, M. Girvan, Finding and evaluating community structure in networks. Phys. Rev. E **69**(2), 026113 (2004)
[9] M.A. Porter, J.-P. Onnela, P.J. Mucha, Communities in networks. Not. Am. Math. Soc. **56**(9), (22 Feb 2009)
[10] U.N. Raghavan, R. Albert, S. Kumara, Near linear time algorithm to detect community structures in large-scale networks. Phys. Rev. E **76**, 036106 (2007)
[11] W. Rand, Objective criteria for the evaluation of clustering methods. J. Am. Stat. Assoc. **66**(336), 846–850 (1971)
[12] E.J. Riedy, H. Meyerhenke, D. Ediger, D.A. Bader, Parallel community detection for massive graphs, in *10th DIMACS Implementation Challenge - Graph Partitioning and Graph Clustering*, 2012
[13] J. Soman, A. Narang, Fast community detection algorithm with GPUs and multicore architectures, in *Proceedings of the 2011 IEEE International Parallel & Distributed Processing Symposium*, 2011
[14] V.A. Traag, P. Van Dooren, Y. Nesterov, Narrow scope for resolution-limit-free community detection. Phys. Rev. E **84**, 016114 (2011)

Multi-scale Modularity and Dynamics in Complex Networks

Renaud Lambiotte

1 Introduction

A broad range of systems are made of elements in interaction and can be represented as networks. Important examples include social networks, the Internet, airline routes, and a wide range of biological networks. The study of networks has emerged in the last decade as one of the fundamental building blocks in the wider study of complex systems [8, 15, 39]. Complex network theory emphasizes that the interactions between the individual components of the system are primordial to understand global emergent behavior and give the possibility to analyze systems of a very different nature within a single framework. Despite the fact that nodes and links have a different nature and are driven by different mechanisms for, e.g., the Web and food webs, the identification of similar structures in various complex systems suggests the existence of generic organization principles. A good example is the omnipresent multi-scale modular organization of complex networks, namely, the fact that they are made of modules, also called communities [62]. The precise mathematical definition of what communities are is still the subject of active research, but most definitions agree on the fact that modules, also called communities, are defined as subnetworks that are locally dense even though the network as a whole is sparse [20, 49]. In many complex systems it seems that modularity does not exist only at a single organizational scale but rather that each module can be further partitioned into a set of sub-modules and within each sub-module there may be sub-sub-modules, etc. In other words, many systems have the fractal property of multi-scale modularity. The detection of such community structure can be of importance for the understanding of the interplay between the structural, dynamical, and functional features of the network. This identification

R. Lambiotte (✉)
Department of Mathematics, University of Namur, Namur, Belgium
e-mail: renaud.lambiotte@fundp.ac.be

A. Mukherjee et al. (eds.), *Dynamics On and Of Complex Networks, Volume 2*,
Modeling and Simulation in Science, Engineering and Technology,
DOI 10.1007/978-1-4614-6729-8_7,

also has the advantage of providing a coarse-grained representation of the system, thereby allowing to sketch its organization and to identify sets of nodes that are likely to have hidden functions or properties in common. The main purpose of this chapter is to present methods uncovering the multi-scale organization of complex networks. In particular, we will emphasize methods finding modules based on the exploration of the network by a random walker at different time scales. The diffusive process allows for multistep transitions exploring further afield the structure of the graph, which results in the detection of community structure across scales, from finer to coarser. Time thus acts as a resolution parameter, allowing to zoom in and out to uncover the multi-scale structure of the system. As we will argue, this dynamic approach has the further advantages of taking into account the constraints imposed by the network structure on flows and of providing a unifying framework for a broad range of community detection methods.

2 Communities: Why Communities?

For many years, researchers, intrigued by the apparent modularity of many social, technological, and biological systems, have searched for mechanisms driving the evolution of systems towards a modular architecture [38]. One of the earliest and most influential ideas was formulated by Simon [62, 63] who argued that a nearly decomposable system, namely, a system of sparsely interconnected modules, allows faster adaptation of the system in response to changing external conditions. Modular systems can evolve by evolution in one module at a time or by duplication and mutation of modules. Well-adapted modules thus represent stable intermediate states whose stability is not jeopardized by evolution in other modules. This robustness represents a major advantage for any system evolving under fluctuating selection criteria, and this may explain the general prevalence of modular architectures across a very wide range of systems. This idea that a hierarchically modular design will be more rapidly and robustly assembled is illustrated by Simon [62] in a parable about two watchmakers, called Hora and Tempus:

> The watches the men made consisted of about 1,000 parts each. Tempus had so constructed his that if he had one partly assembled and had to put it down to answer the phone say it immediately fell to pieces and had to be reassembled from the elements. The better the customers liked his watches, the more they phoned him, the more difficult it became for him to find enough uninterrupted time to finish a watch.
>
> The watches that Hora made were no less complex than those of Tempus. But he had designed them so that he could put together subassemblies of about ten elements each. Ten of these subassemblies, again, could be put together into a larger subassembly; and a system of ten of the latter subassemblies constituted the whole watch. Hence, when Hora had to put down a partly assembled watch in order to answer the phone, he lost only a small part of his work, and he assembled his watches in only a fraction of the man-hours it took Tempus.

This idea has since been explored further. For instance, in [25], it has been argued that modular networks are optimal for performing tasks in a changing environment.

In situations where different tasks share basic sub-functions, evolutionary pressure produces networks where modules specialize in these sub-functions and where each of the tasks is obtained by a rapid recombination of these building functions. Other types of argument have also been put forward, including:

- Modular network topology is associated with a rich nonlinear dynamical behavior. Modular networks tend to produce time-scale separation, i.e., fast intra-modular processes and slow inter-modular processes [3, 45, 46], which helps at preserving coexistence and diversity in the system [27] and at finding a balance between segregated and integrated activity [47, 59]. The high dynamical complexity [64] leads to complex dynamical states such as transient chimera states [1, 60] where synchronization and desynchronization coexist across the network. Hierarchical modularity also enhances dynamical reconnectability [51], as marginally stable networks can be combined or divided while preserving stability.
- Another plausible mechanism for the formation of modules is coevolution, as the network structure and function coevolve [22]. A broad range of models with adaptive rewiring typically incorporate a reinforcement of links between synchronized units and a pruning of links between asynchronized ones. This feedback between structure and dynamics, similar to synaptic plasticity in neuronal dynamics, naturally drives the emergence of inhomogeneities and modules in networks [31].
- Modular networks have the property of small-worldness which is advantageous in a broad range of systems by combining high clustering and global connectivity [70]. For instance, in social systems, small-worldness balances the relative benefits of social cohesion and brokerage [30]. In neuroscience, the high clustering of connections favors locally segregated processing (with low wiring cost) of specialized functions, while the short path length supports globally integrated processing of more generic functions [65].

3 Combinatorial Approaches to Community Detection

3.1 Community Detection

Community detection aims at uncovering the modular organization of large-scale networks in an automatic fashion [20, 35, 49]. Most community detection methods find a partition of the nodes into communities, where most of the links are concentrated within the communities. Each node is assigned to one and only one community, i.e., partitions are not compatible with overlapping communities [2, 16, 44]. At the heart of a partitioning method, there is a mathematical definition for the quality of any partition. Once a quality function has been defined, different types of heuristics can be used in order to find, approximatively, its optimal partition, i.e., to find the partition having the highest value of the quality function. In a majority of cases, this quality function is based on the number of links within and/or across the communities, i.e., this is a combinatorial approach where quality is measured by counting certain motifs in the graph.

Modularity and Its Limitations

The most popular quality function for community detection is Newman–Girvan modularity, which we will describe in detail in this section. In the following, we focus on unweighted networks for the sake of clarity, but the results are directly applicable to weighted networks. Let A be the adjacency matrix of the network, where A_{ij} determines the presence of a link going from j to i. The in-degree and out-degree of node i are defined as $k_i^{\text{in}} \equiv \sum_j A_{ij}$ and $k_i^{\text{out}} \equiv \sum_j A_{ji}$, respectively; $L \equiv \sum_{i,j} A_{ij}$ is the total number of links in the network. If the network is undirected, the adjacency matrix is symmetric $A_{ij} = A_{ji}$, $k_i^{\text{in}} = k_i^{\text{out}} = k_i$, and the number of undirected links $m = L/2$.

Modularity is a function of the adjacency matrix A and of the partition $\mathcal{P}$ of the nodes into communities. It measures if links are more abundant within communities than would be expected on the basis of chance

$$Q = (\text{fraction of links within communities}) - (\text{expected fraction of such links}) \tag{1}$$

and reads

$$Q = \frac{1}{L} \sum_{C \in \mathcal{P}} \sum_{i,j \in C} \left[A_{ij} - P_{ij} \right] \tag{2}$$

where $i, j \in C$ is a summation over pairs of nodes i and j belonging to the same community C of $\mathcal{P}$ and therefore counts intra-community links. The null hypothesis is an extra ingredient in the definition and is incorporated in the matrix P_{ij}. P_{ij} is the expected number of links between nodes i and j over an ensemble of random networks with certain constraints. These constraints correspond to known information about the network organization, i.e., its total number of links and nodes, which has to be taken into account when assessing the relevance of an observed topological feature.

Let us first consider undirected networks. Two standard choices of null model are

$$P_{ij} = \langle k \rangle^2 / 2m, \text{ then } Q \equiv Q_{\text{unif}} \tag{3}$$

where $\langle k \rangle = 2m/N$ is the average degree and the only constraint is thus the total number of links in the network, and

$$P_{ij} = k_i k_j / 2m, \text{ then } Q \equiv Q_{\text{conf}}. \tag{4}$$

where randomized networks now preserve the degree of each node. The latter null model is usually preferred because it takes into account the degree heterogeneity of the network [42]. More complicated null models can in principle be constructed in order to preserve other properties of the network under consideration [5, 43]. For instance, in the case of directed networks, it is common to use the null model

$$P_{ij} = k_i^{in} k_j^{out} / L, \text{ then } Q \equiv Q_{\text{dir}}. \tag{5}$$

preserving the in- and out-degrees of each node.

In the case of undirected networks, again, it is interesting to note that Q_{unif} and Q_{conf} are naturally related to the combinatorial Laplacian $L_{ij}^{(C)} = A_{ij} - k_i \delta_{ij}$ and the (normalized) Laplacian $L_{ij} = A_{ij}/k_j - \delta_{ij}$, respectively,[1] and, more generally, to the dynamics induced by these operators, as we will discuss more in detail in Sect. 4. For Q_{conf}, this relation is particularly clear after expressing modularity in terms of the (right) eigenvectors v_α of L_{ij}, i.e., v_α satisfy $\sum_j L_{ij} v_{\alpha,j} = \lambda_\alpha v_{\alpha,i}$. Without loss of generality, we assume that $\lambda_1 > \lambda_2 \geq \ldots \geq \lambda_\alpha \geq \ldots \geq \lambda_N$. The dominant eigenvector v_1 of eigenvalue $\lambda_1 = 0$ is given by $v_{1;i} = k_i/2m$ and is unique if the network is connected. By using a spectral decomposition of L_{ij}, one finds [13]

$$Q_{\text{conf}} = \sum_{\alpha=2}^{N} \frac{\lambda_\alpha + 1}{2m} \sum_C \sum_{i,j \in C} v_{\alpha;i} v_{\alpha;j} \tag{6}$$

where the contribution of the dominant eigenvector v_1 and the null model have cancelled each other out.

Modularity has become an essential element of a large number of clustering methods for large-scale networks. These methods aim at optimizing the modularity of a graph, i.e., finding the partition having the maximal value of Q. An exhaustive optimization of Q is impossible because of the explosion in the number of ways to partition a graph, when its size increases. It has been shown that the optimization of modularity is an NP-complete problem [9]. For this reason, several heuristics have been proposed to find high-quality partitions [7, 11, 20, 42, 48]. The optimization of modularity has the advantage of being performed without a priori specifying the number of modules nor their size. This procedure has been shown to produce useful and relevant partitions in a number of systems [41]. Unfortunately, it has also been shown that modularity suffers from several limitations [19, 23], partly because modularity optimization produces one single partition, which is not satisfactory when dealing with multi-scale systems. Related to this issue, there is the so-called resolution limit of modularity [19], namely, the fact that modularity is blind to modules smaller than a certain scale. This point originates from the bias of modularity

---

[1] Strictly speaking, the normalized Laplacian of a network is $L'_{ij} = A_{ij}/(k_i^{1/2} k_j^{1/2}) - \delta_{ij}$, but L and L' are equivalent by similarity as $L'_{ij} = k_i^{-1/2} L_{ij} k_j^{1/2}$.

towards modules having a certain scale which might not be compatible with the system architecture [66]. This incompatibility also makes modularity inefficient in practical contexts as it may lead to a high degeneracy of its landscape [21], i.e., the existence of several distinct partitions having a modularity close to the optimum, which implies that approximate solutions of the optimization problem are very dissimilar and that a partition derived from modularity optimization has to be considered with caution.

Multi-scale Methods

Different methods have been proposed to go beyond modularity optimization. A first set of methods looks for local maxima of the modularity landscape in order to uncover partitions at different scales [56]. A good example is the so-called Louvain method, which is a greedy method taking advantage of the hierarchical organization of complex networks in order to facilitate the optimization of modularity [7]. This heuristic performs the optimization in a multi-scale way: by comparing the communities first of adjacent nodes, then of adjacent groups of nodes found in the first round, etc. It has been shown in several examples that modularity estimated by this method is close to the optimal value obtained with slower methods but also that intermediate partitions are meaningful and correspond to communities at intermediate resolutions [37]. This approach has the advantage of being fast, but it lacks theoretical foundations and is not able to uncover coarser partitions than those obtained by modularity optimization. Moreover, it may produce hierarchies even when the system is single scale or, worse, completely random [23] (see [37] for a discussion of how to deal with this issue).

Another class of methods is based on multi-scale quality functions. These quality functions incorporate a resolution parameter allowing to tune the characteristic size of the modules in the optimal partition and aim at uncovering modules at the true scale of organization of a network, i.e., not at a scale imposed by modularity optimization. The two most popular multi-scale quality functions are ad hoc, parametric generalizations of modularity. A first quantity is the parametric modularity introduced by Reichardt and Bornholdt [50]:

$$Q_\gamma = \frac{1}{2m} \sum_{C \in \mathcal{P}} \sum_{i,j \in C} \Big[A_{ij} - \gamma P_{ij} \Big], \tag{7}$$

which is usually defined for the configuration null model $P_{ij} = k_i k_j / 2m$ and mainly consists in changing the effective size of the system $m_{\text{eff}} = m/\gamma$. The optimization of Q_γ leads to larger and larger communities in the optimal partition when γ is decreased. This approach makes use of the size dependence of modularity: because of the factor $1/2m$ in the null model, modularity depends on the total size of the network and not only on its local properties.[2] Decreasing m_{eff} (increasing γ) increases the expected number of links γP_{ij} between i and j, which makes it less advantageous to assign i and j to the same community (because $A_{ij} - \gamma P_{ij}$ decreases).

An alternative approach proposed by Arenas et al. [4] keeps modularity unchanged but modifies the network by adding self-loops to the original network. This approach therefore consists in optimizing

$$Q_r = Q(A_{ij} + rI_{ij}). \tag{8}$$

As expected, increasing r has a tendency to decrease the size of the communities and the optimal partition of Q_∞ is made of single nodes. Even if increasing γ and r has, qualitatively, the same effect on the characteristic size of the communities, one should keep in mind that Q_γ and Q_r are in general optimized by different partitions, except if the network is regular and the resolution parameters verify $\gamma = 1 + r/\langle k \rangle$. It is also interesting to note that the quality function Eq. (8) was first proposed in order to preserve the eigenvectors of the adjacency matrix, as the eigenvectors of $A_{ij} + rI_{ij}$ and A_{ij} are obviously the same. From a partitioning viewpoint, however, the eigenvectors of A_{ij} do not matter as much as the eigenvectors of the combinatorial Laplacian $L_{ij}^{(C)}$ [18] and the normalized Laplacian L_{ij} [61]. Moreover, modularity is related to the eigenvectors of the Laplacian and not of the adjacency matrix; see Eq. (6). These observations suggest to adapt the unfitting quality function Eq. (8) and to optimize the modularity of a modified adjacency matrix preserving the eigenvectors of L_{ij}. This can readily be done by adding degree-dependent self-loops to the nodes

$$A'_{ij} = A_{ij} + r\frac{k_i}{\langle k \rangle}\delta_{ij}, \tag{9}$$

and by optimizing the quality function

$$Q'_r \equiv Q(A_{ij} + r\frac{k_i}{\langle k \rangle}\delta_{ij}). \tag{10}$$

This quality function is equivalent, up to a linear transformation, to Q_γ for any network, i.e., not only for regular networks, with $\gamma = 1 + r/\langle k \rangle$, thereby providing two alternative interpretations to resolution parameters.

Before closing this section, we should emphasize that many other types of methods have also been proposed to detect the multi-scale modular organization

---

[2] In a nutshell, this size dependence originates from a choice of null model where each pair of nodes i and j can be connected, given a certain number of available links m in the system, whatever the distance between i and j in the network. Local null models where pairs of nodes are randomly connected only within a finite radius of interaction are expected to hinder this effect.

of networks, e.g., the hierarchical extension of the Map formalism [55] whose single-scale version is described below, local algorithms [34], or the modeling of the system by hierarchical random graphs [10].

4 Communities: Dynamics and Function

4.1 Linear Dynamics

The behavior of complex systems is determined not only by the topological organization of their interconnections but also by the dynamical processes taking place among their constituents [6]. A faithful modeling of the dynamics is essential because different dynamical processes may be affected very differently by network topology. A full characterization of such systems thus requires a formalization that encompasses both aspects simultaneously, rather than relying only on the topological adjacency matrix [32]. In the simple case of linear dynamics alone, a broad range of qualitatively different processes can be defined on the same graph, e.g., with the same adjacency matrix A_{ij}. Let us consider the class of linear processes defined by the equation

$$x_{i;t+\tau} = \sum_j B_{ij} x_{j;t} \tag{11}$$

where the evolution of a quantity x_i, associated to node i, is driven by B_{ij}, a matrix somehow related to the adjacency matrix A_{ij}.

In the subclass of *random walk processes* alone, modeling the diffusion of some quantity or information between nodes, Eq. (11) takes the form

$$p_{i;t+\tau} = \sum_j T_{ij} p_{j;t} \tag{12}$$

where $p_{j;t}$ is the probability to observe a walker on node i at time t and where T is the transition matrix whose entry T_{ij} represents the probability to jump from j to i in a time interval of duration τ. T_{ij} is a nonnegative matrix verifying the condition $\sum_i T_{ij} = 1$ to ensure the conservation of probability. Except for those constraints, it is arbitrarily defined on a given graph. Important cases include:

- Discrete-time, unbiased random walk, where

$$T_{ij} = A_{ij} / k_j^{out} \tag{13}$$

- Biased random walks, where

$$T_{ij}^{(\alpha)} = \frac{\alpha_i A_{ij}}{\sum_k \alpha_k A_{kj}}, \tag{14}$$

and where α is an attribute biasing the motion of random walkers towards certain nodes.

- Continuous-time random walks, where walkers perform their jumps asynchronously. In the case of exponentially distributed waiting times between jumps, the transition matrix is

$$T_{ij} = \left(e^{-\tau L}\right)_{ij} \tag{15}$$

and relation (12) can be seen as the formal solution of the rate equation

$$\dot{p}_i = \sum_j \left(\frac{A_{ij}}{k_j^{out}} - \delta_{ij} \right) p_j \equiv -\sum_j L_{ij} p_j. \tag{16}$$

A Taylor expansion of (15) in terms of τ clearly shows that the transition matrix accounts for paths of any length on the graph, i.e., each path of length k contributes proportionally to the probability for a walker to perform k jumps in a time interval τ.

It is interesting to note that to each random walk process corresponds a consensus process, in which nodes imitate their neighbors such as to reach a coordinated behavior. In its simplest form, consensus is implemented by the so-called agreement algorithm [67]. Each node i is endowed with a scalar value x_i which evolves as

$$x_{i;t+\tau} = \frac{1}{k_i^{in}} \sum_j A_{ij} x_{j;t}. \tag{17}$$

The matrix driving the dynamics now verifies the constraint $\frac{1}{k_i^{in}} \sum_j A_{ij} = 1$, i.e., the value on a node is updated by computing a weighted average of the values on its neighbors. Similarly to the case of random walks, the process (17) can be generalized either by introducing a bias in the weighted average or by introducing a rate at which nodes update their state.

4.2 Flow-Based Approaches

The multi-resolution quality functions defined in Sect. 3 have been successfully tested on multi-scale benchmark and empirical networks [17, 50, 52]. They have the further advantage of being mathematically very similar to modularity and of being optimized by modularity optimization algorithms with minimum code development. Unfortunately, the introduction of a resolution parameter, γ or r, feels like a trick and lacks theoretical ground. In order to define a resolution parameter in a more satisfying way and, as we will see, to provide a more solid foundation to Q_γ and Q'_r, we look at communities from a different angle; instead from a combinatorial point of view, where intra-community links are counted as in (2), we investigate from a dynamical point of view accounting for the interplay between structure and dynamics. In the following, we describe two such quality functions and emphasize how their dynamical nature helps at clarifying the concept of resolution parameter and at solving some of the issues of the aforementioned combinatorial quality functions. In each case, the starting point is the following: a flow taking place on a network is expected to be trapped for long times in good communities before being able to escape [13, 53, 68]. This argument suggests to measure the quality of a partition in terms of the persistence of flows of random walkers on the network.

From now on, we consider a random walk process at equilibrium. By doing so, we assume that the process is ergodic, i.e., any initial configuration asymptotically reaches the unique stationary solution. If this is not the case, standard tricks, e.g., teleportation, can be introduced to make the dynamics ergodic [33].

Stability

A first quality function, called stability [13], consists in estimating the probability that a random walker stays in a community during a certain time interval

$$
\begin{aligned}
R(\tau) = {} & \text{(probability for a random walker to be in the} \\
& \text{same community initially and at time } \tau) \\
& - \text{(probability for two independent random} \\
& \text{walkers to be in the same community)}
\end{aligned}
\tag{18}
$$

when the system is at equilibrium. To clarify this concept, let us focus on the continuous-time random walk Eq. (15) on an undirected network.[3] By definition, the corresponding stability of a partition is

$$R(\tau) = \sum_C \sum_{i,j \in C} \left[\left(e^{-\tau L}\right)_{ij} \pi_j - \pi_i \pi_j \right], \tag{19}$$

where the probability to find a walker on node i at equilibrium is $\pi_i = k_i/2m$ because of the undirectedness of the links. This expression clearly shows that stability depends on time. The quality of a partition is thus measured differently at different time scales and is, in general, maximized by different partitions when time is tuned, thereby leading to a sequence of optimal partitions.

By looking at limiting values of τ, one can show that time acts as a resolution parameter [13, 28]. As time grows, the characteristic size of the communities is thus adjusted to reveal the possible multi-scale organization of the system. In the limit $\tau \to 0$, keeping linear terms in t in the expansion of $R(\tau)$ leads to

$$R(\tau) \approx (1 - \tau)\, R(0) + \tau\, Q_{\text{conf}} \equiv Q(t), \tag{20}$$

[3]Ergodicity is ensured if the underlying network is non-bipartite and connected.

which is equivalent up to a linear transformation to Q_γ and Q_r' when $P_{ij} = k_i k_j/2m$ (with $\tau = 1/\gamma$, $\tau = \langle k\rangle/(r + \langle k\rangle)$). The latter multi-resolution quality functions can therefore be seen as a simple linear approximation of $R(\tau)$, which provides a physical interpretation to the resolution parameter r and γ, i.e., the inverse of the time used to explore the network. It is also interesting to note that the configuration null model naturally emerges from the definition of stability and from the dynamics (15). Interestingly, other null models, including the uniform null model, are associated to other random walk processes [28]. We should also stress that this connection with modularity only exists for undirected networks and that stability is radically different from modularity when the network is directed. This difference stems from the fact that detailed balance is not verified at equilibrium for directed networks. In the limit $\tau \to \infty$, making use of the spectral decomposition of L, stability simplifies as

$$R(\tau) \approx \frac{1}{2m} e^{\tau\lambda_2} \sum_C \sum_{i,j\in C} v_{2;i} v_{2;j} \tag{21}$$

where it is assumed that the second dominant eigenvalue λ_2 of L is not degenerate and v_2 is its corresponding (right) eigenvector. $R(\tau)$ is therefore maximized by a partition into two communities in accordance with the normalized Fiedler eigenvector [61].

The Map Equation

An alternative way to measure the trapping of walkers in communities is to adopt a coding perspective and to search for a compact description of trajectories on a network in terms of its communities. To do so, the Map equation method [53, 54] relies on a compression of the description length of a random walk inside and between communities. The underlying principle is that for a strongly modular network, the code for the transitions of the random walker can be efficiently compressed by capitalizing on the presence of the community structure. In this formalism, the movement of the walker is described in terms of two features: first, within each community the movement is encoded assigning a unique code word for each node and a particular exit code word for the community. These are stored in an index codebook that is specific to that community. Second, there is intercommunity codebook with unique code words that describe the movements between different communities. The argument is that a walker will rarely leave a good community and a strong community structure leads to a reduction of the code length since short code words can be reused within each community codebook.

Let us focus again on a random walk process in equilibrium, and consider a partition of the network into communities. The probability of leaving a particular community C at equilibrium is denoted by $q_{C\curvearrowright} = \sum_{i\in C}\sum_{j\notin C} T_{ji}\pi_i$, with $\pi_i = k_i/2m$ if the graph is undirected. The Map equation for this partition is

$$L(\mathrm{M}) = q_{\curvearrowright} H(\mathcal{Q}) + \sum_C p_{\circlearrowright}^C H(\mathcal{C}^C), \tag{22}$$

where H is the Shannon entropy. The first term of this equation is the weighted entropy associated with the intercommunity movement of the random walker, where the weighting factor $q_{\curvearrowright} = \sum_C q_{C\curvearrowright}$ is the overall probability of leaving a community. For the Map coding scheme $H(\mathcal{Q})$ is the minimal average per-step code length to describe the transition of the walker between different communities

$$H(\mathcal{Q}) = -\sum_C \frac{q_{C\curvearrowright}}{q_{\curvearrowright}} \log_2 \left(\frac{q_{C\curvearrowright}}{q_{\curvearrowright}} \right). \tag{23}$$

In the second term, each term $p_{\circlearrowright}^C H(\mathcal{C}^C)$ is the weighted average per-step code length needed to describe the movement of the random walker within (and leaving) community C. The entropy $H(\mathcal{C}^C)$ is given analogously by

$$H(\mathcal{C}^C) = -\frac{q_{C\curvearrowright}}{p_{\circlearrowright}^C} \log_2 \left(\frac{q_{C\curvearrowright}}{p_{\circlearrowright}^C} \right) - \sum_{i \in C} \frac{\pi_i}{p_{\circlearrowright}^C} \log_2 \left(\frac{\pi_i}{p_{\circlearrowright}^C} \right), \tag{24}$$

where $p_{\circlearrowright}^C = q_{C\curvearrowright} + \sum_{j \in C} \pi_j$ is the associated weighting factor, describing the probability to use a code word from the codebook of community C. The optimal partition of a network is found by minimizing the Map equation, e.g., by finding the partition minimizing the description length of a random walk on the graph. The Map equation is usually defined for the discrete-time unbiased random walk (13) and is thus based on one-step transitions on the graph. However, it has been shown that the corresponding quality function suffers from some limitations, such as the fact that it is not able to uncover communities in multi-scale networks and that it exhibits a field-of-view limit that can result in undesirable over-partitioning when communities are highly structured [57]. This issue can be addressed by adopting a similar approach as in the previous section, namely, in introducing time explicitly by letting the random walker explore the network over a tunable amount of time [58]. The resulting Markov time sweeping induces a dynamical zooming across scales that can reveal community structure at different scales and circumvent field-of-view limit. In practice, defining the Map equation for the continuous-time random walk (15) leads to time-dependent leaving probabilities $q_{C\curvearrowright}(\tau)$ given as

$$q_{C\curvearrowright}(\tau) = \sum_{i \in C} \sum_{j \notin C} \left(e^{-\tau L} \right)_{ji} \pi_i. \tag{25}$$

With increasing time, this leaving probability and the associated cost for encoding communities increase. Looking at limiting values of τ suggests again that time acts as a resolution parameter [58]. In the limit of vanishing times, when the leaving probabilities go to zero, the Map equation is minimized setting each node in its own community. In the limit $\tau \to \infty$, in contrast, the Map equation finds a partition made of one single module due to the mean-field nature of the dynamics. For intermediate times, the partitions optimizing the Map equation are expected to be made of modules of varying size, as confirmed by numerical simulations [58].

Discussion

As we have discussed in the previous subsections, flow-based approaches have the interesting property to incorporate a natural resolution parameter, time, allowing to explore the graph through paths of different length. However, flow-based approaches offer other advantages that make them an interesting alternative to combinatorial approaches. First, stability is based on flows of probability on the graph and therefore captures how the global structure of the system constrain patterns of flows, while quantities such as modularity focus on pairwise interactions and are blind to such patterns, thereby neglecting important aspects of the network architecture [53]. This difference is particularly marked when the network is directed, when the equilibrium solution of the process depends on the global organization of the process [28]. Flow-based approaches also offer the flexibility to chose a random process properly modeling how, e.g., information or energy flows on the graph and thus to tailor the quality function according to the network nature. For instance, this can be done by using the biased random walk process (14) in systems where unbiased random walks are not realistic models. Important examples include the Internet and traffic networks, where a bias is necessary to account for local search strategies and navigation rules [69].

Partitions at different values of τ are found independently by optimizing either stability or the Map equation, thereby producing a sequence of partitions that are optimal at different scales. However, one expects that only a small number of these partitions are significant, which raises another question: how can one select the most significant partitions, or equivalently the most significant scales of description of the network? It is ironical to note that we are thus confronted with a problem similar to the one that initially led to the definition of modularity.[4] In order to address this problem, it has recently been proposed to look for robust partitions [17, 24, 29, 37]. In practice, the problem is slightly modified, by changing either the resolution parameter, the graph, or the optimization algorithm, and variability among the uncovered partitions is considered. In each case, robustness is related to the ruggedness of the quality function landscape [21]. Lack of robustness corresponds to high degeneracy, namely, to the existence of incompatible partitions that are local maxima of the quality function such that partitions are strongly altered by a slight modification of the optimization problem. Significant partitions are uncovered by identifying values of the resolution parameter where these measures of robustness are significantly low. In the case of the Map equation, an alternative

[4] Modularity was first proposed to find the best partition in a nested hierarchy of possible community divisions [40].

approach is based on an information theoretic indicator for the reliability of the Map equation [58], by measuring the gap between the optimal code compression and the compression achieved by the Map coding strategy. Relevant values of the resolution parameter are then signaled by a low compression gap.

Before concluding, let us mention recent works where the ideas developed in this chapter have been successfully tested on empirical networks of different nature and on benchmark networks [12–14, 28, 29, 57, 58]. We should also stress that the multi-scale methods described here are expected to fail when the system does not exhibit any scale of description, e.g., when the size of the communities is heterogeneously distributed [35]. In that case, other types of methods might reveal more successful, for instance, local methods in contrast with global ones [36].

5 Conclusion

In this chapter, we have focused on the detection of nonoverlapping modules in multi-scale networks. These networks are made of different levels of organization and are typically (but not necessarily) hierarchical, in the sense that the system is made of modules, which themselves are made of sub-modules, etc. In order to account for the presence of multiple levels of organization in the system, multi-scale methods have been introduced that incorporate a resolution parameter allowing to zoom in and out and to focus on the appropriate level of resolution. A class of popular combinatorial quality functions incorporate a resolution parameter to Newman–Girvan modularity in order to adjust the characteristic size of the modules and to uncover the true modular organization of a network. Unfortunately, these multi-resolution quality functions exhibit the same type of limitation as modularity when the resolution parameter is fixed [26]. Moreover, the introduction of a resolution parameter lacks a sound mathematical justification. Here, we argue for the use of flow-based approaches instead of combinatorial ones for several reasons. We have shown that using the trajectories of random walker on the graph defines quality functions with a natural resolution parameter, *time*. As *time* increases, the diffusive process involves longer trajectories and explores further afield the structure of the graph, resulting in the detection of the modular structure across scales. As we have shown, this general scheme can be used in a variety of methods, such as stability and the Map equation. Flow-based approaches have the further advantage of properly taking into account dynamical flows taking place on the graph. This property reveals crucial to provide a faithful characterization of the system whenever complex interdependences between network subunits are generated by patterns of flow, e.g., information in social networks or passengers in airline networks.

Acknowledgements I would like to thank my coauthors involved in the articles described in this chapter, in particular J.-C. Delvenne and M. Barahona for [28], M.T. Schaub for [58], and D. Meunier and E.T Bullmore for [38]. Some of the ideas presented in this chapter are developed in another chapter of this book [14]. I would also like to acknowledge support from FNRS (MIS-2012-F.4527.12) and Belspo (PAI Dysco).

References

[1] D.M. Abrams, R. Mirollo, S.H. Strogatz, D.A. Wiley, Solvable model for chimera states of coupled oscillators. Phys. Rev. Lett. **101**, 084103 (2008)
[2] Y.Y. Ahn, J.P. Bagrow, S. Lehmann, Communities and hierarchical organization of links in complex networks. Nature **466**, 761 (2010)
[3] A. Arenas, A. Díaz-Guilera, C.J. Pérez-Vicente, Synchronization reveals topological scales in complex networks. Phys. Rev. Lett. **96**, 114102 (2006)
[4] A. Arenas, A. Fernández, S. Gómez, Analysis of the structure of complex networks at different resolution levels. New J. Phys. **10**, 053039 (2008)
[5] M.J. Barber, Modularity and community detection in bipartite networks. Phys. Rev. E **76**, 066102 (2007)
[6] M. Batty, K.J. Tinkler, Symmetric structure in spatial and social processes. Env. Plan. B **6**, 3 (1979)
[7] V.D. Blondel, J.-L. Guillaume, R. Lambiotte, E. Lefebvre, Fast unfolding of communities in large networks. J. Stat. Mech. P10008 (2008)
[8] S. Boccaletti, V. Latora, Y. Moreno, M. Chavez, D.-U. Hwang, Complex networks: structure and dynamics. Phys. Rep. **424**, 175–308 (2006)
[9] U. Brandes, D. Delling, M. Gaertler, R. Goerke, M. Hoefer, Z. Nikoloski, D. Wagner, Maximizing modularity is hard. arXiv:physics/0608255
[10] A. Clauset, C. Moore, M.E.J. Newman, Hierarchical structure and the prediction of missing links in networks. Nature **453**, 98–101 (2008)
[11] L. Danon, J. Duch, A. Diaz-Guilera, A. Arenas, Comparing community structure identification. J. Stat. Mech. P09008 (2005)
[12] A. Delmotte, E.W. Tate, S.N. Yaliraki, M. Barahona, Protein multi-scale organization through graph partitioning and robustness analysis: application to the myosin-myosin light chain interaction. Phys. Biol. **8**, 055010 (2011)
[13] J.-C. Delvenne, S. Yaliraki, M. Barahona, Stability of graph communities across time scales. Proc. Natl. Acad. Sci. USA **107**, 12755–12760 (2010)
[14] J.-C. Delvenne, M.T. Schaub, S.N. Yaliraki, M. Barahona, The stability of a graph partition: a dynamics-based framework for community detection, in *Dynamics on and of Complex Networks, vol. 2: Applications to Time-Varying Dynamical Systems*, ed. by A. Mukherjee, M. Choudhury, F. Peruani, N. Ganguly, B. Mitra (Springer, New York, 2013)
[15] T.S. Evans, Complex networks. Contemp. Phys. **45**, 455 (2004)
[16] T.S. Evans, R. Lambiotte, Line graphs, link partitions and overlapping communities. Phys. Rev. E **80**, 016105 (2009)
[17] D.J. Fenn, M.A. Porter, M. McDonald, S. Williams, N.F. Johnson, N.S. Jones, Dynamic communities in multichannel data: an application to the foreign exchange market during the 2007–2008 credit crisis. Chaos **19**, 033119 (2009)
[18] M. Fiedler, A property of eigenvectors of nonnegative symmetric matrices and its application to graph theory. Czech. Math. J. **25**, 619–633 (1975)
[19] S. Fortunato, M. Barthélemy, Resolution limit in community detection. Proc. Natl. Acad. Sci. USA **104**, 36 (2007)
[20] S. Fortunato, Community detection in graphs. Phys. Rep. **486**, 75–174 (2010)
[21] B.H. Good, Y.-A. de Montjoye, A. Clauset, The performance of modularity maximization in practical contexts. Phys. Rev. E **81**, 046106 (2010)
[22] T. Gross, B. Blasius, Adaptive coevolutionary networks: a review. J. R. Soc. Interface **5**, 259–271 (2008)
[23] R. Guimerá, M. Sales, L.A.N. Amaral, Modularity from fluctuations in random graphs and complex networks. Phys. Rev. E **70**, 025101 (2004)
[24] B. Karrer, E. Levina, M.E.J. Newman, Robustness of community structure in networks. Phys. Rev. E **77**, 046119 (2008)

[25] N. Kashtan, U. Alon, Spontaneous evolution of modularity and network motifs. Proc. Natl. Acad. Sci. USA **102**, 13773–13778 (2005)
[26] J.M. Kumpula, J. Saramäki, K. Kaski, J. Kertész, Limited resolution in complex network community detection with Potts model approach. Eur. Phys. J. B **56**, 41–45 (2007)
[27] R. Lambiotte, M. Ausloos, J.A. Holyst, Majority model on a network with communities. Phys. Rev. E **75**, 030101 (2007)
[28] R. Lambiotte, J.-C. Delvenne, M. Barahona, Laplacian dynamics and multiscale modular structure in networks. arXiv:0812.1770
[29] R. Lambiotte, Multi-scale modularity in complex networks, in *Proceedings of the 8th International Symposium on Modeling and Optimization in Mobile, Ad Hoc and Wireless Networks (WiOpt)*, pp. 546–553, 2010
[30] R. Lambiotte, P. Panzarasa, Communities, knowledge creation and information diffusion. J. Informetrics **3**, 180–190 (2009)
[31] R. Lambiotte, J.C. Gonzalez-Avella, On co-evolution and the importance of initial conditions. Phys. A **390**, 392–397 (2011)
[32] R. Lambiotte, R. Sinatra, J.-C. Delvenne, T.S. Evans, M. Barahona, V. Latora, Flow graphs: interweaving dynamics and structure. Phys. Rev. E **84** 017102 (2011)
[33] R. Lambiotte, M. Rosvall, Ranking and clustering of nodes in networks with smart teleportation. Phys. Rev. E **85**, 056107 (2012)
[34] A. Lancichinetti, S. Fortunato, J. Kertész, Detecting the overlapping and hierarchical community structure of complex networks. New J. Phys. **11**, 033015 (2009)
[35] A. Lancichinetti, S. Fortunato, Community detection algorithms: a comparative analysis. Phys. Rev. E **80**, 056117 (2009)
[36] A. Lancichinetti, F. Radicchi, J.J. Ramasco, S. Fortunato, Finding statistically significant communities in networks. PLoS ONE **6**, e18961 (2011)
[37] D. Meunier, R. Lambiotte, A. Fornito, K.D. Ersche, E.T. Bullmore, Hierarchical modularity in human brain functional networks. Front. Neuroinform. **3**, 37 (2009)
[38] D. Meunier, R. Lambiotte, E.T. Bullmore, Modular and hierarchically modular organization of brain networks. Front. Neurosci. **4**, 200 (2010)
[39] M.E.J. Newman, The structure and function of complex networks. SIAM Rev. **45**, 167 (2003)
[40] M.E.J. Newman, M. Girvan, Finding and evaluating community structure in networks. Phys. Rev. E **69**, 026113 (2004)
[41] M.E.J. Newman, Modularity and community structure in networks. Proc. Natl. Acad. Sci. USA **103**, 8577 (2006)
[42] M.E.J. Newman, Finding community structure in networks using the eigenvectors of matrices. Phys. Rev. E **74**, 036104 (2006)
[43] V. Nicosia, G. Mangioni, V. Carchiolo, M. Malgeri, Extending the definition of modularity to directed graphs with overlapping communities. J. Stat. Mech. P03024 (2009)
[44] G. Palla, I. Derenyi, I. Farkas, T. Vicsek, Uncovering the overlapping community structure of complex networks in nature and society. Nature **435**, 814–818 (2005)
[45] R.K. Pan, S. Sinha, Modular networks with hierarchical organization: The dynamical implications of complex structure. Pramana J. Phys. **71**, 331–340 (2008)
[46] R.K. Pan, S. Sinha, Modularity produces small-world networks with dynamical time-scale separation. Europhys. Lett. **85**, 68006 (2009)
[47] R.K. Pan, N. Chatterjee, S. Sinha, Mesoscopic organization reveals the constraints governing Caenorhabditis elegans nervous system. PLoS ONE **5**, e9240 (2010)
[48] P. Pons, M. Latapy, Computing communities in large networks using random walks. J. Graph Algorithm Appl. **10**, 191 (2006)
[49] M.A. Porter, J.-P. Onnela, P.J. Mucha, Communities in networks. Not. Am. Math. Soc. **56**, 1082–1097 (2009)
[50] J. Reichardt, S. Bornholdt, Statistical mechanics of community detection. Phys. Rev. E **74**, 016110 (2006)
[51] P.A. Robinson, J.A. Henderson, E. Matar, P. Riley, R.T. Gray, Dynamical reconnection and stability constraints on cortical network architecture. Phys. Rev. Lett. **103**, 108104 (2009)

[52] P. Ronhovde, Z. Nussinov, Multiresolution community detection for megascale networks. Phys. Rev. E **80**, 016109 (2009)
[53] M. Rosvall, C.T. Bergstrom, Maps of random walks on complex networks reveal community structure. Proc. Natl. Acad. Sci. USA **105**, 1118 (2008)
[54] M. Rosvall, D. Axelsson, C.T. Bergstrom, The map equation. Eur. Phys. J. Spec. Top. **178**, 13–23 (2009)
[55] M. Rosvall, C.T. Bergstrom, Multilevel compression of random walks on networks reveals hierarchical organization in large integrated systems. PLoS ONE **6**, e18209 (2011)
[56] M. Sales-Pardo, R. Guimerá, A. Moreira, L.A.N. Amaral, Extracting the hierarchical organization of complex systems. Proc. Natl. Acad. Sci. USA **104**, 15224 (2007)
[57] M.T. Schaub, J.-C. Delvenne, S.N. Yaliraki, M. Barahona, Markov dynamics as a zooming lens for multiscale community detection: non clique-like communities and the field-of-view limit. PLoS ONE **7**, e32210 (2012)
[58] M.T. Schaub, R. Lambiotte, M. Barahona, Encoding dynamics for multiscale community detection: Markov time sweeping for the map equation. Phys. Rev. E **86**, 026112. Published 21 August 2012
[59] M. Shanahan, Dynamical complexity in small-world networks of spiking neurons. Phys. Rev. E **78**, 041924 (2008)
[60] M. Shanahan, Metastable chimera states in community-structured oscillator networks. Chaos **20**, 013108 (2010)
[61] J. Shi, J. Malik, Normalized cuts and image segmentation. IEEE Trans. Patt. Anal. Mach. Intell. **22**, 888 (2000)
[62] H.A. Simon, The architecture of complexity. Proc. Am. Phil. Soc. **106**, 467–482 (1962)
[63] H.A. Simon, Near-decomposability and complexity: How a mind resides in a brain, in *The Mind, the Brain, and Complex Adaptive Systems*, ed. by H. Morowitz, J. Singer (Addison-Wesley, Reading, 1995), pp. 25–43
[64] O. Sporns, G. Tononi, G. Edelman, Theoretical neuroanatomy: Relating anatomical and functional connectivity in graphs and cortical connection matrices. Cereb Cortex **10**, 127–141 (2000)
[65] O. Sporns, D. Chialvo, M. Kaiser, C.C. Hilgetag, Organization, development and function of complex rain networks. Trends Cognit. Sci. **8**, 418–425 (2004)
[66] V.A. Traag, P. Van Dooren, Y. Nesterov, Narrow scope for resolution-limit-free community detection. Phys. Rev. E **84**, 016114 (2011)
[67] J.N. Tsitsiklis, Problems in decentralized decision making and computation. Ph.D. thesis, MIT (1984)
[68] S. Van Dongen, Graph clustering via a discrete uncoupling process. SIAM. J. Matrix Anal. Appl. **30**, 121–141 (2008)
[69] W.X. Wang, B.-H. Wang, C.-Y. Yin, Y.-B. Xie, T. Zhou, Traffic dynamics based on local routing protocol on a scale-free network. Phys. Rev. E **73**, 026111 (2006)
[70] D.J. Watts, S.H. Strogatz, Collective dynamics of small-world networks. Nature **393**, 440–442 (1998)

Evaluating the Performance of Clustering Algorithms in Networks

Andrea Lancichinetti

1 Introduction

The race towards the ideal clustering method in networks aims at two main goals, i.e., improving the accuracy in the determination of meaningful modules and reducing the computational complexity of the algorithm. The latter is a well-defined objective, while it is more problematic to estimate the accuracy of a method and to compare it with other methods.

A possibility to assess the quality of a clustering algorithm could be to define a set of desirable qualities for a "good" partition. This unsupervised approach has a long tradition in data clustering, where homogeneity, scatter, silhouette metrics, and many others [1–3] have been used for validating the results without requiring any additional information about the groupings.

Here we take the supervised path. We test an algorithm analyzing a network with a well-defined community structure and recovering its communities. Ideally, one would like to have many instances of real networks whose modules are precisely known, but this is unfortunately not the case. Therefore, the most extensive tests are performed on computer-generated networks, with a built-in community structure. This empirical analysis turns out to be a very powerful tool to construct systems where algorithms might give unexpected answers.

Here, we briefly review the most common kind of benchmark graphs in the literature, a class of networks introduced by Girvan and Newman (GN) [4]. We then propose a new class of graphs that accounts for the heterogeneity in the distributions of node degrees and of community sizes and is therefore closer to real properties

A. Lancichinetti (✉)
Howard Hughes Medical Institute (HHMI), Northwestern University, Evanston, IL 60208, USA

Department of Chemical and Biological Engineering, Northwestern University, Evanston, IL 60208, USA
e-mail: andrea.lancichinetti@northwestern.edu

A. Mukherjee et al. (eds.), *Dynamics On and Of Complex Networks, Volume 2*, Modeling and Simulation in Science, Engineering and Technology,
DOI 10.1007/978-1-4614-6729-8_8,

found in real networks. We generalize the new benchmark to include overlapping, weighted, and directed graphs. Finally, we introduce some of the most commonly used community detection algorithms and compare their performances. Most of this subject is covered by [5–7].

2 Girvan and Newman Benchmark Graphs

Although there is no general agreement on the definition of communities, it is well accepted that random networks like Erdös–Rényi (ER) graphs [8] do not present any community structure. Then a good method should partition a graph generated according to this model only in one of the two trivial partitions, grouping all the nodes in one cluster or leaving all of them alone in their own group. This is also true for a network generated via the configuration model [9]. All other splits should not be considered meaningful because nodes would be grouped just according to fluctuations of the model, and the clusters would not be related to any intrinsic property of the nodes.

The *planted ℓ-partition model* [10] is a simple generalization of the ER random graphs to include clusters. In this model one "plants" a partition, consisting of a certain number of groups of nodes. Each node has a probability p_{in} of being connected to nodes of its group and a probability p_{out} of being connected to nodes of different groups. As long as $p_{in} > p_{out}$, the groups might be identified, whereas when $p_{in} = p_{out}$, the network is an ER random graph, without community structure.

Girvan and Newman [4] made some choices for the parameters of this model which became a standard in the community detection literature. Each realization of the GN benchmark graph has 128 nodes, divided into four groups with 32 nodes each. p_{in} and p_{out} are chosen such that the average degree of the network is 16 and the average external degree is k_{out}, where external degree simply means the number of links joining each node to nodes of different groups:

$$p_{in} = (16 - k_{out})/31 \qquad p_{out} = k_{out}/(32 \times 3) \tag{1}$$

When $k_{out} < 8$, each node shares more links with the other nodes of its group than with the rest of the network. When k_{out} is small, the four groups are well-defined communities and a good algorithm should be able to identify them, while when $k_{out} \simeq 12$ one recovers an ER graph (Fig. 1).

This benchmark has been regularly used to test algorithms. However, there are several caveats that one has to consider:

- All nodes of the network have essentially the same degree.
- The communities are all of the same size.
- The network is small.

The first two remarks indicate that the GN benchmark cannot be considered a proxy of a real network with community structure.

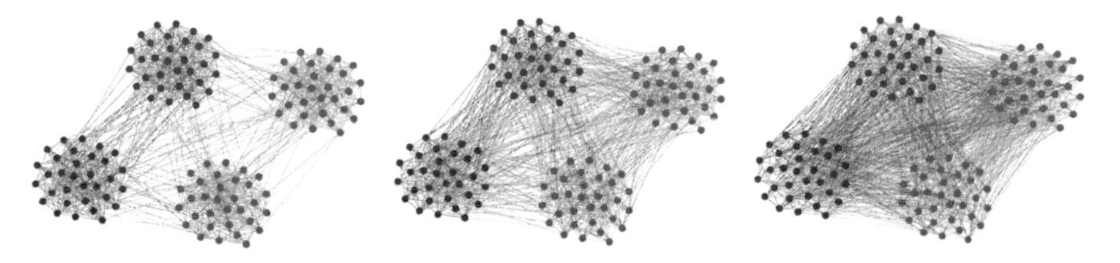

Fig. 1 Three realizations of the GN benchmark model for different values of the average external degree: from left to right, k_{out} was set equal to 2, 6 and 12

Many real networks are characterized by heterogeneous distributions of node degree, whose tails often decay as power laws, and therefore, a good benchmark should have a skewed degree distribution. Likewise, it is not correct to assume that all communities have the same size: the distribution of community sizes of real networks is also broad, with a tail that can be fairly well approximated by a power law [11–14]. A reliable benchmark should include communities of very different sizes. Finally, the GN benchmark was a network of a reasonable size for most existing algorithms at the time when it was introduced. Nowadays, there are methods able to analyze graphs with millions of nodes [14, 15], and it is not appropriate to compare their performances on small graphs. In general, an algorithm should be tested on benchmarks of variable size and average degree, as these parameters may seriously affect the outcome of the method and reveal its limits, as we shall see.

In the next section, we propose a more realistic benchmark for community detection that accounts for the heterogeneity of both degree and community size.

3 LFR Benchmark

We start by presenting the algorithm to build the benchmark for undirected graphs. It is a generalization of the *planted ℓ-partition model* which makes use of the configuration model [9], so that communities are still random graphs (in the sense of [9]) randomly connected. The method includes the possibility to have overlapping modules, i.e., modules which share vertices. This aspect, frequent in certain types of systems, like social networks, has received some attention in the recent years [16–18]. It is then important to test these features against reliable benchmarks too.

In the following sections, we extend the algorithm to the case of weighted and directed graphs. This is very relevant as well, since links of networks from the real world are often directed and carry weights, and both features are essential to understand their function [19, 20]. For these types of benchmark, we will also include the possibility to have overlapping communities.

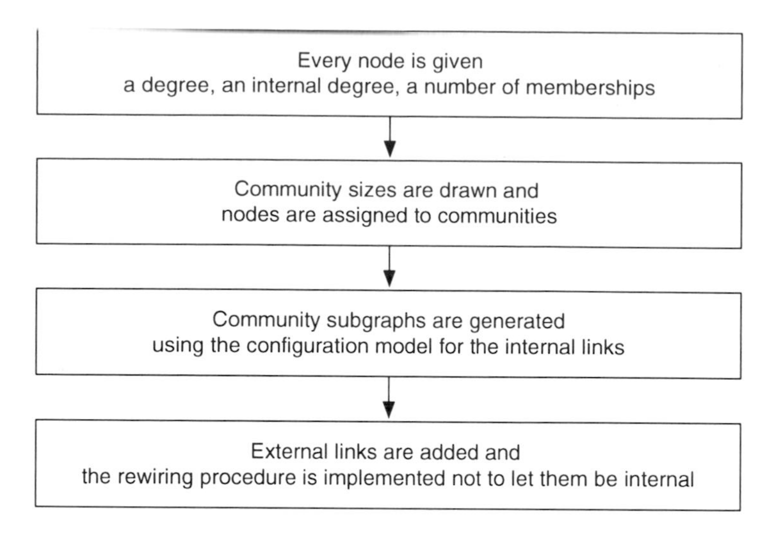

Fig. 2 Flow diagram describing the steps of the algorithm for unweighted networks

3.1 *Unweighted Benchmark with Overlapping Nodes*

The algorithm which generates undirected and unweighted benchmark graphs consists of the following steps (see Fig. 2):

1. We first assign the number ν_i of memberships of node i, i.e., the number of communities the node belongs to. In general we can assign the number of memberships according to a certain distribution. Next, we assign the degrees $\{k_i\}$ by drawing N random numbers from a power law distribution with exponent τ_1 and a hard cutoff k_{max}. We also introduce the *topological mixing parameter* μ_t: $k_i^{(in)} = (1 - \mu_t)k_i$ is the internal degree of the node i, i.e., the number of neighbors of node i which have at least one membership in common with i. In this way, the internal degree is a fixed fraction of the total degree for all the nodes.
2. The community sizes $\{s_\xi\}$ are assigned by drawing random numbers from another power law with exponent τ_2. Naturally, the sum of the community sizes must equal the sum of the node memberships, i.e., $\sum_\xi s_\xi = \sum_i \nu_i$. Furthermore $s_{max} = \max\{s_\xi\} \leqslant N$ and $\nu_{max} = \max\{\nu_i\} \leqslant n_c$, where N is the number of nodes and n_c the number of communities. At this point, we have to decide which communities each node should be included into. This is equivalent to generating a bipartite network where the two classes are the n_c communities and the N nodes; each community ξ has s_ξ links, whereas each node has as many links as its memberships ν_i. The network can be easily generated with the configuration model [9]. To build the graph, it is important to take into account the constraint

$$\sum_{i \to \xi} s_\xi \geqslant k_i^{(in)}, \qquad \forall i, \tag{2}$$

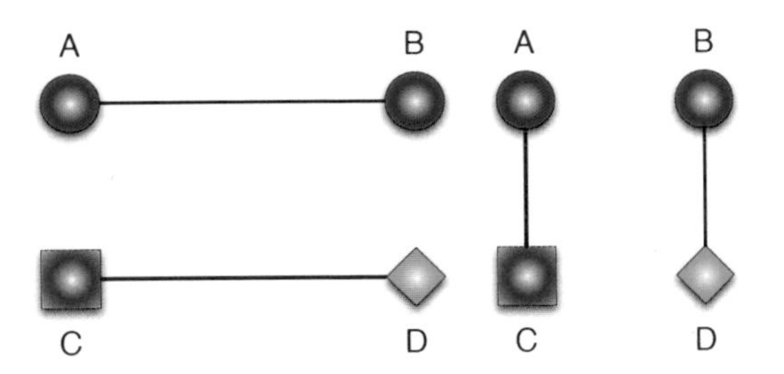

Fig. 3 Scheme of the rewiring procedure necessary to build the graph $\mathcal{G}^{(out)}$, which includes only links between nodes of different communities. (*Top*) If two nodes (A and B) with a common membership are neighbors, their link is rewired along with another link joining two other nodes C and D, where C does not have memberships in common with A, and D is a neighbor of C not connected to B. In the final configuration (*bottom*), the degrees of all nodes are preserved, and the number of links between nodes with common memberships has decreased by one (since A and B are no longer connected), or it has stayed the same (if B and D, which are now neighbors, have common memberships)

where the sum is relative to the communities including node i. This condition means that each node cannot have an internal degree larger than the highest possible number of nodes it can be connected to within the communities it stays in. We perform a rewiring process for the bipartite network until the constraint is satisfied.

So far we assigned an internal degree to each node but it has not been specified how many links should be distributed among the communities of the node. Again, one can follow several recipes; we chose the simple equipartition $k_i(\xi)=k_i^{(in)}/\nu_i$, where $k_i(\xi)$ is the number of links which i shares in community ξ, provided that i holds membership in ξ.

3. Before generating the whole network, we start generating n_c subgraphs, one for each community. In fact, our definition of community ξ is nothing but a random subgraph of s_ξ nodes with degree sequence $\{k_i(\xi)\}$, which can be built via the configuration model, with a rewiring procedure to avoid multiple links. Once each subgraph is built, we obtain a graph divided in components. Note that because of the overlapping nodes, some components may be connected to each other, and in principle, the whole graph might be connected.
4. The last step of the algorithm consists in adding the links external to the communities. To do this, let us consider the degree sequence $\{k_i^{(out)}\}$, where simply $k_i^{(out)} = k_i - k_i^{(in)} = \mu_t k_i$. We like to insert randomly these links in our already built network without changing the internal degree sequences. In order to do so, we build a new network $\mathcal{G}^{(out)}$ of N nodes with degree sequence $\{k_i^{(out)}\}$, and we perform a rewiring process each time we encounter a link between two nodes which have at least one membership in common (Fig. 3), since we are supposed to join only nodes of different communities at this stage.Let us assume

that A and B are in the same community and that they are linked in $\mathcal{G}^{(out)}$; we pick a node C which does not share any membership with A, and we look for a neighbor of C (call it D) which is not a neighbor of B. Next, we replace the links $A - B$ and $C - D$ with the new links $A - C$ and $B - D$. This rewiring procedure can decrease the number of internal links of $\mathcal{G}^{(out)}$ or leaving it unchanged (this happens only when B and D have one membership in common), but it cannot increase it. This means that after a few sweeps over all the nodes, we reach a steady state where the number of internal links is very close to zero (if no node has $k_i \sim N$, the internal links of $\mathcal{G}^{(out)}$ are just a few and one sweep is sufficient). The left panel of Fig. 4 shows how the number of internal links decreases during the rewiring procedure. Finally, we have to superimpose $\mathcal{G}^{(out)}$ on the previous one.

In the benchmarks based on the *planted ℓ-partition model*, some nodes might be misclassified because the internal degree of the nodes fluctuates around its mean value. Here, fluctuations are considerably reduced, and the graphs come as close as possible to the type of structure described by the input parameters.

In fact, our algorithm tries to set the μ-value of each node to the predefined input value. Ideally, the distribution of the μ-values for a given benchmark graph should be close to a Dirac δ-function, but this is not possible especially for nodes of small degree, where the possible values of μ are just a few and clearly separated. Indeed, if the number of internal links of $\mathcal{G}^{(out)}$ goes to zero, the only reason not to have a perfectly sharp function for the distribution of the mixing parameters is a round-off problem, i.e., the problem of rounding integer numbers. Right panel of Fig. 4 shows a bell-shaped curve, with a pronounced peak for several values of the parameters.

We conclude this section with a remark about the complexity of the algorithm. The first two steps take a number of operations proportional to the number of nodes and to the total number of memberships, respectively. The configuration model complexity grows linearly with the number of links m of the network. If the rewiring procedure is successful most of the time, like it happens in most instances (see Fig. 4), it takes a time proportional to the number of external links. Then, the global complexity of the algorithm is $O(m)$ (Fig. 5).

3.2 Weighted Networks

In order to build a weighted network, we first generate an unweighted network with a given topological mixing parameter μ_t and then we assign a positive real number to each link.

To do this we need to specify two other parameters, β and μ_w. The parameter β is used to assign a strength s_i to each node, $s_i = k_i^{\beta}$; such power law relation between the strength and the degree of a node is frequently observed in real weighted networks [20]. The parameter μ_w is used to assign the internal strength $s_i^{(in)} = (1 - \mu_w)\, s_i$, which is defined as the sum of the weights of the links between node i

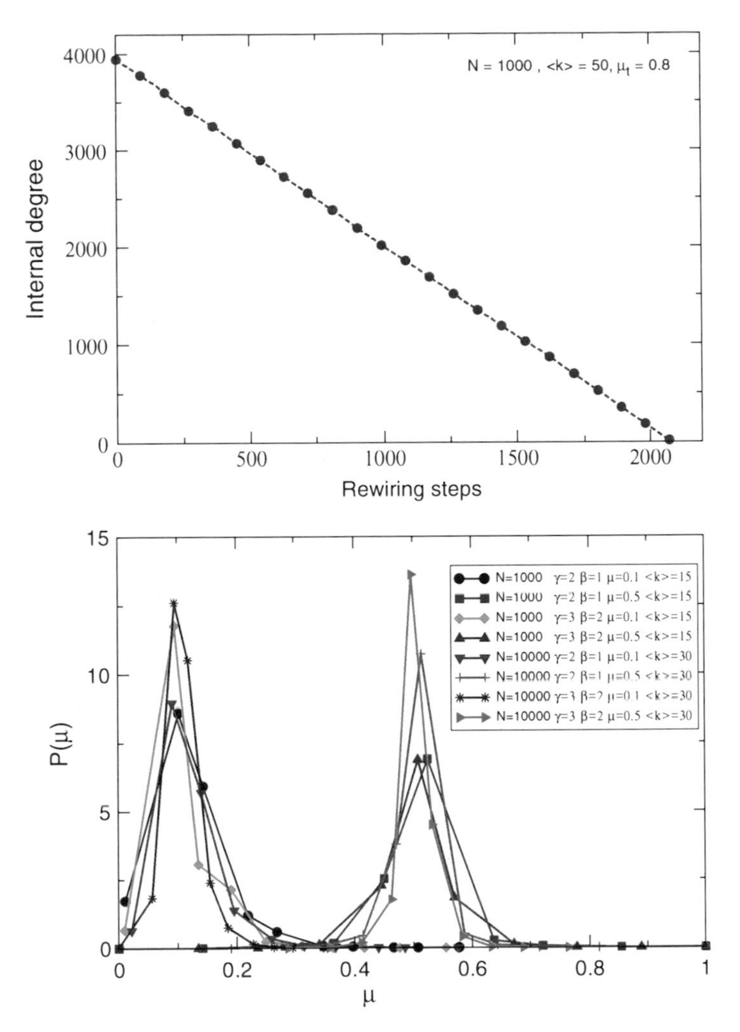

Fig. 4 *Left*. Number of internal links of $\mathcal{G}^{(out)}$ as a function of the rewiring steps. The network has 1,000 nodes and an average degree $\langle k \rangle = 50$. Since the mixing parameter is $\mu_t = 0.8$ and there are 10 equal-sized communities, at the beginning each node has an expected internal degree in $\mathcal{G}^{(out)}$ $k_i^{(in)} = 0.8 * 50 * 1/10 = 4$, so the total internal degree is around 4,000. After each rewiring step, either the internal degree decreases by 2 or it does not change. In this case, less than 2,100 rewiring steps were needed. *Right*. Distribution of the μ-values for benchmark graphs obtained with our algorithm for different choices of the exponents and system size

and all its neighbors having at least one membership in common with i. The problem is equivalent to finding an assignment of m positive numbers $\{w_{ij}\}$ such that the following function is minimized:

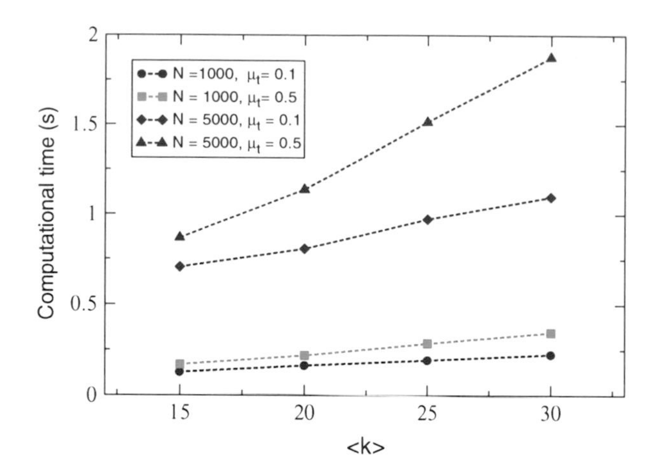

Fig. 5 Computational time to build the unweighted benchmark as a function of the average degree. We show the results for networks of $1,000$ and $5,000$ nodes. μ_t was set equal to 0.1 and 0.5 (the latter requires more time for the rewiring process). Note that between the two upper lines and the lower ones, there is a factor of about 5, as one would expect if complexity is linear in the number of links m

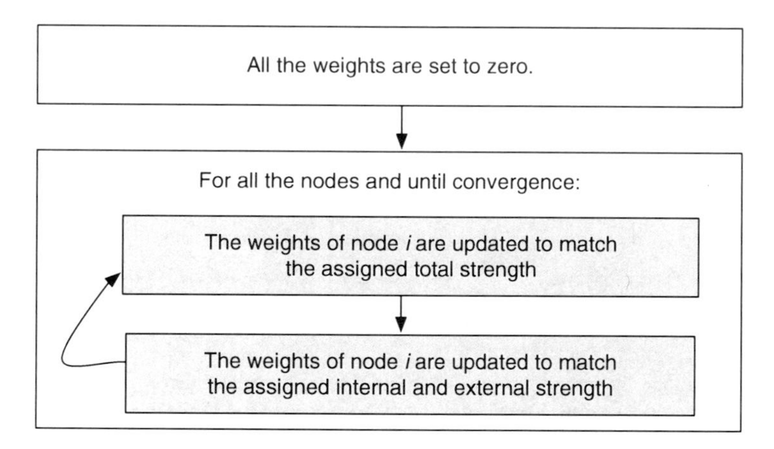

Fig. 6 Flow diagram describing the steps of the algorithm for inserting weights

$$\mathrm{Var}(\{w_{ij}\}) = \sum_i (s_i - \rho_i)^2 + (s_i^{(in)} - \rho_i^{(in)})^2 + (s_i^{(out)} - \rho_i^{(out)})^2 \qquad (3)$$

Here s_i and $s_i^{(in,ext)}$ indicate the strengths which we would like to assign, i.e., $s_i = k_i^{\beta}$, $s_i^{(in)} = (1 - \mu_w)\, s_i$, $s_i^{(out)} = \mu_w\, s_i$; $\{\rho_i^*\}$ are the total, internal, and external strengths of node i defined through its link weights, i.e., $\rho_i = \sum_j w_{ij}$, $\rho_i^{(in)} =$

$\sum_j w_{ij}\ \kappa(i,j)$, $\rho_i^{(out)} = \sum_j w_{ij}\ (1-\kappa(i,j))$, where the function $\kappa(i,j) = 1$ if nodes i and j share at least one membership and $\kappa(i,j) = 0$ otherwise.

We have to arrange things so that s_i and $s_i^{(in,ext)}$ are consistent with the $\{\rho_i^*\}$. For that we need a fast algorithm to minimize $\mathrm{Var}(\{w_{ij}\})$. We found that the greedy algorithm described below can do this job well enough for the cases of our interest (see Fig. 6).

1. At the beginning $w_{ij} = 0, \forall i, j$, so all the $\{\rho_i^*\}$ are zero.
2. We take node i and increase the weight of each of its links by an amount $u_i = \frac{s_i - \rho_i}{k_i}$, where ρ_i indicates the sum of the links' weights resulting from the previous step, i.e., before we increment them. In this way, since initially $\{\rho_i^*\} = 0$, the weights of the links of i after the first step take the (equal for all) value $\frac{s_i}{k_i}$ and $\rho_i = s_i$ by construction, a condition that is maintained along the whole procedure. We update $\{\rho_i^*\}$ for the node i and its neighbors.
3. Still for node i, we increase all the weights w_{ij} by an amount $\frac{s_i^{(in)} - \rho_i^{(in)}}{k_i^{(in)}}$ if $\kappa(i,j) = 1$ and by an amount $-\frac{s_i^{(in)} - \rho_i^{(in)}}{k_i^{(out)}}$ if $\kappa(i,j) = 0$. Again we update $\{\rho_i^*\}$ for the node i and its neighbors. These two steps assure that the contribution of node i to the $\mathrm{Var}(\{w_{ij}\})$ is zero.
4. We repeat steps (2) and (3) for all the nodes. Two remarks are in order. First, we want each weight $w_{ij} > 0$, so we update the weights only if this condition is fulfilled. Second, the contribution of the neighbors of node i in $\mathrm{Var}(\{w_{ij}\})$ will change, and of course, it can increase or decrease. For this reason, we need to iterate the procedure several times until a steady state is reached or until we reach a certain value. With our procedure the value of $\mathrm{Var}(\{w_{ij}\})$ decreases at least exponentially with the number of iterations, consisting in sweeps over all network links (Fig. 7).

Since $\mathrm{Var}(\{w_{ij}\})$ decreases exponentially, the number of iterations needed to reach convergence is weakly dependent on the size of the network, and hence, it does not contribute much to the effective complexity, which remains $O(m)$.

3.3 Directed Networks

It is quite straightforward to generalize the previous algorithms to generate directed networks. Now, we have an indegree sequence $\{y_i\}$ and an outdegree sequence $\{z_i\}$, but we can still go through all the steps of the construction of the benchmark for undirected networks with just some slight modifications. In the following, we list the changes to be made for each of the steps in Sect. 3.1.

1. We decided to sample the indegree sequence from a power law and the outdegree sequence from a δ-distribution (with the obvious constraint $\sum_i y_i = \sum_i z_i$). We need to define the internal indegrees and outdegrees $y_i(\xi)$ and $z_i(\xi)$ with respect

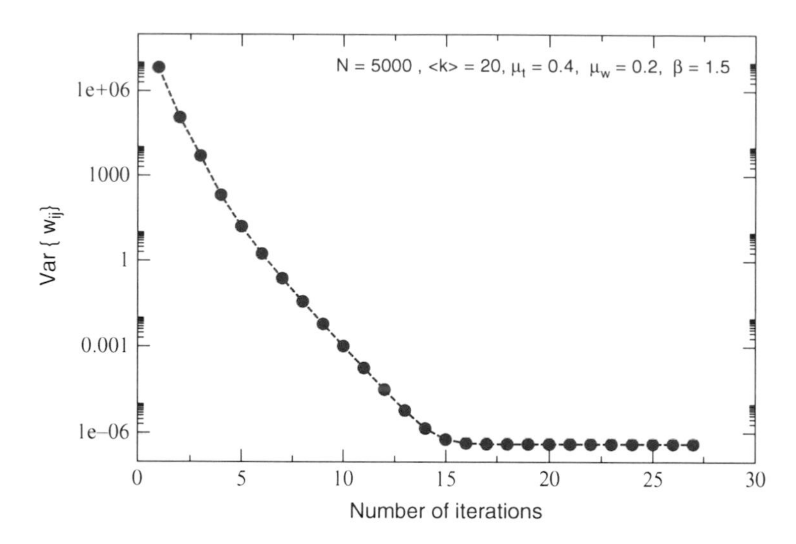

Fig. 7 Value of $\mathrm{Var}(\{w_{ij}\})$ (3) after each update. Each point corresponds to one sweep over all the nodes

to every community ξ, which can be done by introducing two mixing parameters. For simplicity one can set them to the same value.

2. We need to use the configuration model for directed networks, with the condition that $\sum_i y_i(\xi) = \sum_i z_i(\xi)$; because of this condition, it might be necessary to change $y_i(\xi)$ and/or $z_i(\xi)$. We decided to modify only $z_i(\xi)$, whenever necessary.
3. The rewiring procedure can be done by preserving both distributions of indegree and outdegree, for instance, by adopting the following scheme: before rewiring, A points to B and D to C; after rewiring, A points to C and D to B.

In order to generate directed and weighted networks, we use the following relation between the strength s_i of a node and its in- and outdegree: $s_i = (y_i + z_i)^\beta$. Given a node i, one considers all its neighbors, regardless of the link directions (note that i may have the same neighbor counted twice if the link runs in both directions). Otherwise, the procedure to set the weights is equivalent.

4 Tests and Comparative Analysis of Algorithms

In this last section, we compare the performance of some fundamental community detection algorithms in the literature. After discussing how to compare partitions quantitatively, we will describe the algorithms that we examined. We will present the analysis of the algorithms' performance first on the GN benchmark and then on the LFR benchmark, in the undirected unweighted version with no overlapping communities. For a more extensive comparison we refer the reader to [7], where we also considered the issue of whether the algorithms are able to give a null result, i.e., how they handle networks without expected community structure, like random graphs. Our analysis will reveal that there are, at present, algorithms which are fast and reliable in many situations.

4.1 Comparing Partitions

Testing an algorithm on any graph with built-in community structure also implies defining a quantitative criterion to estimate the goodness of the answer given by the algorithm as compared to the real answer that is expected. This can be done by using suitable similarity measures. For reviews of similarity measures, see [21, 22]. In the first tests of community detection algorithms, one used a measure called *fraction of correctly identified nodes*, introduced by Girvan and Newman [4]. However, it is not well defined in some cases (e.g., when a detected community is a merger of two or more "real" communities), so in the recent years, other measures have been used. In particular, measures borrowed from information theory have proved to be reliable. In the following, we will use the definition of normalized mutual information between overlapping partitions provided in [23].

4.2 The Algorithms

We have tested a wide spectrum of community detection methods. We wanted to have a representative subset of algorithms that exploit some of the most interesting ideas and techniques that have been developed recently. We could not by any means perform an analysis of all existing techniques, as their number is huge [21]. Some of them were excluded a priori, if particularly slow, as our tests involve graphs with a few thousand nodes, which old methods are unable to handle. Here is the list of the algorithms we considered:

- Betweenness centrality algorithm of Girvan and Newman (GN) [4, 24]
- Fast greedy modularity optimization by Clauset, Newman, and Moore [14]
- Exhaustive modularity optimization via simulated annealing (Sim. Ann.) [25–28]
- Fast modularity optimization by Blondel et al. [15]
- Algorithm by Radicchi et al. [29]
- CFinder [11]
- Markov cluster algorithm (MCL) [30]
- Structural algorithm by Rosvall and Bergstrom (Infomod) [31]
- Dynamic algorithm by Rosvall and Bergstrom (Infomap) [32]
- Spectral algorithm by Donetti and Munoz (DM) [33]
- Expectation-maximization algorithm by Newman and Leicht (EM) [34]
- Potts model approach by Ronhovde and Nussinov (RN) [35]

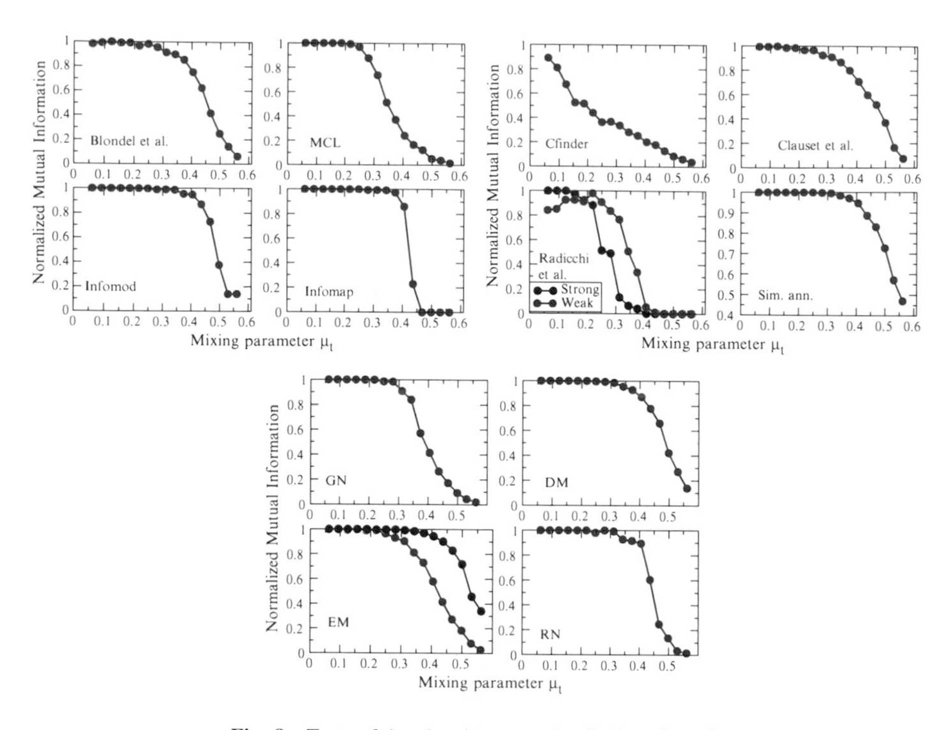

Fig. 8 Tests of the algorithms on the GN benchmark

4.3 Tests on the GN Benchmark

We begin by showing the performance of the algorithms on the GN benchmark. As we have explained in Sect. 2, for the GN benchmark, communities are well defined (in principle) up until a value $3/4 = 0.75$ for the mixing parameter. We will indicate the mixing parameter with the symbol μ_t to mean that we refer to topology. In Fig. 8 we show the results of our analysis. Each point of every curve corresponds to an average over 100 realizations of the benchmark. For the algorithms by Radicchi et al. and by Newman and Leicht (EM), we have put two curves instead of one (likewise in Sect. 4.4). In the first case, we showed the outcome of the method when one uses both possible stopping criteria, corresponding to a partition consisting of strong (black curve) and weak (red curve) communities, respectively. In the case of the EM method, we show the curves delivered by the iterative solution of the EM equations when one starts from a random partition (red) and from the planted partition of the benchmark (black curve). As one can see, results are different in these cases, even if they are solutions of the same equation. This shows how sensitive the solution is to the choice of the initial condition. Moreover, the maximum likelihood achieved when one makes the "intelligent guess" of the real partition is higher compared to the maximum likelihood obtained starting from a random partition. This indicates that the greedy approach to the solution of the EM equations suggested by Newman and Leicht is not an efficient way to maximize the likelihood, as one may expect.

Most methods perform rather well, although all of them start to fail much earlier than the expected threshold of 3/4. The Cfinder fails to detect the communities even when $\mu_t \sim 0$, when they are very well identified. This is due to the fact that, even when μ_t is small, the probe clique that explores the system manages to pass from one group to the other and yields much larger groups, often spanning the whole graph. The method by Radicchi et al. does not have a remarkable performance either, as it also starts to fail for low values of μ_t, although it does better than the Cfinder. The MCL is better than the method by Radicchi et al. but is outperformed by modularity-based methods (simulated annealing, Clauset et al., Blondel et al.), which generally do quite well on the GN benchmark, something that was already known from the literature. The DM and RN methods have a comparable performance as the exhaustive optimization of modularity via simulated annealing. The GN algorithm performs about as well as the MCL. Both methods by Rosvall and Bergstrom have a good performance. In fact, up until $\mu_t \sim 0.4$, they always guess the planted partition in four clusters.

4.4 Tests on the LFR Benchmark

The plots of Fig. 9 illustrate the results of the analysis. As mentioned above, here we only considered unweighted undirected graphs with no overlapping modules, while [7] reports a more extensive analysis.

The average degree is 20, the maximum degree 50, the exponent of the degree distribution is -2, and that of the community size distribution is -1. In each plot, except for the GN and the EM algorithms, we show four curves, corresponding to two different network sizes (1,000 and 5,000 nodes) and, for a given size, two different ranges for the community sizes, indicated by the letters S and B: S (stands for "small") means that communities have between 10 and 50 nodes, and B (stands for "big") means that communities have between 20 and 100 nodes. For the GN algorithm, we show only the curves corresponding to the smaller network size, as it would have taken too long to accumulate enough statistics to present clean plots for networks of 5,000 nodes, due to the high computational complexity of the method. For the EM method we have plotted eight curves as for each set of benchmark graphs we have considered the two outcomes of the algorithm corresponding to the different choices of initial conditions we have mentioned in the previous section, namely, random (bottom curves) and planted partition (top curves). In this case, the difference in the performance of the algorithm in the two cases is remarkable. The fact that, by starting from the planted partition, the final likelihood is actually higher as compared with a random start, as we have seen in the previous section, confirms that the method has a great potential, if only one could find a better way to estimate the maximum likelihood than the greedy approach currently adopted. Nevertheless we remind that the EM also has the big drawback to require to input the number of

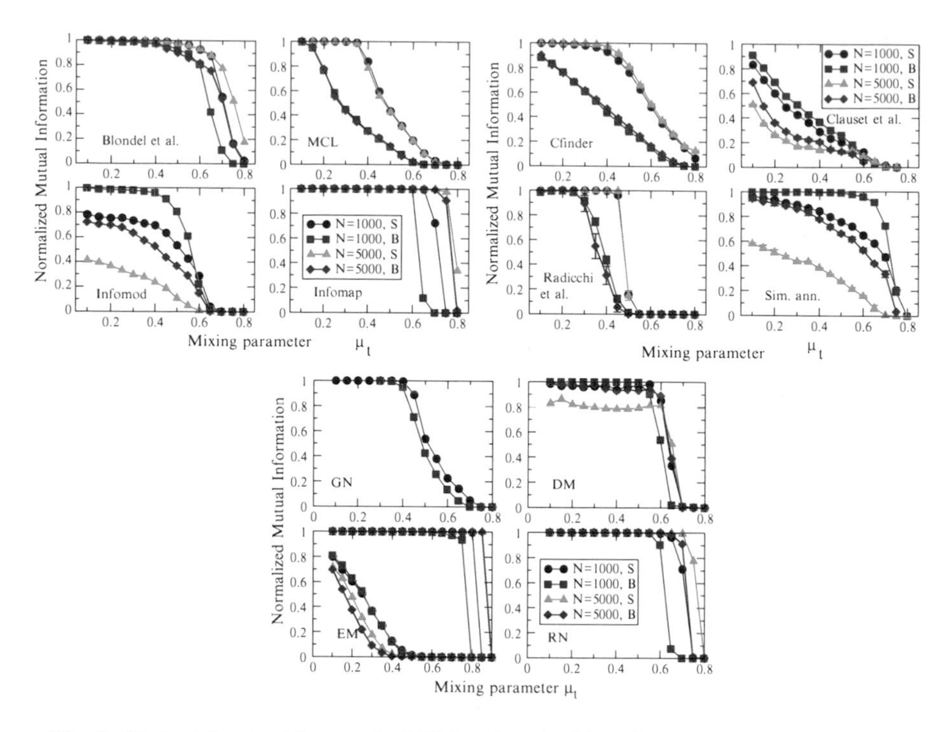

Fig. 9 Tests of the algorithms on the LFR benchmark with undirected and unweighted links

groups to be found, which is usually unknown in applications. As a general remark, we see that the LFR benchmark enables one to discriminate the performances of the algorithms much better than the GN benchmark, as expected. Modularity-based methods have a rather poor performance, which worsens for larger systems and smaller communities, due to the resolution limit of the measure [36]. The only exception is represented by the algorithm by Blondel et al., whose performance is very good, probably because the estimated modularity maximum is not a very good approximation of the real one, which is more likely found by simulated annealing. The Cfinder, the MCL, and the method by Radicchi et al. do not have impressive performances either and display a similar pattern, i.e., the performance is severely affected by the size of the communities (for larger communities, it gets worse, whereas for small communities, it is decent), whereas it looks rather insensitive to the size of the network. The DM has a fair performance, but it gets worse if the network size increases. The same trend is shown by Infomod, where the performance worsens considerably with the increase of the network size. Infomap and RN have the best performances, with the same pattern with respect to the size of the network and of the communities: up to values of $\mu_t \sim 1/2$, both methods are capable to derive the planted partition in the 100% of cases.

We conclude that Infomap, the RN method, and the method by Blondel et al. are the best performing algorithms on the LFR undirected and unweighted benchmark.

5 Conclusions

We have introduced a new model of benchmark graphs whose features are close to those found in real networks. This new test represents a more challenging task for most clustering methods as we have shown carrying out a comparative analysis of the performances of many algorithms. Link direction, weights, and the possibility for communities to overlap have been taken into account in dedicated tests.

We conclude that the Infomap method by Rosvall and Bergstrom [32] is the best performing clustering algorithm on the set of benchmarks we have examined here. In particular, its results on the LFR benchmark graphs as clearly shown by Figs. 8 and 9 are encouraging about the reliability of the method in applications to real graphs. Among the other things, the method can be applied to weighted and directed graphs as well, with excellent performances (see [7]), so it has a large spectrum of potential applications. The algorithms by Blondel et al. [15] and by Ronhovde and Nussinov (RN) [35] also seem to perform competitively and could be used as well. In fact, for a study of the community structure in real graphs, one could think of using all three methods, to be able to extract some algorithm-independent information.

References

[1] E. Pauwels, G. Frederix, Comput. Vis. Image Underst. **75**, 73–85 (1999)
[2] D. Daveis, D. Bouldin, IEEE Trans. Pattern Anal. Mach. Intell. **20**, 53–65 (1979)
[3] P. Rousseeuw, Comput. Appl. Math. **2**, 224 (1986)
[4] M. Girvan, M.E. Newman, Proc. Natl. Acad. Sci. USA **99**, 7821 (2002)
[5] A. Lancichinetti, S. Fortunato, F. Radicchi, Phys. Rev. E **78**, 046110 (2008)
[6] A. Lancichinetti S. Fortunato, Phys. Rev. E **80**, 016118 (2009)
[7] A. Lancichinetti, S. Fortunato, Phys. Rev. E **80**, 056117 (2009)
[8] P. Erdös A. Rényi, Publ. Math. Debrecen **6**, 290 (1959)
[9] M. Molloy, B. Reed, Random Struct. Algorithm **6**, 161 (1995)
[10] A. Condon, R.M. Karp, Random Struct. Algor. **18**, 116 (2001)
[11] G. Palla, I. Derényi, I. Farkas, T. Vicsek, Nature **435**, 814 (2005)
[12] R. Guimerà, L. Danon, A. Díaz-Guilera, F. Giralt, A. Arenas, Phys. Rev. E **68**, 065103 (2003)
[13] L. Danon, J. Duch, A. Arenas, A. Díaz-Guilera, in *Community structure identification. Large Scale Structure and Dynamics of Complex Networks: From Information Technology to Finance and Natural Science*, Singapore: World Scientific. pp. 93–114 (2007)
[14] A. Clauset, M.E. Newman, C. Moore, Phys. Rev. E **70**, 066111 (2004)
[15] V. D. Blondel, J.-L. Guillaume, R. Lambiotte, E. Lefebvre, J. Stat. Mech. **P10008**, (2008)
[16] J. Baumes, M.Goldberg, M. Krishnamoorthy, M. Magdon-Ismail, N. Preston, Finding communities by clustering a graph into overlapping subgraphs, International Conference on Applied Computing (2005)
[17] S. Zhang, R.-S. Wang, X.-S. Zhang, Phys. A **374**, 483 (2007)
[18] T. Nepusz, A. Petróczi, L. Négyessy, F. Bazsó, Phys. Rev. E **77**, 016107 (2008)
[19] E.A. Leicht, M.E.J. Newman, Phys. Rev. Lett. **100**, 118703 (2008)
[20] A. Barrat, M. Barthélemy, R. Pastor-Satorras, A. Vespignani, Proc. Natl. Acad. Sci. USA **101**, 3747 (2004)
[21] S. Fortunato, Community detection in graphs. Phys. Rep. **486**, 75–174 (2010)

[22] M. Meilă, J. Multivariate Anal. **98**, 873 (2007)
[23] A. Lancichinetti, S. Fortunato, J. Kertesz, New J. Phys. **11**, 033015 (2009)
[24] M.E.J. Newman, M. Girvan, Phys. Rev. E **69**, 026113 (2004)
[25] R. Guimerà, M. Sales-Pardo, L.A.N. Amaral, Phys. Rev. E **70**, 025101 (2004)
[26] C.P. Massen, J.P. Doye, Phys. Rev. E **71**, 046101 (2005)
[27] A. Medus, G. Acu na, C.O. Dorso, Phys. A **358**, 593 (2005)
[28] R. Guimerà, L.A.N. Amaral, Nature **433**, 895 (2005)
[29] F. Radicchi, C. Castellano, F. Cecconi, V. Loreto, D. Parisi, Proc. Natl. Acad. Sci. USA **101**, 2658 (2004)
[30] S. van Dongen, Ph.D. thesis, Dutch National Research Institute for Mathematics and Computer Science, University of Utrecht, The Netherlands (2000)
[31] M. Rosvall, C.T. Bergstrom, Proc. Natl. Acad. Sci. USA **104**, 7327 (2007).
[32] M. Rosvall, C.T. Bergstrom, Proc. Natl. Acad. Sci. USA **105**, 1118 (2008)
[33] L. Donetti, M.A. Mu noz, J. Stat. Mech. **P10012**, (2004)
[34] M.E.J. Newman, E.A. Leicht, Proc. Natl. Acad. Sci. USA **104**, 9564 (2007)
[35] P. Ronhovde, Z. Nussinov, Phys. Rev. E **80**, 016109 (2009)
[36] S. Fortunato, M. Barthélemy, Proc. Natl. Acad. Sci. USA **104**, 36 (2007)

Communities in Evolving Networks: Definitions, Detection, and Analysis Techniques

Thomas Aynaud, Eric Fleury, Jean-Loup Guillaume, and Qinna Wang

1 Introduction

A complex network models multiple interactions between components in a complex system. A complex network can be represented by a graph. A diverse range of networks have been studied, for example, the Internet [34], World Wide Web, citation networks [35], coauthorship networks [50], metabolic networks [43], and social networks whose nodes are connected by one or more specific types of relations, like friendship, workplace, and common interest [5, 9]. In the study of these networks, the topic of community structure has attracted much attention in several fields.

A network has an underlying community structure if its nodes can be "naturally" grouped into sets such that each set of nodes is densely connected internally, and these groups of nodes are called *communities*. While other definitions of community exist, not related to the topology but, for instance, to common interests, we refer here to community structure as a network structural property. For instance, in protein–protein interaction networks, communities correspond to specific functions [15]; in the World Wide Web, communities may be related to topics [22]; and in food webs, communities correspond to compartments [51]. Studies in community structure should lead to a better understanding of complex systems.

T. Aynaud (✉) • J.-L. Guillaume
UPMC, CNRS (UMR 7606), France
e-mail: taynaud@gmail.com; jean-loup.guillaume@lip6.fr

E. Fleury
ENS de Lyon (UMR CNRS – ENS de Lyon – UCB Lyon 1 – Inria 5668), France
e-mail: eric.fleury@inria.fr

Q. Wang
Inria (UMR CNRS – ENS de Lyon – UCB Lyon 1 – Inria 5668), France
e-mail: qinna.wang@inria.fr

A. Mukherjee et al. (eds.), *Dynamics On and Of Complex Networks, Volume 2*,
Modeling and Simulation in Science, Engineering and Technology,
DOI 10.1007/978-1-4614-6729-8_9,

In the past, communities in networks have been widely studied. Uncovering community methods proposed usually deal with *static* graphs. Indeed, even if the data observed span a time interval, and not only a specific snapshot, a *static* graph is extracted from the dataset by aggregating interaction behaviors on the whole time interval yielding a static graph, possibly weighted. However, an important property has been largely ignored: networks tend to evolve over time. The interaction between components in a complex system is not stable but changes over time. For example, *Facebook* and *Myspace* social network sites have grown dramatically in recent years, and many people join or post over time. There are also many other dynamic networks with millions of nodes [80, 87]. To learn about how a network evolves over time, it is important to develop methods that allow to study the dynamic aspects of the intrinsic network structure, namely, its community dynamics. In a seminal paper published in 2007, Palla et al. [70] have proposed six possible scenarios that may occur during the evolution of communities: *birth* (a new community appears), *growth* (a growing community), *merging* (merging communities), *contracting* (a shrunken community), *splitting* (a split into communities), and *death* (a community vanishes). In [13], the definition of *change point* is given to represent a significant time point when the system evolves, that is, a major change (or critical event) occurs in the graph structure during a short period.

Several problems and challenges remain while studying community evolution. A first one is that there is no well-established standard definition of the notion itself when dealing with dynamic networks. Some authors are in favor to define a community as a structure that is observable over time. Such communities can be detected by matching observable communities at different time steps. Some authors propose to define a community as a structure that evolves over time. Therefore, communities can be detected by incremental updating. Authors also propose to transform the problem of community detection in dynamic networks into a problem of community detection in static networks. Different approaches are proposed to detect communities and analyze community evolution in dynamic networks.

In this survey, we try to detail current work on community detection in dynamic networks. We review these methods and discuss their performance in analyzing networks. This survey is organized as follows: we first describe the definition of community in static networks (see Sect. 2). In Sect. 3, we present different dynamic models such as network evolution models and community evolution models. These models are the principles of algorithms designed for tracking community evolution. In Sect. 4, we describe different methods for community detection in dynamic networks, discuss how to test them, and visualize their results. Section 5 contains the summary of this survey, along with a discussion about issues raised in tracking community evolution.

2 Communities in Networks

It appears natural to model the topology structure of a complex system by a graph (one can equally use the term network). Many real-world problems (biological, social, web) can be effectively modeled by graphs where nodes represent entities of interest and edges mimic the interactions or relationships among them.

In the study of networks, such as computer networks, information networks, social networks, or biological networks, uncovering the underlying community structure is essential. Social networks often include community groups based on common location, interests, hobbies, etc. Metabolic networks have communities based on modular functions [76]. Citation networks form communities by research topics. In each context, communities in a network can be described by dense groups of nodes, with more edges inside groups than edges linking the rest of the network.

In the following, we introduce the definition of community which depends on the context. Social network analysts have devised many definitions of communities with various degrees of internal cohesion among nodes [46, 80]. Many other definitions have been introduced by computer scientists and physicists. We distinguish three main classes of definitions: local, global, and based on vertex similarity. We review the notion of community structure. We also discuss the definition of the modularity function, derived to measure the quality of a graph partition into communities.

2.1 Preliminaries, Notations, and Definitions

A graph $G = (V, E)$ consists of two sets V and E, where $V = \{v_1, v_2, \dots, v_n\}$ are the nodes (or vertices, or points) of the graph G and $E \subseteq V \times V$ are its links (or edges, or lines). The number of elements in V and E are denoted by $n = |V|$ and $m = |E|$, respectively.

In the context of graph theory, an adjacency (or connectivity) matrix $\mathbf{A}$ is often used to describe a graph G. Specifically, the adjacency matrix of a finite graph G on n vertices is the $n \times n$ matrix $\mathbf{A} = [A_{ij}]_{n \times n}$, where an entry A_{ij} of $\mathbf{A}$ is equal to 1 if the link $e_{ij} = (v_i, v_j) \in E$ exists, and zero otherwise.

A *partition* is a division of a graph into disjoint communities, such that each node belongs to a unique community. A division of a graph into overlapping (or fuzzy) communities is called a *cover*. We use $\mathcal{P} = \{\mathcal{C}_1, \dots, \mathcal{C}_{n_c}\}$ to denote a partition, which is composed of n_c communities. In $\mathcal{P}$, the community to which the node v belongs to is denoted by σ_v. By definition we have $V = \cup_1^{n_c} \mathcal{C}_i$ and $\forall i \neq j, \mathcal{C}_i \cap \mathcal{C}_j = \emptyset$. We denote by $\mathcal{S} = \{S_1, \dots, S_{n_c}\}$ a cover composed of n_c communities. In $\mathcal{S}$, we may find a pair of community S_i and S_j with $i \neq j$ such that $S_i \cap S_j \neq \emptyset$.

Given a community $\mathcal{C} \subseteq V$ of a graph $G = (V, E)$, we define the internal degree $k_v^{\text{int}} = |\{e = (v, u) \mid u \in \mathcal{C}\}|$ (respectively the external degree $k_v^{\text{ext}} = |\{e = (v, u) \mid u \notin \mathcal{C}\}|$) of a node $v \in \mathcal{C}$, as the number of edges connecting v to other nodes u belonging to $\mathcal{C}$ (respectively to the rest of the graph). If $k_v^{\text{ext}} = 0$, the only neighbors of node v are within $\mathcal{C}$: assigning v to the current community $\mathcal{C}$ is likely to be a good choice. If $k_v^{\text{int}} = 0$ instead, the node is disjoint from $\mathcal{C}$ and it should be better to assign it to a different community. Classically, we note $k_v = k_v^{\text{int}} + k_v^{\text{ext}}$ the degree of node v. The internal degree k^{int} of $\mathcal{C}$ is the sum of the internal degrees of its nodes: $k^{\text{int}} = \sum_{v \in \mathcal{C}} k_v^{\text{int}}$. Likewise, the external degree k^{ext} of $\mathcal{C}$ is the sum of the external degrees of its nodes. The total degree $k_{\mathcal{C}}$ is the sum of the degrees of the nodes of $\mathcal{C}$. By definition, $k_{\mathcal{C}} = k_{\mathcal{C}}^{\text{int}} + k_{\mathcal{C}}^{\text{ext}}$.

2.2 Definitions of a Community

When studying community structure, authors often analyze structural properties of communities in the networks. The notion of communities can be formalized based on statistical properties. Furthermore, we can distinguish three types of community definition: local definition, global definition, and the definition based on node similarity. There, definitions are used to automatically detect community structure in networks.

Local Definitions

Communities are parts of the graph (group of nodes), within which the connections are dense and between which the connections are sparse. In some specific systems or applications, they can be considered as separate entities with their own autonomy, which do not depend on the whole graph. For instance, in [59], communities are defined in a very strict sense and require that all pairs of nodes are connected. In other words, this corresponds to a clique, that is, a subset of nodes such that every two vertices in the subset are connected by an edge. However, such criteria are too strict. A relaxable extended definition is *k-clique community*, which is the basis of CPM (Clique Percolation Method) [71]. A *k-clique community* is a sequence of adjacent cliques, where two k-cliques are *adjacent* if they share k-1 nodes.

Another criterion for community cohesion is the difference between the internal and external cohesion of the community. This idea is also used to define communities. For instance, Radicchi et al. [75] proposed the definitions of *strong communities* and *weak communities*. A set of nodes is a community in a strong sense if the internal degree of each node is greater than its external degree. This definition seems too strict. Its relaxable definition is the community in a weak sense: the internal degree of the community (sum of all its node internal degree) should exceed its external degree. Note that a community in a strong sense is also a weak community, while the converse is not generally true.

Global Definitions

Communities can be defined with respect to the graph as a whole. This seems to be reasonable when the community structure is exactly the division of the graph into several groups of nodes. In such a context, many global criteria are used to identify communities, which are all based on the intrinsic idea that a graph offers a community structure if its structure is far from a random graph. Random networks such as Erdös–Renyi's graphs do not display community structure. Indeed, as any pair of nodes are independently linked with the same probability, there should be no preferential wiring involving special groups of nodes. Therefore, one may define a *null model*, that is, a random graph that shares some structural properties of the original graph such as its degree distribution. The null model is the basic element in the conception of the notion of quality function named *modularity*. The modularity evaluates the quality of a graph partition into disjoint communities. The most popular modularity is proposed by Newman and Girvan [65], which compares the number of edges inside the community to the expected number of internal edges in the null model. A series of algorithms using modularity maximization heuristics [28, 66] for finding communities are proposed and developed.

Definitions Based on Node Similarity

It seems also natural to assume that communities are groups of nodes similar to each other. One can compute the similarity between each pair of nodes with respect to some reference properties. An important class of node similarity measures is based on properties of random walks on graphs, such as *commute time*. The *commute time* between a pair of nodes is the average number of steps needed for a random walker, starting at either node, to reach the other node for the first time and to come back to the starting node. Saerens et al. [79] have studied and used the commute time as a similarity measure: the larger the commute time is, the less similar the nodes are.

2.3 Modularity

One may want to measure the quality of a partition through a *quality function*, which assigns a score to each partition of a graph. In this way, partitions can be ranked based on their score given by the quality function. Partitions with high scores are *"good"*, so the one with the highest score is by definition the best.

The widest accepted quality function is the modularity introduced by Newman and Girvan [65, 68]. Let e_{ij} be the fraction of edges in the network that connect nodes in community i to those in community j, and $a_i = \sum_j e_{ij}$. The modularity measure is defined as

$$Q = \sum_i \left(e_{ii} - a_i^2\right). \tag{1}$$

This quantity measures the fraction of the within-community edges in the network minus the expected value in a network with the same community division but when connections between nodes are random. If the number of within-community edges is less than the expected number of edges in a random graph, we will get $Q = 0$. Values approaching $Q = 1$, which is the maximum, indicate networks with strong community structure. In practice, values for real networks typically fall in the range from 0.3 to 0.7. Higher values are rare.

Suppose we have a division of a network into communities. Let σ_i be the community to which node i is assigned. The fraction of the edges in the graph that fall within communities, that is, that connect nodes that both lie in the same community, is

$$\frac{\sum_{ij} A_{ij}\delta(\sigma_i, \sigma_j)}{\sum_{ij} A_{ij}} = \frac{1}{2m}\sum_{ij} A_{ij}\delta(\sigma_i, \sigma_j)$$

where the function $\delta(\sigma_i, \sigma_j)$ is 1 if $\sigma_i = \sigma_j$ and 0 otherwise. At the same time, the expected number of edges between nodes i and j, if edges are drawn at random, is $k_i k_j / 2m$, where k_i and k_j are the degrees of the nodes and m is the total number of edges in the network. Thus, the modularity [64], as defined above, is given by

$$Q = \frac{1}{2m}\sum_{i \neq j}\left(A_{ij} - \frac{k_i k_j}{2m}\right)\delta(\sigma_i, \sigma_j) \tag{2}$$

Note that the modularity is always smaller than one but can be negative as well. For instance, the partition where each node represents a single community is always negative. When considering the whole graph as a single community, the modularity is zero as the two terms are equal and cancels each other out. There are also other types of modularity, some of which are motivated by specific classes of clustering problems or graphs. The goal here is not to list exhaustively all algorithms which were built upon the modularity. The interested reader should turn to Fortunato's extensive review of the field [32], which does not only cover the modularity but community detection as a whole.

Modularity has been employed as quality function in many algorithms, like some division algorithms [66] which give a trade-off between high accuracy and low complexity. In addition, modularity optimization is the most popular method for community detection. Heuristic proposed in [11] runs fast and handles very large-scale networks. Modularity also allows to assess the stability of partitions [60].

However, the applicability and reliability of modularity for the problem of graph clustering may be limited. An important issue concerning the limits of modularity is raised by Fortunato and Barthelemy [33]. The study shows that a large value for the maximum modularity does not necessarily mean that a graph has a clear community structure. In a random graph, such as the Erdös–Rényi model, the distribution of edges among the nodes is highly homogeneous. For instance, the distribution of the number of neighbors of a node, or degree, is binomial, so most nodes have equal or similar degree. The random graph is supposed to have no community structure, as the link probability between nodes is either constant or a function of the node degrees, so there is no bias a priori towards special groups of nodes. Still, random graphs may have partitions with large modularity values [42, 77]. This is due to fluctuations in the distribution of edges in the graph, which determine concentrations of links in some subsets of the graph, which then appear as communities.

Moreover, Fortunato and Barthelemy [33] have found that modularity optimization has a resolution limit. It may prevent from detecting communities which are comparatively small with respect to the graph as a whole. Given two communities $\mathcal{A}$ and $\mathcal{B}$, with a total degree $k_{\mathcal{A}}$ and $k_{\mathcal{B}}$, respectively, and where the number of edges connecting $\mathcal{A}$ and $\mathcal{B}$ is $l_{\mathcal{AB}}$, the difference of modularity determining the merger of two communities with respect to the whole graph partition is

$$\Delta Q = \left[\frac{k_{\mathcal{A}}^{int} + k_{\mathcal{B}}^{int} + 2l_{\mathcal{AB}}}{2m} - \left(\frac{k_{\mathcal{A}} + k_{\mathcal{B}}}{2m} \right)^2 \right] - \left[\frac{k_{\mathcal{A}}^{int} + k_{\mathcal{B}}^{int}}{2m} - \left(\frac{k_{\mathcal{A}}}{2m} \right)^2 - \left(\frac{k_{\mathcal{B}}}{2m} \right)^2 \right]$$

If $l_{\mathcal{AB}} = 1$, that is, there is a single edge joining $\mathcal{A}$ to $\mathcal{B}$, we expect that the two communities should be separated. If $k_{\mathcal{A}} k_{\mathcal{B}} / 2m^2 < \frac{1}{m}$, we have $\Delta Q_{\mathcal{AB}} > 0$. For simplicity, let us suppose that $k_{\mathcal{A}} \sim k_{\mathcal{B}} = k$, that is, that the two subgraphs have roughly the same number of edges. We conclude that when $k < \sqrt{2m}$ and the two communities $\mathcal{A}$ and $\mathcal{B}$ are connected, then the modularity is higher if they are in the same cluster [33]. So, if the partition with maximum modularity includes clusters with total degree of the order of $\mathcal{O}(\sqrt{m})$ (or smaller), one cannot know a priori whether the clusters are composed of single communities or are in fact a combination of smaller weakly interconnected communities. This resolution problem may have important impacts in practical applications.

3 Community Evolution in Dynamic Networks

In complex networks, the interactions between entities dynamically evolve over time [7]. Let us take Facebook[1] as an example: users add or delete "friends" [26]. Similarly, new forms of social contacts can be observed in phone calls, e-mail exchanges [58], or other communications on the Internet.

As mentioned previously, traditional analysis treats networks as *static* graphs, which is derived either from an aggregation of data over the whole network life (experiment measure) or from a snapshot of data at a particular time step. Although

[1] http://www.facebook.com/

this study provides meaningful results, the dynamic features are neglected. Dynamic features are however crucial in order to better understand complex networks.

During the last decade, the avalanche of data footprint provided by the traceability of many social activities represents a major scientific, economic, and social revolution. The availability of large dataset (thanks to Open Data initiative), the optimized rating of computing facilities, and the development of powerful and reliable data analysis tools have constituted a better and better machinery to explore the topological properties of several networked systems from the real world. This has allowed to study the topology of the dynamic interactions in a large variety of Big Data [19] as diverse as communication [72, 73], social [24, 67], and biological systems [12, 48].

In the following, we discuss network evolution models in Sect. 3.1. Next, we introduce the definitions and notations of community evolution in dynamic networks in Sects. 3.2 and 3.3. Moreover, we discuss how to evaluate community evolution. Evaluating community evolution depends on the definition about community evolution and the chosen similarity measures. There is a similarity measure proposed for measuring the quality of the found community structure at each time step. It is called *α-cost*. There are also many other similarity measures such as *matching metrics*. Matching metrics are used for matching communities at different time steps. We introduce the quality function: α-cost (see Sect. 3.4) and discuss matching metrics (see Sect. 3.5). In addition, we discuss the definition of community dynamics, which are based on matching metrics (see Sect. 3.6).

3.1 Network Evolution Models

When networks tend to gradually evolve over time, a first class of models considering networks as dynamical systems can be derived. A first common class of evolving network models is based on two ingredients: *growth* and *preferential attachment*. The growth hypothesis suggests that networks continuously expand through the arrival of new nodes and new links between existing nodes, while the preferential attachment hypothesis states that strongly connected nodes increase their connectivity faster than less connected nodes. In [49], the authors have measured different networks. Results show different attachment rate functions: the attachment rate in some systems depends linearly on the node degree, while the dependence of other systems follows a sublinear power law.

In order to mine dynamic properties of networks, Leskovec et al. [57] have studied a wide range of real networks from several domains. Empirical observations revealed that most networks are becoming denser over time with the increasing average degree and decreasing effective diameters.

As one may expect, the evolution of real networks is complicated. It is possibly related to community structure. For instance, Backstrom et al. [5] investigated an online network [2]. They computed the probability of joining a community as a function of internal friends (who are already in the community). The empirical studies showed that the probability of joining a community increases with the number of internal friends but is very noisy. Moreover, their studies also compare the effects between different friends in attracting new community members: the number of friends who know each other provides a stronger effect than the friends who do not know each other.

Asur et al. [3] have measured *sociability index* in testing the DBLP coauthor dataset. The *sociability index* gives high scores to nodes that are involved in interactions with different groups. Their analysis showed that the sociability index could be used to predict future co-occurrences of nodes in clusters.

Therefore, current statistical models based on growth and preferential attachment only capture one part of dynamic behaviors of real networks and fail to capture many other dynamic structural properties such as the evolution of community structure. Community evolution is important for analyzing dynamic structural properties of real networks.

3.2 *Community Evolution Models*

There is no accepted standard definition of community in dynamic networks. A community can be defined as a structure that is observable over time. Such communities can be detected by matching observable communities at different time steps. In [44], Hopcroft et al. have detected the partition for each snapshot graph by hierarchical clustering [45] and then matched communities at different time steps through the notion of *natural communities*. They define *natural communities* as groups of nodes having high stability against perturbations of their interactions. When analyzing citation networks, natural communities can be used to denote topics of communities. Tracking natural community evolution allows them to understand the history of topics, such as the emergence of new topics. The idea of detecting *time-independent* communities at different time steps and then matching them becomes the basis for several algorithms. They are called *two-stage approaches* (see Sect. 4.1). Each *time-independent* community is detected independent of the results at other time steps.

Another definition of community in dynamic networks is a structure that evolves over time. For instance, a community is defined as the neighborhood of a chosen node in [27]. Therefore, each community is detected by incrementally updating the neighborhood of the chosen node corresponding to the evolution of the graph structure over time. Such methods that propose to detect *time-dependent* temporal clusters are called *evolutionary clustering* (see Sect. 4.2). The principle of *evolutionary clustering* [13] is to simultaneously optimize two potentially conflicting criteria:

[2]LiveJournal (LJ)http://www.livejournal.com/

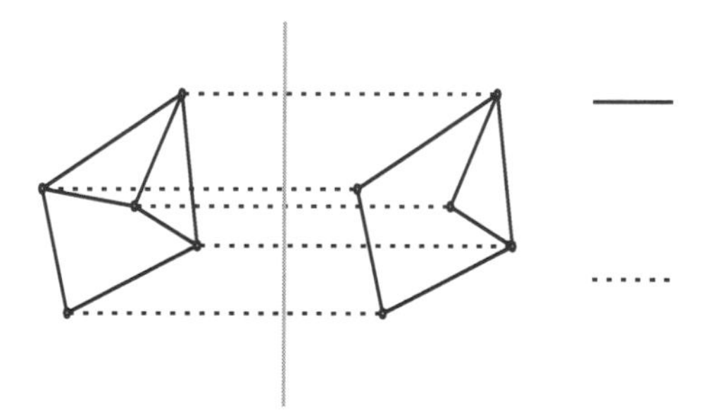

Fig. 1 An example of a coupling graph, where graphs at different time steps are connected through couplings. The real interactions between nodes are shown in *solid lines*, while the coupling interactions are denoted by *dotted lines*. The figure is gained from [47]

(i) first, the clustering at any time step should remain faithful to the current data as much as possible *(ii)* and second, the clustering should not shift dramatically from one time step to the next time step.

There are also many other models for capturing community structure in dynamic networks. The *coupling graph clustering* is a framework which detects community structure of a *coupling graph* (see Sect. 4.3). A *coupling graph* is a graph linking a sequence of graphs over several time steps by adding coupling edges between the same nodes at different time steps (see Fig. 1). Given a coupling graph, a subgraph which describes all interactions at a specific time step is called a *slice*.

3.3 Notations of Community Evolution

A *dynamic graph* $\mathcal{G}(V, \mathcal{E})$ on a finite time sequence $1 \ldots \Delta$ is a sequence of graph snapshots $\{G(1), \ldots, G(\Delta)\}$. There is a set $V = \{v_1, \ldots, v_n\}$ of nodes. Each node $v_i \in V$ appears at least one during the dynamic graph lifetime, *that is*, $\exists t \; s.t. \; v_i \in G(t)$.

At each time step t where $1 \leq t \leq \Delta$, the corresponding snapshot $G(t)$ describes interactions between active nodes at time t, where the edges of a snapshot graph are a set of active dynamic links. $G(t)$ is partitioned into a set of *temporal clusters* $\mathcal{P}(t) = \{C_1(t), \ldots, C_{n_c^t}(t)\}$, where n_c^t denotes the number of temporal clusters in $G(t)$. In some definitions of communities in dynamic networks [30, 31], the number of temporal clusters may be not equal to the number of communities at the same time step t . One community $\mathcal{C}_i$ at time step t is possibly represented by a set of temporal clusters such that $\mathcal{C}_i(t) = \{C_1(t), \ldots\}$.

The problem of tracking community evolution can be resolved by the identification of a set of *community evolution paths* (or *community evolution traces* [92], *dynamic communities* [39]).

Definition 1 (Community evolution path). For a given time window $[\delta_0, \delta_0 + \Delta]$, an evolution path $\text{Evol}(\mathcal{C}_i)$ is a time series of temporal clusters: $\text{Evol}(\mathcal{C}_i) := \{C_i(\delta_0), \ldots, C_i(\delta_0 + \Delta)\}$ where each temporal cluster $C_i(t) \in \text{Evol}(\mathcal{C}_i), t \in [\delta_0, \delta_0 + \Delta]$ is the observation of the community $\mathcal{C}_i$.

In the definition of Wang et al. [92], the observation of the community C_i at time t can be the union of several temporal clusters. When a community appears for the first time, it should be a unique temporal cluster.

3.4 Quality Function: α-Cost

Reliable algorithms are supposed to provide results having a high-quality value. In the case of community detection in dynamic networks, a famous function named α-cost has been used by several algorithms [52, 83, 89] for measuring the quality of the found dynamic communities. This α-cost is motivated by the principle of clustering evolution: the community structure at each time step is the evolution of the community structure at the previous time step. Therefore, one can see the evolution as a combination of a *snapshot cost* and a *past history cost*. The parameter α controls the relative weight of recent and past history:

$$\text{cost} = \alpha\,\mathcal{CS} + (1-\alpha)\,\mathcal{CT} \tag{3}$$

where the snapshot cost $\mathcal{CS}$ measures how a community structure fits the graph interactions at time t and the past history cost $\mathcal{CT}$ qualifies how consistent the community structure is with the past history community structure at time $t-1$.

Let X represent the current community structure, Y represent the community structure at the previous time step, W denote current graph interaction, Λ be a nonnegative diagonal matrix, and $D(\bullet)$ be the function for measuring the cost such that $D(\bullet)$ computes the similarity between the network structure and the community structure and the similarity between the current community structure and the previous community structure.

In [89], authors defined $D(\bullet)$ as a KL-divergence between two objects such that $\mathcal{CS} = D(W \parallel X\Lambda X^T)$ and $\mathcal{CT} = D(Y \parallel X\Lambda)$. Given two objects A and B, $D(A \parallel B) = \sum_{ij}\left(a_{ij}\log\frac{a_{ij}}{b_{ij}} - a_{ij} + b_{ij}\right)$. Through this cost definition, the snapshot cost is high when the approximate community structure fails to fit the graph interactions at time t, while the past history cost is high when there is a dramatic change of community structure from time $t-1$ to t.

There exist also other definitions of $D(\bullet)$. In [52], two definitions of the cost are introduced: one is the distance between all pairs of objects in an agglomerative hierarchical clustering, and the other is associated with the centroid of the community in k-means clustering [10]. In k-means clustering, community memberships are measured by the membership degrees of nodes, that is, the distance between the node to the centroid of its community. Then, in the cost of the community structure of a dynamic graph, the snapshot cost is associated with the distance between the node and the centroid of its community, and the past history cost is computed by the difference between the current community centroid and the community centroid at the previous time step.

In the case of multimode networks, Tang et al. [84] have suggested the resolution by transforming the problem in multimode networks into the problem of two mode. Most of existing work concentrates on *one-mode network*. That is, there is only one type of social actors (nodes) involved in the network and the ties (interactions) between actors are all of the same type. This is common in a broad sense such as friendship network, the Internet, and phone call network. However, some applications such as web mining, collaborative filtering, and online-targeted marketing involve more than one type of actors and multiple heterogeneous interactions between different types of actors. Such a network is called *multimode network* [93].

Given an m-mode network, for each mode i, let $\mathbb{X}_i$ denote this mode of nodes, such as $\mathbb{X}_i = \{x_1^i, \ldots, x_{n_i}^i\}$, where n_i is the number of nodes for $\mathbb{X}_i$. Then, for each pair of modes, we use $\mathbf{R}_{ij}^t \subseteq \mathbb{X}_i \times \mathbb{X}_j$ to represent interactions between two modes of nodes $\mathbb{X}_i, \mathbb{X}_j$ at time t. Ideally, the interaction between nodes can be approximated by

$$\mathbf{R}_{ij}^t \approx \mathbf{C}_i^t \mathbf{A}_{ij}^t (\mathbf{C}_j^t)^T$$

where $\mathbf{C}_i^t$ is the cluster membership for $\mathbb{X}_i$ at time t and A_{ij}^t represents the group interaction. The group interaction is computed by $A_{ij}^t = (\mathbf{C}_i^t)^T \mathbf{R}_{ij}^t \mathbf{C}_j^t$. Therefore, for each temporal m-mode graph at time t, its snapshot cost $\mathcal{CS}$ can be formulated as

$$\mathcal{CS} = \sum_{1 \leqslant i < j \leqslant m} w_a^{(i,j)} D\left(\mathbf{R}_{ij}^t \parallel \mathbf{C}_i^t \mathbf{A}_{ij}^t (\mathbf{C}_j^t)^T\right)$$

and its history cost $\mathcal{CT}$ is expressed as

$$\mathcal{CT} = \sum_{1 \leqslant i \leqslant m} w_b^i D\left(\mathbf{C}_i^t \parallel \mathbf{C}_i^{t-1}\right)$$

where w_a^{ij} is an importance factor for every pair of modes i and j, and w_b^i is a relative importance factor for each mode i.

3.5 *Matching Metrics*

A matching metric is a similarity function, which measures how similar two communities are. It is often used in two-stage approaches (see Sect. 4.1) to connect similar communities. One can measure the similarity between two temporal clusters at different time steps and naturally obtain how one community evolves from one time step to the following time steps.

Hopcroft et al. [44] defined a *match* function. Let C and C' be two clusters; their match value is written as follows:

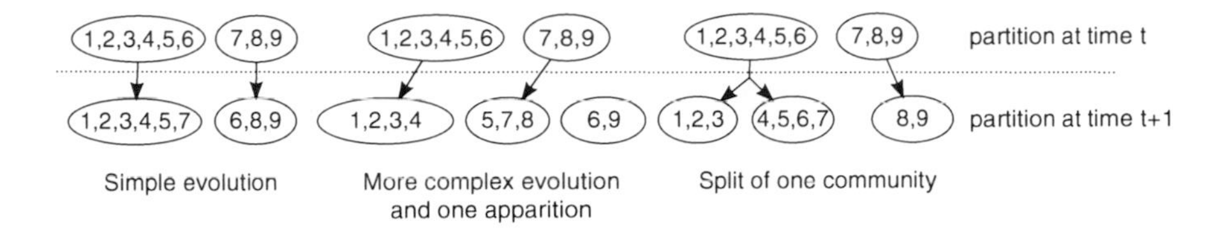

Fig. 2 Examples of community evolution in a time period $[t, t+1]$. We match clusters at time t to the clusters at time $t+1$. Given a cluster at time t, it remains stable if it is matched to a unique cluster at time $t+1$; it splits if it is matched to more than one cluster at time $t+1$. In addition, one new cluster appears at time $t+1$, if no cluster at time t is matched to it. The figure is obtained from [4]

$$\text{match}(C, C') = \min\left(\frac{|C \cap C'|}{|C|}, \frac{|C \cap C'|}{|C'|}\right) \tag{4}$$

The definition ensures that a high matching value (close to 1) occurs when two clusters have many common nodes and are roughly of the same size. The best match value for C at time t is the highest match(C, C') value for any cluster C' at time t (Fig. 2).

Palla et al. [70] defined *relative overlap*, which is a Jaccard index. The relative overlap value between two communities X and Y is written as follows:

$$J(X, Y) = \frac{|X \cap Y|}{|X \cup Y|} \tag{5}$$

By definition, the cluster $C(t+1)$ at time $t+1$ is matched to the cluster $C(t)$ which has the largest overlap at time t.

Another bipartite mapping metric is *dynamic Jaccard's index*, whose definition is

$$JacD'(X, Y) = \frac{J(X, Y)}{|t - t'|} \tag{6}$$

where $|t - t'|$ represents the time interval duration between communities X and Y. It allows a temporal cluster to be matched to an older one ($|t - t'| > 1$) which may have disappeared during several time steps.

Two communities are matched if they share the highest matching value. The matching metric is a natural resolution to connect temporal clusters over time. So, it is often used in two-stage methods [44, 70, 86]. Its other advantage is to characterize community dynamics. However, there is no standard definition of matching metric. In Hopcroft et al. [44]'s *match* function (4), the minimum size of communities is important for the comparison. Instead, the size of the union of communities is essential in the relative overlap (5). Furthermore, a minimum intersection size threshold needs to be set, that is, the minimum number of common nodes shared by the matching communities.

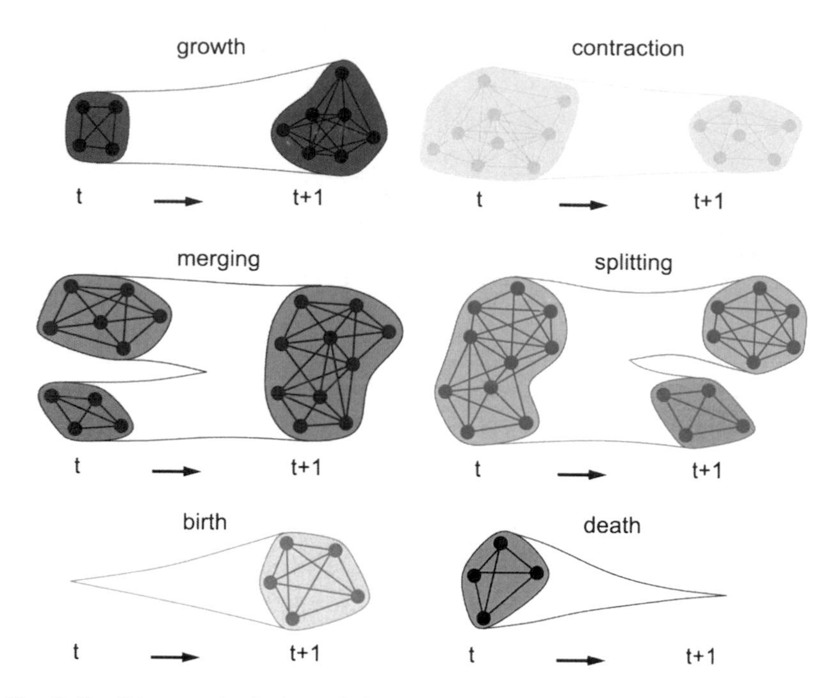

Fig. 3 Possible scenarios in the evolution of communities. The figure is gained from [70]

3.6 *Definition of Community Dynamics*

When we track community evolution, one problem is to characterize community dynamics. How a community changes over time? Chakrabarti et al. [13] proposed the definition of *change point* to describe a significant change in community structure. In the following, we describe them in details. Moreover, Palla et al. have introduced the main phenomena occurring during the lifetime of a community (see Fig. 3): birth, growth, contraction, merging, splitting, and death.

Change Point

Chakrabarti et al. [13] have detected *change points*, which represent the significant time points when the system evolves, that is, a major change (or critical event) occurs in the graph structure during a short period. The approach called GraphScope [13] applied the MDL (Minimum Description Length) principle [40] to compute the encoding cost of assigning nodes into communities. A *segment* presents a sequence of graphs without any change in its community structure. So the graphs of each segment are characterized by the same partition with the lowest encoding cost. If the cost for encoding a graph into the existing segment is higher than the cost for encoding the graph into a new segment, a significant change of community structure occurs. The change point offers one important benefit of detecting community evolution using information theory.

Community Changes

We show six community changes in Fig. 3, which are used to describe the main events occurring in dynamic graphs. In order to identify them, matching metrics are often used such as the definition given in [3].

Definition 2. Let $G(t)$ and $G(t+1)$ be two snapshots of $\mathcal{G}$ at two consecutive time steps. Let the cluster $C_i(t)$ and $C_i(t+1)$ denote the observations of the community $\mathcal{C}_i$ at time step t and $t+1$, respectively.

Continue: $C_i(t+1)$ is the continuation of $C_i(t)$ if $C_i(t+1)$ is the same as $C_i(t)$:

$$C_i(t) = C_i(t+1)$$

κ–Merge: Two clusters $C_i(t)$ and $C_j(t)$ merge into $C_i(t+1)$ if $C_i(t+1)$ contains at least $\kappa\%$ of nodes belonging to the union of $C_i(t)$ and $C_j(t)$ and the renewal of $C_i(t)$ and $C_j(t)$ is at least 50%:

$$\frac{|(C_i(t) \cup C_j(t)) \cap C_i(t+1)|}{\max(|C_i(t) \cap C_j(t)|, |C_i(t+1)|)} > \kappa$$

$$|C_i(t) \cap C_i(t+1)| > |C_i(t)|/2$$

$$|C_j(t) \cap C_i(t+1)| > |C_j(t)|/2$$

κ–Split: $C_i(t)$ is split into $C_i(t+1)$ and $C_j(t+1)$ if $\kappa\%$ of nodes belonging to $C_i(t)$ are in two different clusters at time $t+1$, such as

$$\frac{|(C_i(t+1) \cup C_j(t+1)) \cap C_i(t)|}{|\max(|C_i(t+1) \cap C_j(t+1)|, |C_i(t)|)} > \kappa$$

$$|C_i(t) \cap C_i(t+1)| > |C_i(t+1)|/2$$

$$|C_i(t) \cap C_j(t+1)| > |C_j(t+1)|/2$$

Emerge: A new cluster $C_i(t+1)$ emerges at time $t+1$ if none of the nodes in the cluster $C_i(t+1)$ are grouped together at time t, that is, $\nexists\ C_i(t)$, such that $|C_i(t) \cap C_i(t+1)| > 1$.

Disappear: $C_i(t)$ disappears if none of the nodes in the cluster $C_i(t)$ are grouped at time $t+1$, that is,$\nexists\ C_i(t+1)$, such that $|C_i(t) \cap C_i(t+1)| > 1$.

This definition has several limits. First, the definition of one continuation is so strict that almost all communities do not have any continuation at the next time step. Second, the value of κ needs to be set to determine when a community is merged or when a community is splited. Varying κ may lead to different results. Finally, the definition of emerging community or disappearing community has weakness. Some clusters may be generated only by the fluctuation of degree distribution. These artificial clusters will not share a strong common interest. For the disappearance, the process may be too slow: a community may lose its core nodes but still have node attached to it. In this case, the observed community does not share a strong common interest anymore. It is difficult to determine whether a community exists.

Another definition of community dynamics is based on community predecessor/successor relationship [91].

Definition 3 (Community predecessor and successor). Given a temporal cluster $C_i(t)$ at time t, if the temporal cluster $C_j(t-1)$ has the maximum overlap size among all temporal clusters at time $t-1$, we define that $C_j(t-1)$ is the predecessor of $C_i(t)$. If the temporal cluster $C_k(t+1)$ has the maximum overlap size among all temporal clusters at time $t+1$, we define that $C_k(t+1)$ is the successor of $C_i(t)$.

In the following, given a pair of temporal clusters (X, Y), $X \rightarrow Y$ is used to denote that Y is Xs successor and $X \leftarrow Y$ represents that X is Ys predecessor.

The relationship between one community and its successor (or its predecessor) may be asymmetrical. That is, for one community and its successor, this community may not be the predecessor of its successor. Similarly, for one community and its predecessor, it is possible that the community is not the successor of its predecessor. This asymmetrical property is used to characterize community dynamics.

Definition 4. Let $G(t)$ and $G(t+1)$ be snapshots of $\mathcal{G}$ at two consecutive time steps with the temporal partition $\mathcal{P}(t)$ and $\mathcal{P}(t+1)$ denoting the community structure of $\mathcal{G}$ at time step t and $t+1$, respectively.

Survive. $C_j(t+1)$ is the continuation of $C_i(t)$, if and only if $C_i(t)$ is the predecessor of $C_j(t+1)$ and $C_j(t+1)$ is the successor of $C_i(t)$, such that

$$C_i(t) \leftarrow C_j(t+1) \wedge C_i(t) \rightarrow C_j(t+1)$$

This relationship is denoted by $C_i(t) \leftrightarrows C_j(t+1)$. When a community survives from t to $t+1$, it is a growing community if it has an increasing number of community members; otherwise, it is a shrinking community.

Emerge. $C_j(t+1)$ is a creation if and only if $C_j(t+1)$ has no predecessor such that

$$\nexists C_i(t) \in \mathcal{P}(t) | \left(C_i(t) \rightarrow C_j(t+1)\right)$$

Merge. $C_j(t+1)$ is a fusion if and only if $C_j(t+1)$ is the successors of several clusters at time step t such that

$$\exists \{C_i(t), C_k(t)\} \subseteq \mathcal{P}(t) | \left(C_i(t) \rightarrow C_j(t+1) \wedge C_k(t) \rightarrow C_j(t+1)\right)$$

where $i \neq k$.

Split. $C_j(t+1)$ is a split if and only if $C_j(t+1)$ is not the successor of its predecessor such that

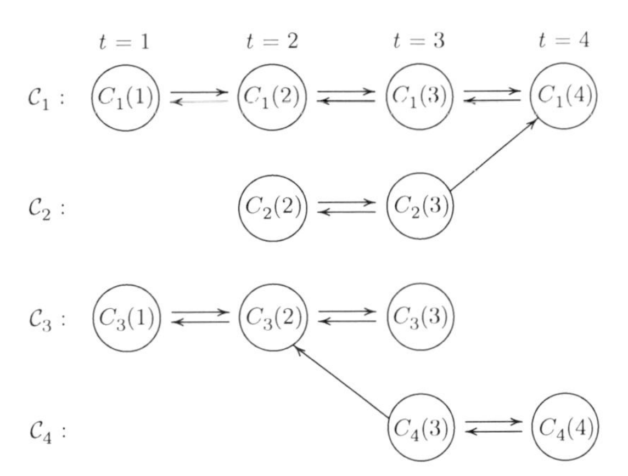

Fig. 4 Diagrams of four communities observed during four time steps, featuring continuation, creation, disappearance, fusion, and split

$$C_i(t) \leftarrow C_j(t+1) \wedge C_i(t) \nrightarrow C_j(t+1)$$

Disappear. A community disappears at time $t+1$ if and only if its observation $C_i(t)$ at time step t has no successor such that

$$\nexists C_j(t+1) \in \mathcal{P}(t+1) | \left(C_i(t) \nrightarrow C_j(t+1)\right)$$

Diagrams in Fig. 4 show several cases illustrating community dynamics which can be featured by continuation, creation, disappearance, fusion, and split. For better understanding of community evolution, their evolution paths (Def. 1) will be shown. For each community $\mathcal{C}$, its evolution path is $\text{Evol}(\mathcal{C}) := \{C(1), \ldots, C(\Delta)\}$, where each element $C(i)$ $(1 \le i \le \Delta)$ represents its observation at time step $t = i$.

In the example illustrated by the Fig. 4, four communities have the evolution paths:

- $\text{Evol}(\mathcal{C}_1) := \{C_1(1), C_1(2), C_1(3), C_1(4)\}$
- $\text{Evol}(\mathcal{C}_2) := \{C_2(2), C_2(3)\}$
- $\text{Evol}(\mathcal{C}_3) := \{C_3(1), C_3(2), C_3(3)\}$
- $\text{Evol}(\mathcal{C}_4) := \{C_4(3), C_4(4)\}$

Nearly all types of community changes are observed:

- Community $\mathcal{C}_2$ is created at $t = 2$ as it has no predecessor at $t = 1$.
- Community $\mathcal{C}_3$ disappears at $t = 4$ as it has no successor at $t = 4$.
- Community $\mathcal{C}_2$ is merged into $\mathcal{C}_1$ at $t = 4$ since its successor at $t = 4$ is $C_1(4)$ whose predecessor is not $C_2(3)$.
- Community $\mathcal{C}_4$ is split from $\mathcal{C}_2$ since $t = 2$ as its predecessor at $t = 2$ is $C_3(2)$ whose successor is not $C_4(3)$.

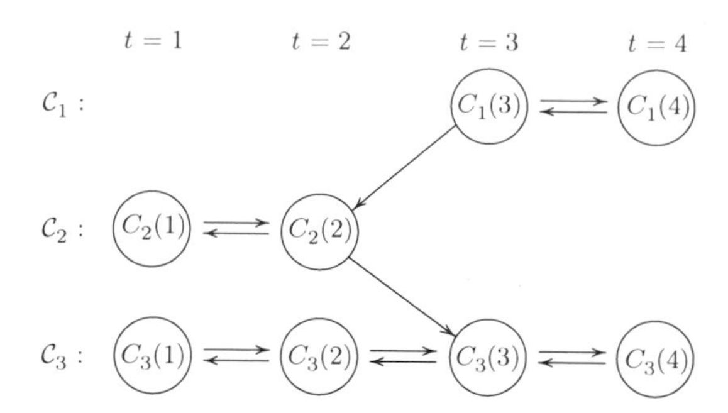

Fig. 5 Diagram of four clusters observed over 4 time steps, featuring fusion and split community events

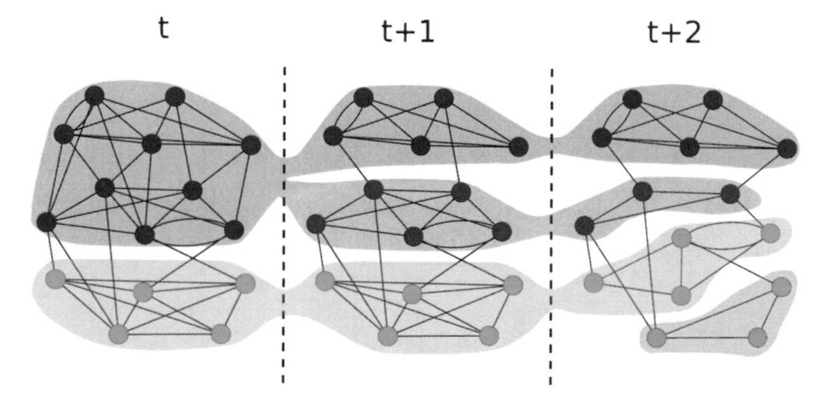

Fig. 6 Examples of community evolution over three snapshot graphs by matching temporal clusters to dynamic communities. We observe 4 dynamic communities, indicated by colors: C_1 in *dark blue*, C_2 in *red*, C_3 in *green*, and C_4 in *light blue*. During their evolution, we observe the community C_1 is split into C_1 and C_2 between t and $t + 1$

Community C_1 is observable during all the observation window (only four time steps on this toy example). At time step $t = 4$, community C_2 joins it. This community fusion event seems to be more an event related to C_2 rather than to C_1.

A more complex diagram is displayed in Fig. 5. We observe the changes of communities from time step $t = 2$ to $t = 3$. At time step $t = 3$, community C_2 partially merges with C_3, while its split $C_1(3)$ starts a new community C_1.

There are also other types of definitions [16,37,39]. For example, Chen et al. [16] characterize community dynamics by tracking community core evolution. Greene et al. [39] use the definition of dynamic communities described above but require that if several dynamic communities share the same temporal cluster at time t, then these dynamic communities should merge.

In Fig. 6, we have shown examples of community evolution. There are four dynamic communities over the total three time steps, whose evolution paths are expressed as following:

$$\mathrm{Evol}(\mathcal{C}_1) \leftarrow \{C_1(t), C_1(t+1), C_1(t+2)\}$$
$$\mathrm{Evol}(\mathcal{C}_2) \leftarrow \{C_2(t+1), C_2(t+2)\}$$
$$\mathrm{Evol}(\mathcal{C}_3) \leftarrow \{C_3(t), C_3(t+1), C_3(t+2)\}$$
$$\mathrm{Evol}(\mathcal{C}_4) \leftarrow \{C_4(t+2)\}$$

Through these evolution paths, we observe two new communities appearing during network evolution: the community $\mathcal{C}_2$ is the branch of $\mathcal{C}_1$ and the community $\mathcal{C}_2$ emerges at time $t = 2$.

This is an example to illustrate the relationship between community dynamics and community evolution paths. We conclude that the problem of identifying and characterizing community dynamics can be revealed by community evolution paths, whereas the problem of tracking community evolution in dynamic networks can be reformulated as a problem of constructing community evolution paths across one or more time steps.

4 Tracking Community Evolution

Tracking community evolution is a key problem for which many algorithms have been proposed and developed. In the following, we will review them in details. Being able to benchmark proposed heuristics appears to be an important challenge too. Indeed, once an algorithm is designed, it is mandatory to test its performance. A natural solution is to design benchmark graphs (see Sect. 4.4). Benchmark graphs should offer an a priori known community structure. When an algorithm is reliable and efficient, it has to perform well when applied to benchmark graphs. Another important challenge is to design adapted tools for visualizing and representing how community structure evolves over time (see Sect. 4.5).

4.1 Two-Stage Approaches

The basic idea of two-stage approaches is to detect temporal clusters at each time step and then establish relationships between clusters for tracking community evolution over time. Figure 6 illustrates the result of applying a two-stage approach to a dynamic network across three time steps. In a first phase, clusters at each time step are detected: at time t, there are two clusters, then there are three clusters at time $t + 1$ and four clusters at time $t + 2$. In a second phase, the relationship between clusters at different time steps is established, which is shown by colors. Through the above results, we learn how the community structure of this graph evolves from the time step t to the time step $t + 2$. For the first phase, we apply a graph clustering algorithm [36]. For the second phase, we can use a matching metric (see Sect. 3.5). However, it may lead to noisy results where some nodes often change their community memberships. Therefore, many advanced resolutions are proposed to resolve this general matching problem.

Core-Based Methods

If a partition is significant, it will be recovered even if the structure of the graph is modified, as long as the modification is not too extensive. Instead, if a partition is not significant, we may observe that a minimal perturbation of the graph is enough to disrupt its group memberships. Since most stochastic community detection are searching for local optima due to computational costs, the detection results can be different simply if the ordering of the nodes is modified, without any modification of the topology. This is often referred as the consistency problem [53, 54, 81]. A *significant cluster*, that is, a significant group of nodes, is often defined as a *community core*. We can reduce noisy results by matching community cores. This is the main principle of core-based methods. The matching metric (see Sect. 3.5) is often applied. Two temporal clusters are matched if their community cores share the highest similarity value.

Hopcroft et al. have proposed the concept of natural communities, which are significant clusters that have high stability against modification of graph structure. Given a temporal graph, by applying 5% of perturbations, a set of modified graphs are produced, each of which has 95% of core nodes. Each natural community is identified by the partitions corresponding to these modified graphs, which has the best match value with clusters in those partitions.

Rosvall et al. [78] used a bootstrap method [25] to detect significance of clusters. The bootstrap method assesses the accuracy of an estimate by resampling from the empirical distribution of observations. Each graph can be resampled by assigning to each edge a weight taken from a Poisson distribution with mean equal to the original edge weight. A graph clustering method is applied to the original graph and the samples. For each community in the original graph's partition, they define its largest subset of nodes that are classified in the same community in at least 95% of all bootstrap samples, as the significant cluster.

In some methods, core nodes are identified through their roles within their communities. Given a community, there are core nodes and peripheral nodes. Guimerá and Amaral [41] have classified community members into different roles according to intra- and inter-module connection patterns. With respect to core node identification, Wang et al. [92] defined core nodes, where each core node v satisfies $\sum_{u \in \text{neighbours}} (k_v - k_u) > 0$. In [8], k-cores nodes [1] are detected with a threshold k where k-core decomposition is used for filtering out peripheral nodes.

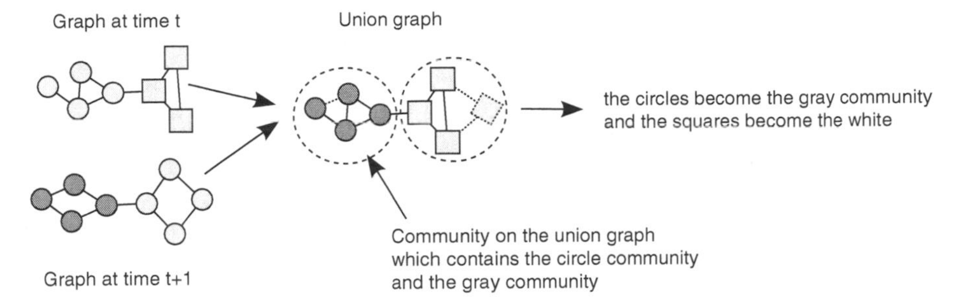

Fig. 7 An example of a union graph which is constructed by jointing two graphs at time t and $t+1$. The figure is obtained from [70]

Although core-based approach can smooth variances caused by peripheral nodes, its results still suffer from some limits such as the parameters used in matching metrics. In additional, if we only track evolution of community cores, there is a risk of missing important structural changes which are related to peripheral nodes.

Union-Graph-Based Methods

Another important early work [70] for detecting community evolution is related to the *union graph*. Each union graph merges two graphs (union of their links) present at contiguous time steps. Let $G(t, t+1)$ denote the union graph resulting from the union of two graphs at time t and $t+1$. We have $E_{t,t+1} = E_t \cup E_{t+1}$. Figure 7 gives an example of a union graph. Any community present at t or $t+1$ is contained in exactly one community in the union graph. Thus, communities in the union graph provide a natural connection between communities at t and $t+1$. If a community in the joined graph contains a single community from t and a single community from $t+1$, then they are matched. If the joined group contains more than one community from both time steps, the communities are matched in decreasing order of their relative node overlap (5). The technique is validated by applying it to two social systems: a graph of phone calls between customers of a mobile phone company over one year and a collaboration network between scientists spanning a period of 142 months.

The union graph smooths the changes between every pair of consecutive time steps. This property can reduce fluctuations caused by noisy data. In addition, the union graph allows us to directly determine the links between temporal clusters at consecutive time steps. It simplifies the problem of tracking community evolution.

The main disadvantage of this technique is that the CPM algorithm used only detects communities in certain contexts, that is, CPM algorithm fails to detect community structure of networks with few cliques. In addition, some parameters are used to determine how community changes due to the application of similarity metric.

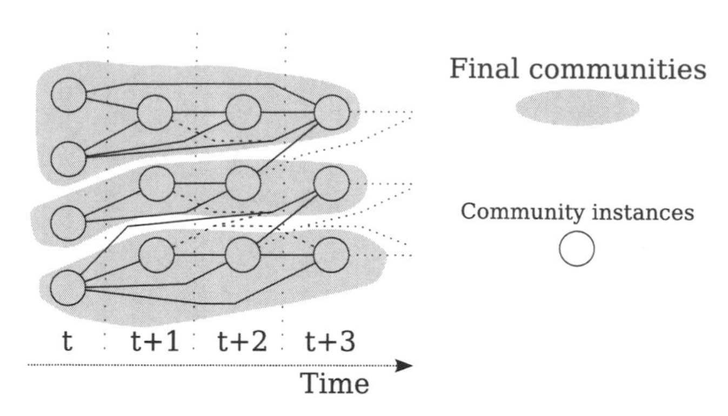

Fig. 8 An example of dynamic networks with community instances (nodes) and final communities (*in grey*). At each time step, we may match several community instances to the same temporal community

---

Algorithm 4 Hierarchical edge betweenness clustering

Input: $G = (V, E)$
Output: A dendrogram
  repeat
    Compute edge betweenness for all edges
    Remove edge with highest betweenness
  until no more edges in graph
  Return a dendrogram // *The dendrogram is produced from a top down approach: the network is split into different communities with successive removals of links. The leaves of the dendrogram are individual nodes.*

Survival-Graph-Based Methods

Given a dynamic graph, its *community survival graph* is constructed by representing community instances as nodes which are linked via edges based on their similarity. One can divide this community survival graph into final communities. Each final community groups a set of temporal clusters and spans several time steps as shown in Fig. 8.

The first approach associated with survival graph is proposed by Falkowski et al. [30, 31]: first cluster each temporal graph to find community instances at each time step, then construct a community survival graph, and finally cluster the community survival graph to find final communities by using a hierarchical edge betweenness clustering [36].

To construct a community survival graph, a time window is set to compare the similarity between community instances and connect the similar community instances with edges. In another words, this time window size is the largest time distance between every pair of connected community instances in a community survival graph. The applied hierarchical edge betweenness clustering (see Algorithm 4) contains an iteration, which eliminates edges to separate subgraphs. In Falkowski et al.'s method, a parameter k is applied to determine the number of iterations. The conncсted subgraphs retained after k iterations correspond to the final communities. A connected subgraph consists of similar community instances.

Chi et al. [17] have detected final communities through a soft clustering [2], after detecting community instances [66, 82, 94] at each time step. At a time step i, the graph interaction is denoted by $\mathbf{A}^i \subseteq V \times V$ with l_i *basis subgraphs* $\mathcal{B}^i = [\mathbf{B}^i_1, \ldots, \mathbf{B}^i_{l_i}]$. Each *basis subgraph* describes interactions between nodes within a community instance. Across a time window $[1, \ldots, \Delta]$, graph interactions can be denoted by a three-dimensional tensor: $\mathbb{A} = [\mathbf{A}^1, \ldots, \mathbf{A}^\Delta] \in \mathcal{R}^{n \times n \times \Delta}$. For the total $N_c = \sum_{i=1}^{\Delta} l_i$ *basis subgraphs*, another three-dimensional tensor is defined: $\mathbb{B} = [\mathbf{B}^1_1, \ldots, \mathbf{B}^\Delta_{l_\Delta}] \in \mathcal{R}^{n \times n \times N_c}$.

Then, the final communities are obtained by minimizing the objective function: $D(\mathbb{A} \parallel \mathbb{B}\mathbf{U}\mathbf{V}^T)$. The matrices $\mathbf{U} = [u_{kj}]_{N_c \times n_c}$ and $\mathbf{V} = [v_{ij}]_{\Delta \times n_c}$ are the solution of the optimization problem. For each dynamic community j, u_{kj} is a vector of weights on kth basis subgraph. At each time step i, v_{ij} is a community intensity for the jth final community.

In this method, the size of time windows and basis subgraphs are an issue. A good size value of time windows allows us to group small community instances into a final community, if these small community instances have high frequency grouped together. The size of basis subgraphs is related to insignificant subgraphs (e.g., a subgraph with only a couple of nodes), as insignificant subgraphs are removed for the computation. The larger size threshold of basis subgraphs is, the less iterations are used for computing $\mathbf{U}$ as less number of N_c. Therefore, the computation time can be optimized by increasing the size threshold of basis subgraphs.

For the number of communities n_c, they try different values to compare the reconstruction error and then choose one that is reasonably small and, at the same time, explains data reasonably well.

In [86], authors use a similar approach which tracks community evolution by connecting community instances, but they use another notion of final community. A quality function called *node cost* is defined to determine the community membership for each node over time. This function is the sum of two costs: the cost of one node to keep its community membership and the cost of one node to change its community membership. Therefore, final community detection is transformed into the problem of optimizing this function. Optimizing this function is shown to be an NP-complete problem. Another solution with an approximate factor is proposed in [85]. In their proposed node cost function, the importance of different costs is predefined. Giving a high importance to cost of a node, to keep its community membership, makes node membership stable for a long time duration. Giving a high importance to the cost of a node, to change its community member, makes node membership to fit to current snapshot structure.

Survival-graph-based method gives results about how dynamic communities evolve over time directly. It simplifies the problem of tracking community evolution. Compared to other two-stage approaches, which track community evolution by identifying observations at each time step, this technique is more practical. However, some issues arise: How to choose the *time window size*? How to choose *the number of clustering iterations* in [31]? How to choose *the size threshold of basis subgraphs* and *the number of final communities* in [17]? And how to choose *the importance value* in [86]?

Conclusion

Methods presented above are two-stage-like approaches:

1. Clusters are detected at each time step independently of the results at any other time step.
2. Relationships between clusters at different time steps are inferred successively.

Such natural process often producces significant variations between partitions that are close in time, especially when the datasets are noisy. Since the first phase is independent of the past history, smooth transitions are impossible. Such approach may produce artifacts if the data are noisy and variations between partitions may also be generated by the community detection algorithm itself. Such artifacts yield to artificial community dynamics rather than the real graph evolution. For each graph, let $\mathcal{O}(P)$ denote the partition detection time and $\mathcal{O}(M)$ represent the computation time for the matching problem. The total time complexity of a two-stage approach on a time window of length $\mathcal{T}$ is in $\mathcal{O}((P + M)\mathcal{T})$.

4.2 Evolutionary Clustering

An evolutionary clustering approach follows a principle of detecting community structure based on the current graph topology information at a given time t and on the community structure at previous time steps. The quality function used for dynamic community structure is α-cost (see Eq. (3)). By assuming that a good community structure has a high α-cost value, many optimization methods are proposed and are applied to real dynamic networks. For instance, Lin et al. [88, 89] used a probabilistic model to capture community evolution by maximizing α-cost. On one hand, proposed frameworks called *community model* usually search the optimal community structure for modeling the sequence of graphs by encompassing interactions of the whole graphs. On the other hand, *incremental/online algorithms* only consider interaction changes such as link insertion or link deletion which also make sense in detecting structural changes. In the following, we will review these evolutionary clustering methods.

Community Model

Community evolution can be modeled by a sequence of graphs based on a probabilistic model, which assumes that:

1. The interactions of the graph at each time step follow a certain distribution.
2. The community structure follows a certain distribution that is determined by the community structure at the previous time step.

The first attempt has been done by Lin et al. [88, 89] through α-cost function optimization. Let $\mathbf{W}^t$ denote a graph structure at time t and $\mathbf{X}^t\mathbf{\Lambda}^t$ represent its community structure. By defining $\mathbf{Z}^t = \mathbf{X}^t\mathbf{\Lambda}^t(\mathbf{X}^t)^T$, the authors have devised an α-cost (3):

$$\text{cost} = \alpha\, D(W^t \parallel \mathbf{Z}^t) + (1-\alpha)\, D(\mathbf{Z}^{t-1} \parallel \mathbf{Z}^t)\,.$$

Consequently, they estimate $\mathbf{X}^t$ and $\mathbf{\Lambda}^t$ for optimizing the cost. The problem of community detection at each time step becomes a problem in terms of maximum a posteriori (MAP) estimation. An EM algorithm for solving the MAP problem is given in [88, 89] with a low complexity when the graph structure is sparse.

This technique enables to detect overlapping community structure and track community evolution directly. So it is a good resolution for the problem of community detection in dynamic graphs. However, determining a priori the value of α is a drawback.

Yang et al. [95] also used a dynamic stochastic block model (DSBM) for finding communities and their evolutions in a dynamic social network. In their study, they have applied a Bayesian treatment for parameter estimation that computes the posterior distributions for all the unknown parameters.

Let $\mathbf{W}^t \in \mathbb{R}^{n\times n}$ denote a graph structure at time t and $\mathbf{Z}^t \in \mathbb{R}^{n\times n_c}$ is its community structure. For each node i, it is assigned into community k with a probability π_k, such as $\Pi = [\pi_1, \ldots, \pi_{n_c}] \in \mathbb{R}^{n_c}$. For a pair of nodes i and j whose community memberships are k and l, respectively, the link connecting them is assumed to follow a Bernoulli distribution with parameter P_{kl}, such as $w_{ij}^t \sim \text{Beronulli}(\bullet \mid \text{P}_{\text{kl}})$, that is, $\mathbf{W}^t \sim \Pr(\mathbf{W}^t \mid \mathbf{P}, \mathbf{Z}^t)$, where $\mathbf{P} = [P_{kl}]_{n_c \times n_c}$. For a community matrix $\mathbf{Z}^{t-1}$, a transition matrix $\mathbf{B} \in \mathbb{R}^{n\times n}$ is assumed to model $\mathbf{Z}^t$, such as $\mathbf{Z}^t \sim \Pr(\mathbf{Z}^t \mid \mathbf{Z}^{t-1}, \mathbf{B})$. So we write the likelihood for the DSBM model as follows:

$$\Pr(\mathbf{W}^t, \mathbf{Z}^t \mid \Pi, \mathbf{P}, \mathbf{B})\,.$$

With the Bayesian Model, a posteriori probability $\Pr(\mathbf{Z}^t \mid \mathbf{W}^t)$ is computed with an inference algorithm.

There is no parameter in this technique. However, the authors only provide performances of the applications to networks with nearly ten time steps and a few hundred nodes. For large networks such as millions nodes and hundreds of time steps, the performance of this technique is not clear.

The community model captures community evolution by modeling the sequence of graphs. It performs well when applied to stable evolving graphs. However, it suffers from scalability problems due to an expensive matrix computation and storage cost.

Incremental/Online Algorithms

The incremental spectral clustering [69] is one of the early incremental algorithms that update matrices like the degree matrix or the Laplacian matrix according to changes of graph interactions [90]. In traditional spectral clustering, community detection is transformed into the eigenvalue problem of $L\mathbf{q} = \lambda D\mathbf{q}$, where L is the Laplacian matrix, $\mathbf{q}$ is the cluster indicator, λ is the eigenvalues, and D is the degree matrix. Using incremental computation yields to a lower computational cost than the standard spectral clustering. Incremental computation only takes into account changes; thus, the computation matrix is sparse. In addition, a tunable threshold τ is used to balance the computational cost and the accuracy. One drawback is that errors are accumulated after several steps, and when the dataset grows or changes frequently, the associated cost becomes expensive.

Modularity optimization is the most popular method for community detection. It is extended to detect community evolution, *for example*, the modularity-driven clustering proposed by Gorke et al. [38]. Their basic idea is to detect community structure by starting from a pre-clustering obtained from a standard modularity optimization heuristic. Then, they proposed and discussed heuristics based on global greedy algorithms or on local greedy algorithms. They pass a pre-clustering to the global version to adapt it to the dynamic case (dGlobal). Similarly, the local version remembers its old results: roughly speaking, the dynamic local version (dLocal) starts by letting all free (elementary) nodes reconsider their cluster. Then it lets all those (super-)nodes on higher levels reconsider their cluster, whose content has changed due to lower-level revisions. Similarly, Dinh et al. [21] proposed another method extended from community optimization.

The community detection based on node similarity such as DBSCAN [28] is also extended for detecting dynamic community evolution [27]. DBSCAN considers a community as a core node and a *neighborhood*. For each core node, its community must consist of at least η nodes within a radius distance ε. In IncrementalDB-SCAN [27], each community updates its neighborhood if its community members have changed their neighbors. Similarly, DENGRAPH [29] detects community evolution according to the core nodes and their neighborhoods. Instead of a distance radius ε, a different distance function is proposed to compute core nodes and their neighborhoods.

Incremental or online method can detect dynamic communities and save time by avoiding computations on subgraphs where there is no change. However, all above approaches need predefined parameters.

Conclusion

There exist many other evolutionary clustering approaches. For instance, information theory has been used to detect community evolution in dynamic graphs. Sun et al. applied the MDL to find the minimum encoding cost to describe a time sequence of graphs and their partitions into communities. The basic principle of this method is to encode the graph topology into a compression information with the minimum cost of the description. This method enables to provide meaningful information on community evolutions. However, one drawback is the problem called *relevant variable*, which is the variance between real data and data compression. To what extent is information theory able to capture community structures? To our knowledge, we are still far from a precise definition of community, while modularity (defined by Eq. (2)) is the widest accepted quality function.

As opposed to two-stage approaches, evolutionary clustering does not encounter the matching problem. However, most methods are using parameters. Furthermore, we stress that evolutionary clustering results are generally too strongly correlated with community history which may occult structural changes.

4.3 Coupling Graph Clustering

Coupling graph clustering approach is based on a coupling graph as shown in Fig. 1. The underlying idea is once the coupling graph built (encompassing the time dimension as edges) to use an efficient standard static community detection heuristic. The first attempt is [47] where authors built a temporal graph and then used the classical community detection algorithm Walktrap [74]. The community evolution can be traced through group memberships over time.

Another method is proposed by Mucha et al. [63]. They detected dynamic communities by optimizing a modified modularity, which is motivated by α-cost (3). The modified modularity balances the contribution of community memberships to each slice and the cost for changing community memberships. The major advantage of this algorithm is to smooth community evolution. However, its results rely on the parameter α and the relative weight of coupling.

This idea of coupling graph clustering simplifies the problem of detecting community evolution. However, it introduces the problem about how to construct coupling graphs: how to add *the weight on coupling edges*? what is *the length of coupling windows* (i.e., the longest time interval between nodes connected by coupling edges)? For the length of coupling windows, we illustrate examples in Fig. 9. This figure is taken from [63], where each snapshot graph is called a *slice*. Two different lengths of coupling windows are given: (*a*) couplings between neighboring slices such that the length is two time steps and (*b*) all-to-all inter-slice couplings such that the length is the total time steps.

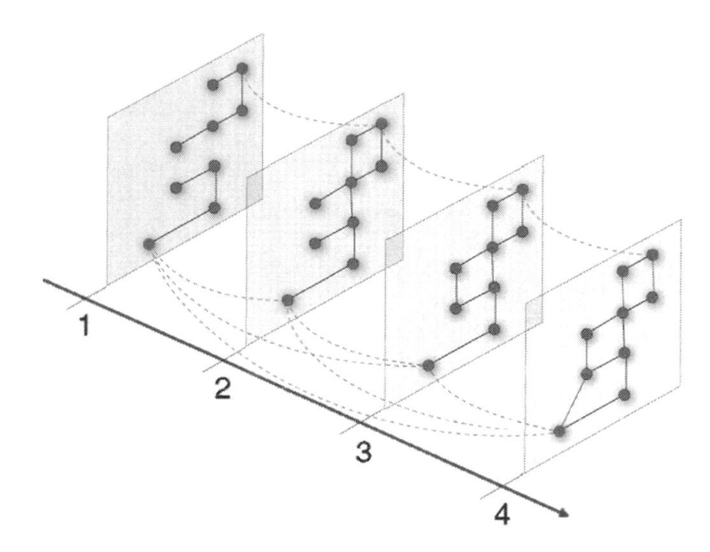

Fig. 9 Schematic of a multislice (couplings) network. Four slices $s = \{1, 2, 3, 4\}$ represented by adjacencies A_{ijs} encode intra-slice connections (*solid*). Inter-slice connections (*dashed*) are encoded by C_{jrs} , specifying coupling of node j to itself between slices r and s. For clarity, inter-slice couplings are shown for only two nodes and depict two different types of couplings: (**a**) coupling between neighboring slices, appropriate for ordered slices, and (**b**) all-to-all inter-slice coupling, appropriate for categorical slices. The figure is gained from [63]

4.4 Benchmarks

When designing a new algorithm, it is necessary to stress it through series of simple benchmark graphs, artificial or from the real world, for which the community structure is known. If the algorithm provides results agreeing with the ground truth, we may consider that the algorithm is reliable and can be used in applications. In this section, we firstly describe current benchmarks for testing dynamic community detection algorithms and secondly review measures for comparing the similarity between computed modular structure and a ground truth.

Computer-Generated Graphs

Computer-generated graphs try to build random graphs that have natural partitions. The simplest model of this form is for the graph bisection problem. This is the problem of partitioning the vertices of a graph into two equal-sized sets while minimizing the number of edges bridging the sets. To create an instance of the planted bisection problem, we first choose a partition of the vertices into equal-sized sets V_1 and V_2. We then choose probabilities $p_{\text{in}} > p_{\text{out}}$ and place edges between vertices with the following probabilities: the expected number of edges crossing between V_1 and V_2 will be $p_{\text{out}}|V_1|\,|V_2|$. If p_{in} is sufficiently larger than p_{out}, then every other bisection will have more crossing edges. There have been many analyses of the generalization of planted partition models to more than 2 partitions [18, 61]. The number of subgraphs is equal to the number of predefined communities, and nodes within the same community are connected with a probability of p_{in} and connect to the rest with a probability of p_{out}. In addition, each subgraph is modeled by an Erdös–Rényi's model, which assigns equal probability to all graph edges. The model is motivated by the idea that vertices (or general items) belong to certain categories and that vertices in the same categories are more likely to be connected. Such models also arise in the analysis of clustering algorithms. However, it is not clear that these models represent practice very well.

Lin et al. [88] have proposed a computer-generated benchmarks for testing their evolutionary clustering framework called FacetNet. They use the model of Newman [68] similar to the previous model as a basis (4 clusters of 32 nodes). They generate different graphs for each time steps. In each time step, dynamic is introduced as follows: from each community, they randomly select 3 members to leave their original community and to join randomly the other three communities. Edges are added randomly with a higher probability p_{in} for within-community edges and a lower probability p_{out} for between-community edges. The average degree for nodes is set to 16.

Another similar benchmark is proposed in [23]. To introduce change points, a sequence of graphs are separated into *segments*. Each *segment* is a sequence of graphs sharing the same community structure. The average degree of nodes and the internal and external connection probability are fixed. The edge weights are integers randomly chosen from 1 to 10 for intra-community edges and from 1 to 6 for inter-community edges.

All benchmarks for dynamic community detection extended from the planted partition model, used by Newman et al. have two main drawbacks: (*a*) all nodes have the same expected degree and (*b*) all communities have equal size. These features are unrealistic, as complex networks are known to be characterized by heterogeneous distributions of degree and community sizes.

Greene and Doyle [39] proposed a set of benchmarks based on Lancichinetti and Fortunato's technique [56]. Lancichinetti and Fortunato assumed that the distributions of degree and community size are power laws, with exponents τ_1 and τ_2, respectively. Each node shares a fraction $1-\mu$ of its edges with the other nodes of its community and a fraction μ with the rest of the graph; μ is a mixing parameter in range of [0, 1]. Greene and Doyle contracted four different synthetic networks for four different event types, covering $15,000$ nodes over 5 time steps. In each of the four synthetic datasets, 20% of node memberships were randomly permuted at each step to simulate the natural movement of users between communities over time. Subsequently, community dynamic events were added as follows:

Intermittent communities: At each time step, 10% of communities are unobserved from time $t = 2$ onwards.

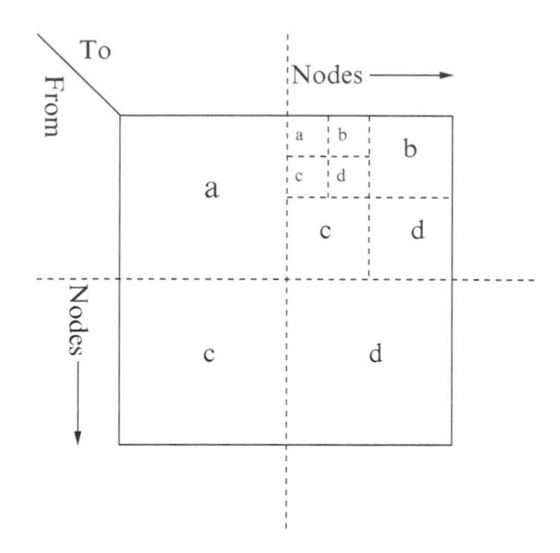

Fig. 10 The R-MAT model. The figure is gained from [14]

Expansion and contraction: At each time step, 40 randomly selected communities expand or contract by 25% of their previous size.

Birth and death: At each time step, 40 additional communities are created by removing nodes from other existing communities and randomly removing 40 existing communities.

Merging and splitting: At each time step, 40 temporal clusters of communities split, together with 40 cases where two existing communities were merged.

Chen et al. [16] constructed benchmark graphs using GTgraph [6] based on a recursive matrix graph model (R-MAT) [14]. The R-MAT model follows the preferential attachment idea (growing model where new nodes prefer to connect to existing nodes with higher degrees). In order to build a graph, the R-MAT recursively subdivides the adjacency matrix into four equal-sized partitions and assigns edges within these partitions with unequal probabilities:

1. Starting with an empty adjacency matrix, which represents a subgraph for edge assignment.
2. Assign edges into the matrix with probabilities a, b, c, d, respectively (see Fig. 10).

The chosen partition is again subdivided into four smaller partitions, and the above procedure is repeated until the chosen partition is composed of a simple cell such as a single node. In Chen et al.'s method, they define some nodes as graph-dependent nodes. These graph-dependent nodes play the role of core nodes and are used to identify communities. The community dynamics can be revealed by the community member changes, where these communities are mapped through graph-dependent nodes.

The main drawback of above computation-generated benchmarks is that the evolution of a dynamic network corresponds to a fixed probability. We may expect that in real networks, communities may experience heterogeneous changes such as bursty node insertion probability, node deletion probability, link insertion probability, or link deletion probability.

Real Networks

Real networks are also used to show performances of algorithms, such as Karate, Football, Dolphins, and Neural. When dealing with real data, the main issue is generally the ground truth or a fine and precise expertise on the datasets. Real networks are released by Newman and can be downloaded from http://www-personal.umich.edu/~mejn/netdata/.

Mucha et al. [63] performed simultaneous community detection across multiple resolutions (scales) in the well-known Zachary Karate Club network, which encoded the friendships between 34 members of a 1970s university karate club [96]. Keeping the same unweighted adjacency matrix across *slices* (each *slice* represents a graph at a time step), the resolution associated to each slice is dictated by a specified sequence of γ_Δ parameters, such as $\gamma_\Delta = \{0.25, 0.5, 0.75, \ldots, 4\}$. In other words, given a sequence of slices $\mathcal{A}_{ij\Delta} = \{A_{ij}(1), \ldots, A_{ij}(\Delta)\}$, these slices share the same unweighted adjacency matrix such as $\forall\ t_r, t_s\ , A_{ij}(t_r) = A_{ij}(t_s)$. Figure 11 depicts the community assignments obtained for coupling strengths $\omega = \{0, 0.1, 1\}$ between each neighboring pair of the 16 ordered slices. These results simultaneously probe all scales, including the partition of the Karate Club into four communities at the default resolution of modularity. Additionally, nodes that have an especially strong tendency to break off from larger communities are identified.

The previous definition for building benchmark graphs does not change interactions between nodes. Community structure changes observed are caused by tuning the resolution (scale) of the networks. Therefore, we cannot use it to test the reliability of community dynamic detecting algorithms. Its other drawback is that the algorithm should use the same resolution parameter; otherwise, it fails to test the performance of the algorithm in smoothing community evolution.

Comparing Partitions

To measure the similarity between the built-in modular structure of a benchmark and the one delivered by an algorithm, several similarity measurements are possible. The most used similarity measurement is the *normalized mutual information*, which is based on information theory [20]. The idea is that, if two community structures are similar to each other, only little information is used to infer one community structure by given the other one.

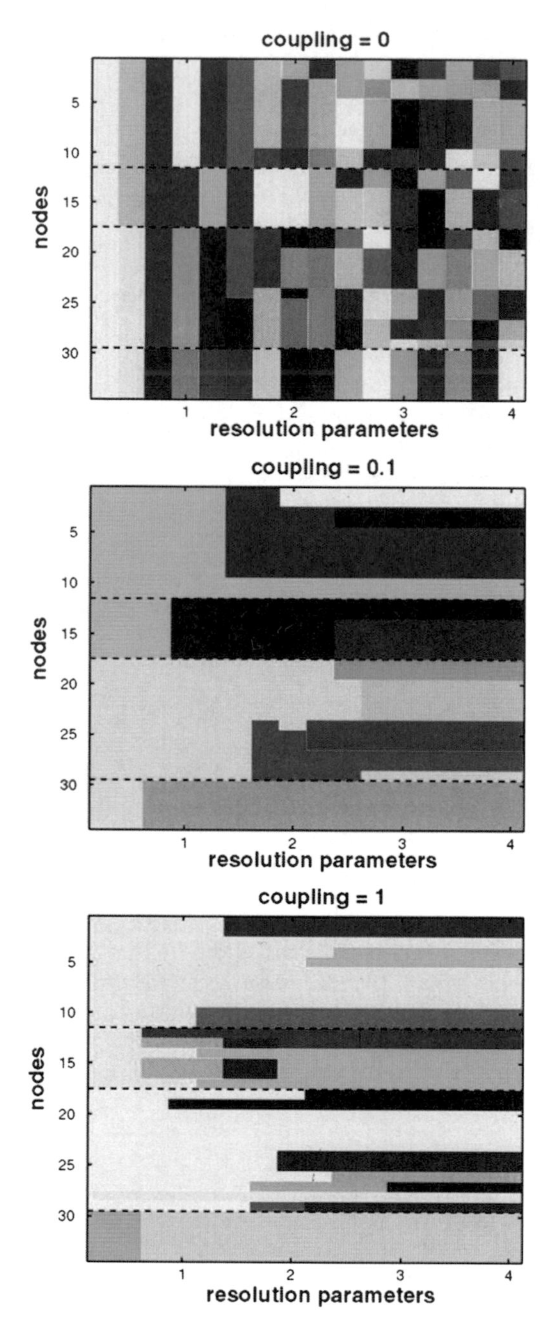

Fig. 11 Multislice community detection of the Zachary Karate Club network [96] across multiple resolutions. Colors depict community assignments of the 34 nodes in each of the 16 slices (with resolution parameters $\gamma_\Lambda = \{0.25, 0.5, \ldots, 4\}$), for $\omega = 0$ (*top*), $\omega = 0.1$ (*middle*), and $\omega = 1$ (*bottom*). Dashed lines bound the communities obtained using Newman–Girvan modularity [68]. The figure is gained from [63]

The normalized mutual information is based on the *mutual information*. The mutual information for two random variables X, Y is denoted by $\mathrm{I}(X,Y)$ and is defined as

$$\mathrm{I}(X,Y) = \sum_x \sum_y P(x,y) \log \frac{P(x,y)}{P(x)P(y)}$$

where $P(x)$ indicates the probability that $X = x$ (similarly for $P(y)$) and $P(x,y)$ is the joint probability of X and Y, that is, $P(x,y) = P(X = x, Y = y)$. Actually, $\mathrm{I}(X,Y) = \mathrm{H}(X) - \mathrm{H}(X|Y)$, where $\mathrm{H}(X)$ is the Shannon entropy of X and $\mathrm{H}(X|Y)$ is the entropy of X conditional on Y.

Danon et al. [20] defined the normalized mutual information (NMI) for comparing the similarity between two partitions: $\mathcal{P}_x$ and $\mathcal{P}_y$. Let n_x and n_y denote the number of communities in the partition $\mathcal{P}_x$ and $\mathcal{P}_y$, respectively. The normalized mutual information is defined as

$$\mathrm{NMI} = \frac{2\mathrm{I}(\mathcal{P}_x, \mathcal{P}_y)}{\mathrm{H}(\mathcal{P}_x) + \mathrm{H}(\mathcal{P}_y)} \,. \tag{7}$$

Let

$$\mathrm{I}(\mathcal{P}_x, \mathcal{P}_y) = \sum_{i=1}^{n_x} \sum_{j=1}^{n_y} P(\mathcal{C}_i, \mathcal{C}'_j) \log \frac{P(\mathcal{C}_i, \mathcal{C}'_j)}{P(\mathcal{C}_i)P(\mathcal{C}'_j)}$$

$$\mathrm{H}(\mathcal{P}_x) = -\sum_{i=1}^{n_x} P(\mathcal{C}_i) \log P(\mathcal{C}_i)$$

where n_x and n_y denote the number of communities in two partitions $\mathcal{P}_x$ and $\mathcal{P}_y$, respectively, $P(\mathcal{C}_i) = \frac{|\mathcal{C}_i|}{n}$ and $P(\mathcal{C}_i, \mathcal{C}'_j) = \frac{|\mathcal{C}_i \cap \mathcal{C}'_j|}{n}$.

Danon et al.'s normalized mutual information can be directly written in

$$\mathrm{NMI} = \frac{-2 \sum \sum N_{ij} \log \frac{N_{ij} N}{N_{i\bullet} N_{\bullet j}}}{\sum_{i=1}^{n_x} N_{i\bullet} \log \frac{N_{i\bullet}}{N} + \sum_{j=1}^{n_y} N_{\bullet j} \log \frac{N_{\bullet j}}{N}} \,, \tag{8}$$

where N_{ij} represents the size of overlaps in communities i and community j, $N_{i\bullet}$ is the sum of ith row in matrix N_{ij}, and $N_{\bullet j}$ is the sum of jth column. The normalized mutual information is equal to 1 if the partitions are identical, whereas it has an expected value of 0 if the partitions are independent.

This normalized mutual information is extended for comparing covers in [55]. The normalized mutual information for covers $\mathcal{S}_x$ and $\mathcal{S}_y$ is denoted by $\mathrm{N}(\mathcal{S}_x|\mathcal{S}_y)$ and is defined as

$$\mathrm{N}(\mathcal{S}_x|\mathcal{S}_y) = 1 - \frac{1}{2}\left[\mathrm{H}(\mathcal{S}_x|\mathcal{S}_y)_{\mathrm{norm}} + \mathrm{H}(\mathcal{S}_y|\mathcal{S}_x)_{\mathrm{norm}}\right] \tag{9}$$

where the normalized conditional entropy of $H(\mathcal{S}_x|\mathcal{S}_y)_{norm}$ (similarly to $H(\mathcal{S}_y|\mathcal{S}_x)_{norm}$) of the cover $\mathcal{S}_x$ with respect to $\mathcal{S}_y$ is defined as

$$H(\mathcal{S}_x|\mathcal{S}_y)_{norm} = \frac{1}{n_x}\sum_{i=1}^{n_x}\frac{H(S_i|\mathcal{S}_y)}{H(S_i)}\text{ , where } S_i \in \mathcal{S}_x \text{ , } n_x = |\mathcal{S}_x|$$

The conditional entropy of S_i with respect to all the components of $\mathcal{S}_y$ is defined by

$$H(S_i|\mathcal{S}_y) = \min_{S'_j \in \mathcal{S}_y} H(S_i|S'_j) \tag{10}$$

where $H(S_i|S'_j)$ denotes the conditional entropy of a community S_i by given a community S'_j.

As Eq. (10) only counts the minimum $H(S_i|S'_j)$, this extended normalized mutual information suffers from the following problem: some communities sharing few common nodes may not be taken into account. Moreover, this normalized mutual information is not ideal: given two covers $\mathcal{S}_x, \mathcal{S}_y$, if only one community of $\mathcal{S}_x$ is divided into several small ones in $\mathcal{S}_y$, while all the other communities stay identical, the normalized mutual information is low because some communities have very low conditional entropy.

The main drawback of the above similarity measurements is that they are proposed for static graphs, and they do not consider the community dynamics. Therefore, we propose to measure the similarity between the found community structure and the ground truth of dynamic graphs by counting the similarity between every pair of communities' evolution paths. We can write NMI (7) by setting

$$P(\mathcal{C}_i) = \frac{\sum_{t=1}^{\Delta}|C_i(t)|}{\sum_{t=1}^{\Delta}n(t)}$$

$$P(\mathcal{C}_i, \mathcal{C}_j) = \frac{\sum_{t=1}^{\Delta}|C_i(t) \cap C_j(t)|}{\sum_{t=1}^{\Delta}n(t)}$$

where $n(t)$ represents the nodes assigned to the partition in time t and $C_i(t)$ represents the observation of community $\mathcal{C}_i$ at time t (similarly for $C_j(t)$).

4.5 *Visualizing Dynamics in Communities*

In early work, several tools such as SoNIA [62] and TeCFlow visualize dynamic networks by creating graph movies, where nodes move as a function of changes in relations. However, these tools fail to indicate a changing behavior of community memberships and community dynamics. In [63], matrix is used (Similar as Fig. 11), whose element represents the community membership of a node at a time step. Each

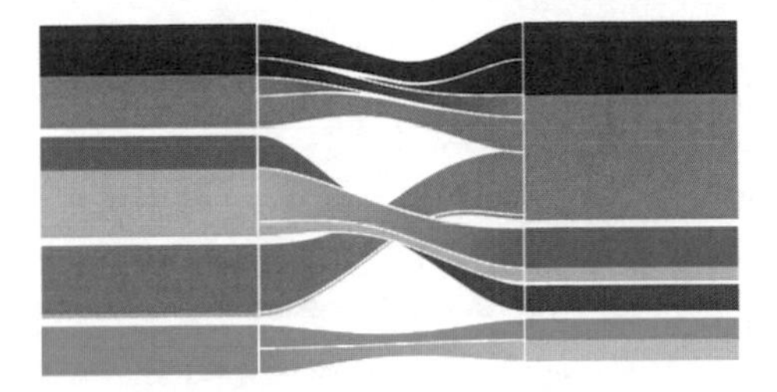

Fig. 12 *Example of mapping between communities.* In the bottom networks, the *darker colors* represent nodes that are clustered together in at least 95% of the 1,000 bootstrap networks. The alluvial diagram highlights and summarizes the structural changes between the time 1 and time 2 significance clusters. The height of each block represents the volume of flow through the cluster. The clusters are ordered from bottom to top by their size, with mutually nonsignificant clusters placed together and separated by a third of the standard spacing. The *orange module* merges with the *red module*, but the nodes are not clustered together in 95% of the bootstrap networks. The *blue module* splits, but the significant nodes in the *blue and purple modules* are clustered together in more than 5% of the bootstrap networks. The figure is obtained from [78]

node occupies a column. Colors are used to depict communities. We can observe how a node changes its community membership through the color change in the corresponding column. The drawback is that we do not directly observe how one community emerges, merges, splits, or disappears.

An example of a graph with dynamic communities is depicted in Fig. 12. The evolution path of a dynamic community is depicted by a diagram occupying a column. Each diagram represents a community as a block and shows relationships between preceding and succeeding clusters through horizontally connected stream fields. This result is obtained by the algorithm of bootstrap [25] in [78]. It enables to show community dynamics. For example, we observe the orange module merges with the red module in Fig. 12. In addition, in this case, we are also able to observe the significance of clusters, which is shown by dark color.

Another visualization tool illustrates community evolution through *lineage diagrams* [91]. Each lineage represents a separate evolutionary path and occupies a column. Each cluster is shown by a circle whose size is proportional to its number of nodes. A lineage tie is added between two clusters if they share a successor or predecessor relationship. Therefore, if a circle has a link to another column, it indicates a community change. For example, in Fig. 13, we observe a violet cluster having a link with blue color which connects it and a blue cluster. The link color is given by the predecessor if there exists a predecessor relationship; otherwise, the color is given by the successor. Therefore, we easily identify community dynamics. In this case, we say that the blue cluster is a split of the violet community. Moreover, the blue cluster has a link with green color which links it to a green cluster. We say that the green community merges into the blue community.

Although the layout displaying the community evolution in a movie seems suitable to analyze the dynamic activities of networks and communities, it is less suitable to analyze community dynamics. Displaying the community dynamics is

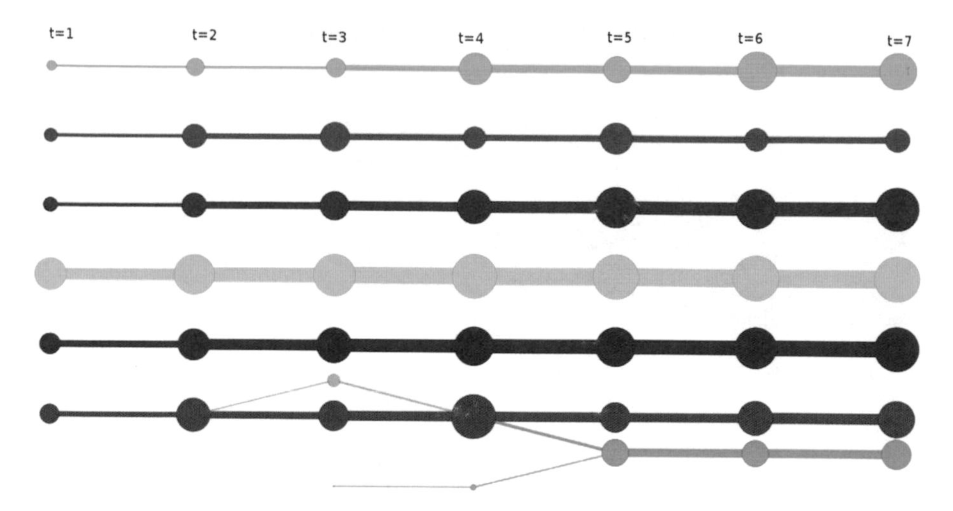

Fig. 13 Example of lineage diagrams. Between $t = 2$ and $t = 3$, an orange cluster is split from the violet community. At $t = 3$, a new green cluster is emerged. Between $t = 3$ and $t = 4$, the orange cluster is merged into the violet community. Between $t = 4$ and $t = 5$, a blue cluster is split from the violet cluster and it is merged with the green community simultaneously

very meaningful to understand the "history" of community structure in networks. There are several visualization tools which provide an overview of community dynamics. Their main drawback is that the roles and dynamic behaviors of nodes are unobservable for analysis. For example, how overlapping nodes evolve is still a problem in the visualization. Therefore, many effects need to make particularly in visualizing roles and behaviors of nodes in dynamic aspect.

5 Conclusion

In this chapter, we have reviewed current research about community detection in dynamic networks. From our exposition, we have seen a great number of clustering techniques. Two-stage algorithms are nature and simple, but the matching problem is hard to resolve. The results also suffer from temporal variance and/or algorithm-generated influence. Techniques based on the principle of evolutionary clustering also have many limits such as the tunable parameter value for community model. However, a method without faults will never be found. Thus, it is better to find a general method that gives good results for network analysis. Moreover, a number of important open issues are left in current research, such as benchmark graphs, quality function, and overlapping community evolution.

Various algorithms have been proposed and developed. Which algorithm has the best performance? To resolve this problem, we need reliable benchmarks to test algorithms. We have reviewed benchmarks [16, 23, 39]. The existing computer-generated benchmarks correspond to a fixed probability. Oppositely, communities in real networks may evolve with heterogeneous changes such as bursty node insertion probability, node deletion probability, link insertion probability, or link deletion probability. Moreover, to oppose the current real network benchmarks which are static networks, we lack real network benchmarks whose interactions change over time. The benchmark graph is a crucial issue in the area of community detection in dynamic networks.

After applying algorithms to benchmarks, we need to compare their performances to find the best one. At this moment, we cannot determine since no quality function is accepted to measure the goodness of the found dynamic communities. The α-cost function (3) depends on a tunable parameter with no a priori knowledge. There is a quality function without any parameter. It is proposed in [13] by transforming the problem of community detection into a problem of message decoding within information theory. The problem is whether information theory can be used to define communities. To our knowledge, we are still far from a precise definition of communities, while modularity (2) is the widely accepted quality function. Nonetheless, information theory study represents an important research line for the future of the field: (*a*) respect community history to better understand communities and (*b*) capture significant changes which represent crucial changes in community structure of networks.

Most algorithms in the literature deal with communities without overlapping nodes. We know that real networks may be overlapped like interdisciplinary which combines two or more academic fields in science, multiple citizenship which allows a person to be a citizen of more than one state, and proteins playing the intermediate role between several functions in biology networks. The frameworks [16] are proposed to trace overlapping community evolution. However, it is under one assumption that communities are clique-like objects, which is only suitable to detect communities in some specific networks. To detect overlapping communities in different types of networks, we need another resolution which can capture the general characteristics of overlapping nodes.

Finally, the main motivation encouraging us for community detection is to mine the relationship between the algorithmic communities with the reality. Why do communities split, or merge, or disappear? What are the effects of overlapping nodes? To answer these questions, we study features behind graph evolution and hope to learn more information.

Acknowledgement This work is supported in part by the French National Research Agency contract DynGraph ANR-10-JCJC-0202.

References

[1] J.I. Alvarez-hamelin, A. Barrat, A. Vespignani, Large scale networks fingerprinting and visualization using the k-core decomposition, in *Advances in Neural Information Processing Systems 18* (MIT Press, 2006), pp. 41–50
[2] S. White, P. Smyth, A spectral clustering approach to finding communities in graphs, in *SIAM International Conference on Data Mining* (2005)
[3] S. Asur, S. Parthasarathy, D. Ucar, An event-based framework for characterizing the evolutionary behavior of interaction graphs, in *Proceedings of the 13th ACM SIGKDD International Conference on Knowledge Discovery and Data Mining* (KDD, 2007), 978-1-59593-609-7, San Jose, CA, pp. 913–921. http://doi.acm.org/10.1145/1281192.1281290
[4] T. Aynaud, J.-L. Guillaume, Long range community detection, *LAWDN - Latin-American Workshop on Dynamic Networks*, INTECIN - Facultad de Ingeniería (U.B.A.) - I.T.B.A, p. 4 (2010). http://hal.inria.fr/inria-00531750/PDF/lawdn2010_submission_5.pdf
[5] L. Backstrom, D. Huttenlocher, J. Kleinberg, X. Lan, Group formation in large social networks: membership, growth, and evolution, in *Proceedings of the 12th ACM SIGKDD International Conference on Knowledge Discovery and Data Mining*, p. 54, 2006
[6] D.A. Bader, K. Madduri, Gtgraph: A synthetic graph generator suite (2006). http://www.cse.psu.edu/~madduri/software/GTgraph/index.html
[7] Y. Bar-Yam, *Dynamic of Complex Systems*. Addison-Wesley stydies in nonlinearity. The Advanced Book Program (Westview Press, Boulder, 1997), p. 848., ISBN 0813341213, 9780813341217
[8] M. Beiró, J. Busch, Visualizing communities in dynamic networks, in *Latin American Workshop on Dynamic Networks*, vol. 1, 2010
[9] R. Bekkerman, A. Mccallum, G. Huang, Automatic Categorization of Email into Folders: Benchmark Experiments on Enron and SRI Corpora. Center for Intelligent Information Retrieval, Department of Computer Science, University of Massachusetts Amherst, Technical Report IR (2004) pp. 4–6
[10] J.C. Bezdek, *Pattern Recognition with Fuzzy Objective Function Algorithms* (Kluwer Academic, Norwell, MA, 1981)
[11] V.D. Blondel, J.L. Guillaume, R. Lambiotte, E. Lefebvre, Fast unfolding of communities in large networks. J. Stat. Mech. Theory Experiment **2008**(10), P10008 (2008). http://stacks.iop.org/1742-5468/2008/i=10/a=P10008
[12] J. Camacho, R. Guimerá, L.A.N. Amaral, Robust patterns in food web structure. Phys. Rev. Lett. **88**(22), 228102 (2002)
[13] D. Chakrabarti, S. Papadimitriou, D.S. Modha, C. Faloutsos, Fully automatic cross-associations, in *Proceedings of the Tenth ACM SIGKDD International Conference on Knowledge Discovery and Data Mining*, KDD '04 (ACM, New York, NY, 2004), pp. 79–88
[14] D. Chakrabarti, Y. Zhan, C. Faloutsos, R-mat: A recursive model for graph mining, in *In SDM*, 2004
[15] J. Chen, B. Yuan, Detecting functional modules in the yeast protein-protein interaction network. Bioinformatics **22**(18), 2283–2290 (2006)
[16] Z. Chen, K.a. Wilson, Y. Jin, W. Hendrix, N.F. Samatova, Detecting and tracking community dynamics in evolutionary networks, in *2010 IEEE International Conference on Data Mining Workshops*, pp. 318–327, Dec. 2010
[17] Y. Chi, S. Zhu, X. Song, J. Tatemura, B.L. Tseng, Structural and temporal analysis of the blogosphere through community factorization, in *Proceedings of the 13th ACM SIGKDD International Conference on Knowledge Discovery and Data Mining - KDD '07*, p. 163, 2007
[18] A. Condon, R.M. Karp, Algorithms for graph partitioning on the planted partition model. Random Struct. Algorithm **18**, 116–140 (1999)

[19] K. Crawford, Presented at Oxford Internet Institute's A Decade in Internet Time: Symposium on the Dynamics of the Internet and Society, Six provocations for big data. Social Science Research Network Working Paper Series (13 September 2011) Computer 1–17 (2011)
[20] L. Danon, A. Díaz-Guilera, J. Duch, A. Arenas, Comparing community structure identification. J. Stat. Mech. Theor Exp. **2005**(09), P09008 (2005)
[21] T. Dinh, I. Shin, N. Thai, M. Thai, A general approach for modules identification in evolving networks. Dynam. Inform. **40**(4), 83–100 (2010)
[22] Y. Dourisboure, F. Geraci, M. Pellegrini, Extraction and classification of dense communities in the web, in *Proceedings of the 16th International Conference on World Wide Web*, WWW '07 (ACM, New York, NY, 2007), pp. 461–470
[23] D. Duan, Y. Li, Y. Jin, Z. Lu, Community mining on dynamic weighted directed graphs, in *Proceeding of the 1st ACM International Workshop on Complex Networks Meet Information and Knowledge Management* (ACM, 2009), pp. 11–18
[24] H. Ebel, L.-I. Mielsch, S. Bornholdt, Scale-free topology of e-mail networks. Phys. Rev. E Stat. Nonlin. Soft. Matter. Phys. **66**(3 Pt 2A), 035103 (2002)
[25] B. Efron, R.J. Tibshirani, *An Introduction to the Bootstrap* (Chapman & Hall, New York, 1993)
[26] N.B. Ellison, C. Steinfield, C. Lampe, The benefits of facebook friends: Social capital and college students use of online social network sites. J. Comput. Mediat. Comm. **12**(4), 1143–1168 (2007)
[27] M. Ester, H.-P. Kriegel, J. Sander, M. Wimmer, X. Xu, Incremental clustering for mining in a data warehousing environment, in *Proceedings of 24rd International Conference on Very Large Data Bases, VLDB'98* ed. by A. Gupta, O. Shmueli, J. Widom (Morgan Kaufmann, New York City, New York, 1998), pp. 323–333
[28] M. Ester, H.-P. Kriegel, J. Sander, X. Xu, A density-based algorithm for discovering clusters in large spatial databases with noise in *Conference on Knowledge Discovery and Data Mining* (AAAI Press, 1996), pp. 226–231
[29] T. Falkowski, A. Barth, M. Spiliopoulou, Studying community dynamics with an incremental graph mining algorithm, in *Proc. of the 14 th Americas Conference on Information Systems (AMCIS 2008)*, pp. 1–11, 2008
[30] T. Falkowski, M. Spiliopoulou, Data mining for community dynamics. Kunstliche Intelligenz **3**, 23–29 (2007)
[31] T. Falkowski, M. Spiliopoulou, Users in volatile communities: Studying active participation and community evolution. Lect. Note Comput. Sci. **4511**, 47 (2007)
[32] S. Fortunato, Community detection in graphs. Phys. Rep. **486**(3–5), 75–174 (2010)
[33] S. Fortunato, M. Barthelemy, Resolution limit in community detection. Proc. Natl. Acad. Sci. USA **104**(1), 36–41 (2007)
[34] E. Gabrilovich, S. Markovitch, Computing semantic relatedness using wikipedia-based explicit semantic analysis, in *Proceedings of The Twentieth International Joint Conference for Artificial Intelligence*, pp. 1606–1611, Hyderabad, India, 2007
[35] L.C. Giles, K. Bollacker, S. Lawrence, Citeseer: an automatic citation indexing system, in *The Third ACM Conference on Digital Libraries*, (ACM Press, Pittsburgh, 1998), pp. 89–98
[36] M. Girvan, M.E.J. Newman, Community structure in social and biological networks. Proc. Natl. Acad. Sci. USA **99**, 7821–7826 (2002)
[37] P. Gongla, C.R. Rizzuto, Where did that community go? - communities of practice that disappear, in *Knowledge Networks: Innovation Through Communities of Practice*, ed. by P. Hildreth, C. Kimble (Idea Group, Hershey, PA, 2004)
[38] R. Görke, P. Maillard, C. Staudt, Modularity-driven clustering of dynamic graphs. In: *Proceedings of the 9th International Symposium on Experimental Algorithms (SEA'10)*, Lecture Notes in Computer Science, vol. 6049 (Springer, 2010), pp. 436–448
[39] D. Greene, D. Doyle, P. Cunningham, Tracking the evolution of communities in dynamic social networks, in *2010 International Conference on Advances in Social Networks Analysis and Mining (ASONAM 2010)*, 9–11 August 2010, Odense, Denmark, N. Memon, R. Alhajj. The Institute of Electrical and Electronics Engineers, Inc. (IEEE, 2010), pp. 9–11. ISBN: 978-1-4244-7787-6

[40] P.D. Grnwald, I.J. Myung, M.A. Pitt, *Advances in Minimum Description Length: Theory and Applications (Neural Information Processing series)* (MIT Press, 2005), 444 p. ISBN 0262072629, 9780262072625
[41] R. Guimerá, L.A.N. Amaral, Cartography of complex networks: modules and universal roles. J. Stat. Mech. **2005**(P02001), nihpa35573 (2005)
[42] R. Guimerá, M. Sales-Pardo, L.A.N. Amaral, Modularity from fluctuations in random graphs and complex networks. Phys. Rev. E Stat. Nonlin. Soft. Matter. Phys. **70**(2 Pt 2), 025101 (2004)
[43] L.H. Hartwell, J.J. Hopfield, S. Leibler, A.W. Murray, From molecular to modular cell biology. Nature **402**(6761), 47 (1999)
[44] J. Hopcroft, O. Khan, B. Kulis, B. Selman, Tracking evolving communities in large linked networks. Proc Natl Acad Sci USA **101**(1), 5249–5253
[45] A. Jain, R. Dubes, *Algorithms for Clustering Data* (Prentice-Hall, Upper Saddle River, NJ, 1988)
[46] M. James, D.R, White, Structural cohesion and embeddedness: A hierarchical concept of social groups. Am. Socio. Rev. **68**(1), 103–127 (2003)
[47] M.B. Jdidia, C. Robardet, E. Fleury, Communities detection and analysis of their dynamics in collaborative networks, in *Dynamic Communities: from Connectivity to Information Society Workshop Co-located with ICDIM'07* (IEEE, 2007), pp. 11–17. http://liris.cnrs.fr/publis/?id=3258
[48] H. Jeong, S.P. Mason, A.L. Barabási, Z.N. Oltvai, Lethality and centrality in protein networks. Nature **411**(6833), 41–42 (2001)
[49] H. Jeong, Z. Néda, A.L. Barabási, Measuring preferential attachment in evolving networks. Europhys. Lett. **61**(4), 567–572 (2003)
[50] E.M. Jin, M. Girvan, M.E. Newman, Structure of growing social networks. Phys. Rev. E Stat. Nonlin. Soft. Matter. Phys. **64**(4 Pt 2), 046132 (2001)
[51] A.E. Krause, K.A. Frank, D.M. Mason, R.E. Ulanowicz, W.W. Taylor, Compartments revealed in food-web structure. Nature **426**(6964), 282–285 (2003)
[52] R. Kumar, A. Tomkins, D. Chakrabarti, Evolutionary clustering, in *Proc. of the 12th ACM SIGKDD Conference*, 2006
[53] H. Kwak, Y. Choi, Y.-H. Eom, H. Jeong, S. Moon, Mining communities in networks: a solution for consistency and its evaluation, in *Proc. of the 9th ACM SIGCOMM Conference on Internet Measurement Conference*, pp. 301–314, 2009
[54] A. Lancichinetti, S. Fortunato, Consensus clustering in complex networks. Sci. Rep. **2**, 336 (2012)
[55] A. Lancichinetti, S. Fortunato, Benchmarks for testing community detection algorithms on directed and weighted graphs with overlapping communities. Phys. Rev. **E80**, 016118 (2009)
[56] A. Lancichinetti, S. Fortunato, F. Radicchi, Benchmark graphs for testing community detection algorithms. Phys. Rev. **E78**, 046110 (2008)
[57] J. Leskovec, J. Kleinberg, C. Faloutsos, Graphs over time: densification laws, shrinking diameters and possible explanations, in *Proceedings of the Eleventh ACM SIGKDD International Conference on Knowledge Discovery in Data Mining* KDD'05, Chicago, IL (ACM, New York, 2005), pp. 177–187. ISBN 1-59593-135-X, http://doi.acm.org/10.1145/1081870.1081893
[58] J. Leskovec, K.J. Lang, M. Mahoney, Empirical comparison of algorithms for network community detection, in *Proceedings of the 19th International Conference on World Wide Web*, WWW '10 (ACM, New York, NY, 2010), pp. 631–640
[59] R.D. Luce, A.D. Perry, A method of matrix analysis of group structure. Psychometrika **14**(2), 95–116 (1949)
[60] C.P. Massen, J.P.K. Doye, Thermodynamics of community structure. Arxiv preprint cond-mat/0610077 (2006)
[61] F. McSherry, Spectral partitioning of random graphs, in *Proceedings of the 42nd IEEE Symposium on Foundations of Computer Science*, FOCS '01 (IEEE Computer Society, Washington, DC, 2001), p. 529

[62] J. Moody, D.A. McFarland, S. Bender-DeMoll, Dynamic network visualization: methods for meaning with longitudinal network movies. Am. J. Sociol. **110**, 1206–1241 (2005)
[63] P. Mucha, T. Richardson, K. Macon, M.A. Porter, Community structure in time-dependent, multiscale, and multiplex networks. Science **876**, 10–13 (2010)
[64] M.E.J. Newman, Analysis of weighted networks. Phys. Rev. E **70**, 056131 (2004)
[65] M.E.J. Newman, Fast algorithm for detecting community structure in networks. Phys. Rev. E Stat. Nonlin. Soft. Matter. Phys. **69**(6 Pt 2), 066133 (2004)
[66] M.E.J. Newman, Modularity and community structure in networks. Proc. Natl. Acad. Sci. USA **103**(23), 8577–8582 (2006)
[67] M.E.J. Newman, A.L. Barabási, D.J. Watts (eds.), *The Structure and Dynamics of Networks* (Princeton University Press, Princeton, 2006)
[68] M.E.J. Newman, M. Girvan, Finding and evaluating community structure in networks. Phys. Rev. E **69**(2), 26113 (2004)
[69] H. Ning, W. Xu, Y. Chi, Y. Gong, T. Huang, Incremental spectral clustering with application to monitoring of evolving blog communities, in *SIAM International Conference on Data Mining*, 2007
[70] G. Palla, A.-L. Barabasi, T. Vicsek, Quantifying social group evolution. Nature **446**, 664–667 (2007)
[71] G. Palla, I. Derenyi, I. Farkas, T. Vicsek, Uncovering the overlapping community structure of complex networks in nature and society. Nature **435**, 814 (2005)
[72] R. Pastor-Satorras, A. Vázquez, A. Vespignani, Dynamical and correlation properties of the internet. Phys. Rev. Lett. **87**(25), 258701 (2001)
[73] R. Pastor-Satorras, A. Vespignani, *Evolution and Structure of the Internet: A Statistical Physics Approach* (Cambridge University Press, New York, NY, 2004)
[74] P. Pons, M. Latapy, Computing communities in large networks using random walks. J. Graph Algorithm Appl. **10**, 191–218 (2006)
[75] F. Radicchi, C. Castellano, F. Cecconi, V. Loreto, D. Parisi, Defining and identifying communities in networks. Proc. Natl. Acad. Sci. USA **101**(9), 2658–2663 (2004)
[76] E. Ravasz, A.L. Somera, D.A. Mongru, Z.N. Oltvai, A.L. Barabási, Hierarchical organization of modularity in metabolic networks. Science **297**(5586), 1551–1555 (2002)
[77] J. Reichardt, S. Bornholdt, Statistical mechanics of community detection. Phys. Rev. E **74**(1), 016110 (2006)
[78] M. Rosvall, C.T. Bergstrom, Mapping change in large networks. PLoS ONE **5**(1), e8694 (2010). doi:10.1371/journal.pone.0008694
[79] M. Saerens, F. Fouss, L. Yen, P. Dupont, The principal components analysis of a graph, its relationships to spectral clustering, in *Proceedings of the 15th European Conference on Machine Learning (ECML 2004)*. Lecture Notes in Artificial Intelligence (Springer, 2004), pp. 371–383
[80] J. Scott, *Social Network Analysis: A Handbook*, 2nd edn. (SAGE Publications, 2000), 240 p.
[81] M. Seifi, J.-L. Guillaume, I. Junier, J.-B. Rouquier, S. Iskrov, Stable community cores in complex networks, in *Proceedings of the 3rd Workshop on Complex Networks (CompleNet 2012)*, 2012
[82] J. Shi, J. Malik, Normalized cuts and image segmentation. IEEE Trans. Pattern Anal. Mach. Intell. **22**, 888–905 (1997)
[83] J. Sun, C. Faloutsos, S. Papadimitriou, P. Yu, Graphscope: parameter-free mining of large time-evolving graphs, in *Proceedings of the 13th ACM SIGKDD International Conference on Knowledge Discovery and Data Mining* (ACM, New York, NY, 2007), pp. 687–696
[84] L. Tang, H. Liu, J. Zhang, Z. Nazeri, Community evolution in dynamic multi-mode networks, in *International Conference on Knowledge Discovery and Data Mining*, p. 8, 2008
[85] C. Tantipathananandh, T. Berger-Wolf, Constant-factor approximation algorithms for identifying dynamic communities, in *Proceedings of the 15th ACM SIGKDD International Conference on Knowledge Discovery and Data Mining*, KDD '09 (ACM, New York, NY, 2009), pp. 827–836

[86] C. Tantipathananandh, T. Berger-Wolf, D. Kempe, A framework for community identification in dynamic social networks, in *Proceedings of the 13th ACM SIGKDD International Conference on Knowledge Discovery and Data Mining - KDD '07*, p. 717, 2007
[87] M. Toyoda, M. Kitsuregawa, Extracting evolution of web communities from a series of web archives, in *Proceedings of the Fourteenth ACM Conference on Hypertext and Hypermedia* (ACM, New York, NY, 2003), pp. 28–37
[88] B.L. Tseng, Y.-R. Lin, Y. Chi, S. Zhu, H. Sundaram, Facetnet: a framework for analyzing communities and their evolutions in dynamic networks. Soc. Network 685–694 (2008)
[89] B.L. Tseng, Y.-R. Lin, Y. Chi, S. Zhu, H. Sundaram, Analyzing communities and their evolutions in dynamic social networks. ACM Trans. Knowl. Discov. Data **3**(2), 1–31 (2009)
[90] U. von Luxburg, A tutorial on spectral clustering. Stat. Comput. **17**(4), 395–416 (2007)
[91] Q. Wang, Détection de communautés recouvrantes dans des réseaux de terrain dynamiques. PhD thesis, Docteur de l'Ecole Normale Supérieure de Lyon, 2012
[92] Y. Wang, B. Wu, N. Du, Community evolution of social network: feature, algorithm and model. Sci. Tech. (60402011) (2008). arXiv:0804.4356
[93] S. Wasserman, K. Faust, *Social Network Analysis: Methods and Applications (Structural Analysis in the Social Sciences)*. Structural analysis in the social sciences, vol. 8, 1st edn. (Cambridge University Press, Cambridge, 1994)
[94] S. White, P. Smyth, A spectral clustering approach to finding communities in graphs, in *SDM*, pp. 43–55, 2005
[95] T. Yang, Y. Chi, S. Zhu, Y. Gong, R. Jin, A bayesian approach toward finding communities and their evolutions in dynamic social networks, in *SIAM Conference on Data Mining (SDM)*, 2009
[96] W.W. Za chary, An information flow model for conflict and fission in small groups. J. Anthropologica **1**(33), 452–473 (1977)

Clustering Hypergraphs for Discovery of Overlapping Communities in Folksonomies

Abhijnan Chakraborty* **and Saptarshi Ghosh**

1 Introduction

Online social systems have become very popular in today's Web, where millions of users form social relationships with one another and generate and share various forms of contents. Among these websites, some are specifically designed for content sharing which are known as *social tagging systems* or *folksonomies*. Here, users share various types of contents or resources—such as URLs in Delicious (www.delicious.com), images in Flickr (www.flickr.com), music files in LastFm (www.last.fm) and movie reviews in MovieLens (www.movielens.org)—and collaboratively annotate resources with descriptive keywords (known as "tags") in order to facilitate search and retrieval of interesting resources.

With the growing popularity of folksonomies, a tremendous amount of resources are being uploaded to these sites, and it has become practically impossible for users to discover on their own interesting resources and people having common interests. Hence it is important to develop algorithms for personalized search [36] as well as resource and friend recommendation [20]. One approach to these tasks is to group the entities (resources, tags, users) into communities or clusters which are typically groups of entities having more or better interactions or similarity among themselves than with entities outside the group.

*This work was completed when the author was in IIT Kharagpur.

A. Chakraborty (✉)
Microsoft Research India, Bangalore – 560001, India
e-mail: t-abcha@microsoft.com

S. Ghosh
Department of Computer Science and Engineering Indian Institute of Technology Kharagpur, Kharagpur – 721302, India
e-mail: saptarshi@cse.iitkgp.ernet.in

A. Mukherjee et al. (eds.), *Dynamics On and Of Complex Networks, Volume 2*, Modeling and Simulation in Science, Engineering and Technology,
DOI 10.1007/978-1-4614-6729-8_10,

Folksonomies are modelled in the Complex Networks literature as *tripartite hypergraphs* [7,36], which include user, resource and tag nodes, where a hyperedge (u, t, r) indicates that the user u has assigned the tag t to the resource r. Detecting communities from such hypergraphs is a challenging problem. On the other hand, this not only helps in efficient search and recommendation of resources or friends to users but also in the organization of the vast amount of the resources present in folksonomies into different semantic categories.

Several algorithms have been proposed for detecting communities in hypergraphs [4,25,27,28,34] (details in Sect. 2). However, most of the prior approaches do not consider an important aspect of the problem—they assign a single community to each node, whereas in reality, nodes in folksonomies frequently belong to *multiple overlapping communities*. For instance, users have multiple topics of interest, and thus they link to resources and tags of many different semantic categories. Similarly, the same resource is frequently associated with semantically different tags by users who appreciate different aspects of the resource. Considering multiple overlapping communities enables more complete knowledge about the characteristics of the users and the resources and hence can lead to better recommendations.

To the best of our knowledge, only two prior studies have addressed the problem of identifying overlapping communities in folksonomies—(i) Wang et al. [35] proposed an algorithm to detect overlapping communities considering only the user-tag relationships (i.e. the user-tag bipartite projection of the hypergraph), and (ii) Papadopoulos et al. [32] detected overlapping *tag* communities by taking a projection of the hypergraph onto the set of tags. It can be noted that these approaches work on bipartite and unipartite projections of the tripartite hypergraph, respectively.

Taking projections result in loss of some of the information contained in the original tripartite network, and it is known that qualities of the communities obtained from projected networks are not as good as those obtained from the original network [18]. Another important drawback of these algorithms is that none of them consider the resource nodes. However, it is necessary to detect overlapping communities of users, resources and tags simultaneously for personalized recommendation of resources to users. The goal of the presented "Overlapping Hypergraph Clustering" algorithm is to detect overlapping communities, utilizing the complete tripartite structure of folksonomies. It achieves this goal by using the concept of *link clustering*, which is explained next.

Though a node in a network can be associated to multiple semantic topics, a *link* (or edge, the terms are used interchangeably) is usually associated with only one semantics [1]. For instance, a user can have multiple topical interests, but each link created by the user is likely to be associated with exactly one of his interests. *Link clustering* algorithms utilize this notion to detect overlapping communities, by clustering *links* instead of the more conventional approach of clustering nodes. Although each link is placed in exactly one link cluster, this automatically associates multiple overlapping communities with the nodes since a node inherits membership of all the communities into which its links are placed.

Link clustering algorithms have recently been proposed for unipartite [1, 11] and bipartite networks [35]. However, to our knowledge, "Overlapping Hypergraph Clustering" is the first link clustering algorithm for tripartite hypergraphs. Initial versions of this algorithm were presented in [13] and [9]. In this chapter, we present the "Overlapping Hypergraph Clustering" algorithm and its analysis in more detail. We compare the performance of this algorithm with the existing algorithms by Papadopoulos et al. [32] and Wang et al. [35]. Section 3 details the working of the algorithm. Extensive experiments on synthetically generated hypergraphs (Sect. 4) as well as on real data from three popular folksonomies—Delicious, MovieLens and LastFm (Sect. 5)—show that the "Overlapping Hypergraph Clustering" algorithm outperforms both these algorithms. Section 6 concludes the chapter by summarizing the contributions and potential applications of the presented algorithm.

2 Related Work

As networks are increasingly being used to model complex systems [1, 3, 8, 14, 33], several algorithms have been proposed for finding communities in networks. Girvan and Newman proposed one of the initial algorithms for community detection [14], which iteratively removes edges based on their betweenness centrality and, thus, splits the network into disconnected communities. Later, they introduced the notion of *modularity* as a measure of the quality of community structure in a network [15]. Subsequently, several community detection algorithms based on modularity maximization have been proposed, such as techniques based on agglomerative hierarchical clustering [10]. The reader is referred to [12] for a detailed survey of community detection algorithms for graphs.

A large majority of the proposed algorithms assign unique communities to nodes. However, as stated earlier, nodes in social networks (including folksonomies) typically belong to multiple overlapping communities; for example, a user is usually a part of multiple communities of family members, colleagues, schoolmates, college mates and so on. Next we discuss some of the algorithms which have been proposed to identify overlapping communities.

Overlapping Community Detection in Graphs: One of the earliest methods to find overlapping communities was by Baumes et al. [2], which find subsets of nodes whose induced subgraph locally optimizes a given metric based on the edge density of the subgraph. As different overlapping subsets may all be locally optimal, nodes may belong to multiple communities. Clique Percolation Method (CPM) by Palla et al. [31] is possibly the most well-known overlapping community detection technique. CPM finds all k-cliques with a fixed constant k, and each community is formed by merging a maximum set of such k-cliques if they share $k - 1$ nodes. One node may belong to multiple disconnected k-cliques and hence to multiple communities. Among other methods, Lancichinetti et al. [21] proposed a local clustering algorithm which optimizes a fitness function defined using the internal and external degrees of the computed clusters. By varying the parameters in the fitness function, both overlapping and hierarchical community structures can be obtained using the algorithm. Gregory [16] proposed an algorithm which works in multiple stages. First, the nodes with highest split betweenness centrality are identified (these are the nodes which may potentially belong to multiple communities) and are split into multiple nodes connected by edges. The original graph is thus transformed into a larger graph including these node sets, on which any nonoverlapping clustering technique can be applied. Finally, the communities are mapped back into the original graph. The well-known modularity metric has also been extended to the overlapping community scenario [30], using which any modularity maximization algorithms can be applied to detect overlapping communities.

Some of the recent algorithms [1, 11] adopt the methodology of *link clustering* for detecting overlapping communities: that is, they group "similar" links (edges) unlike conventional attempts to group similar nodes. Link clustering strategies build upon the idea that even though actors can belong to multiple groups, a link is mostly associated with a single category. Evans et al. [11] considered a modified random walk on the line graph of a given graph along with other diffusion processes. Ahn et al. [1] proposed to cluster links with an agglomerative hierarchical clustering technique and, thus, identify overlapping communities for nodes. The advantage of these algorithms is that while overlapping communities of nodes are indeed discovered (since a given node inherits membership of all communities that contain the edges associated with the node), these algorithms are much simpler and more efficient than the ones which directly find overlapping groups of nodes. Hence in the present study, we adopt the link clustering methodology to propose an algorithm for overlapping community detection in tripartite hypergraphs. In the next section, we discuss existing community detection algorithms for hypergraphs.

Community Detection in Hypergraphs: Several algorithms have been proposed for detecting communities in hypergraphs. Vazquez [34] proposed a Bayesian formulation of the problem of finding hypergraph communities—starting from a statistical model on hypergraphs, they use a mean field approximation to identify communities. Bulo et al. [5] proposed a game-theoretic approach to hypergraph clustering. They show that the hypergraph clustering problem can be converted into a non-cooperative multiplayer clustering game, where the notion of a cluster is equivalent to a classical game-theoretic equilibrium concept. Zhou et al. [37] generalized spectral clustering techniques to hypergraphs.

Approaches based on tensor decomposition [19] have also been proposed—for instance, Lin et al. [25] proposed an efficient multi-tensor factorization method for detecting hypergraph communities. Neubauer et al. [28] proposed a modularity maximization technique to extract communities from hypergraphs. The original k-partite hypergraph is decomposed into $k(k+1)/2$ bipartite projections. The algorithm tries to optimize a joint modularity measure, which is based on the average bipartite modularity in the individual bipartite graphs, in a brute-force, greedy bottom-up fashion. Later, Murata defined a tripartite modularity metric and proposed an algorithm to detect communities from hypergraphs using tripartite modularity maximization principle [27]. Other possible approaches include mapping each hyperedge into a multidimensional space and then applying standard clustering algorithms (e.g. the ROCK algorithm [17]) to the point space.

Overlapping Community Detection in Folksonomies: All the hypergraph community detection algorithms stated above assign a single community to each node. To our knowledge, only two studies have addressed the problem of overlapping community detection in folksonomies.

Wang et al. [35] proposed an edge clustering methodology to detect overlapping communities that uses only user-tag subscription information. In effect, they consider the projection of the given tripartite hypergraph onto a user-tag bipartite graph. Their algorithm is a k-mean variant which maximizes intra-cluster similarity. The network is considered in an edge-centric view where, for determining each centroid, only a small set of edges are compared against each other. Though this is a computationally fast algorithm, it requires the number of communities as an input which is difficult to predict in case of real-world folksonomies.

Papadopoulos et al. [32] proposed an algorithm to detect overlapping communities of *tags*. This algorithm extracts a resource-tag association graph from the tripartite hypergraph, transforms it to a tag co-occurrence network and then finds overlapping tag communities. The proposed scheme initially searches for core sets (densely connected groups) in the tag co-occurrence network and then successively expands the identified cores by maximizing a local subgraph quality measure.

Both the above approaches actually work on unipartite or bipartite projections of the given tripartite hypergraph. Although taking projections reduces the complexity associated with hypergraphs, it loses a significant part of the information contained in the original tripartite network. As an example, let us suppose that two users u_1 and u_2 annotate a common resource r with two different tags t_1 and t_2. In the user-tag bipartite projection (as considered in [35]), u_1 will be linked with t_1 and u_2 will be linked with t_2. However, the information that both these annotations were applied on the same resource will be lost. Since the tags applied on the same resource are likely to be semantically related, this information could have been useful for community discovery. Moreover, Guimera et al. [18] have shown that the quality of communities obtained from projected networks is usually worse than those obtained from the original network. "Overlapping Hypergraph Clustering" algorithm, which is detailed in the next section, does not lose this information as it detects overlapping communities in folksonomies considering the complete tripartite hypergraph structure.

3 Overlapping Hypergraph Clustering Algorithm

This section details the "Overlapping Hypergraph Clustering" algorithm for detecting overlapping communities in tripartite hypergraphs. "Overlapping Hypergraph

Clustering" is abbreviated to "OHC", and both these names are used interchangeably in the remaining part of the chapter.

As discussed earlier, a folksonomy is modelled as a tripartite hypergraph, more specifically a 3-uniform tripartite hypergraph, denoted as $G = (V, E)$, where V is the set of nodes and E is the set of hyperedges. V is composed of three partitions (i.e. three types of vertices) V^X, V^Y and V^Z, and each hyperedge in E connects triples of nodes (a, b, c) where $a \in V^X, b \in V^Y, c \in V^Z$.

3.1 Basic Idea of the Algorithm

For a given hypergraph G, the algorithm converts G to the *weighted line graph* G' which is a *unipartite graph* in which the hyperedges in G are the nodes. Two nodes e_1 and e_2 in G' are connected by an edge if the hyperedges e_1 and e_2 are "*similar*" in G, and the weight of the edge (e_1, e_2) in G' is the similarity between the two hyperedges. Section 3.2 details different ways to compute the similarity between the hyperedges.

Once the *weighted line graph* G' is constructed from the given tripartite hypergraph G, any conventional (nonoverlapping) community detection algorithm for simple graphs can be used to cluster the nodes in G' (i.e. the hyperedges in G).[2] Section 3.3 discusses about the various options and the selection of the community detection algorithm for the line graph.

As we get the node communities in G', each hyperedge in G gets placed into a single *link community*. This automatically assigns multiple overlapping communities to the *nodes* in G since a node inherits membership of all those communities into which the hyperedges connected with this node are placed.

3.2 Calculating Similarity Between Hyperedges

Similarity between a pair of hyperedges can be computed in various ways. For instance, hyperedges can be expressed as feature vectors, on which vector-based similarity measures can be used. Another way of measuring similarity is by considering the neighbourhood of the end vertices of hyperedges.

Expressing Hyperedges as Vectors: For a hypergraph having n nodes, one can express each hyperedge as a vector of size n, where an element of the vector represents the amount of participation of a particular node in that hyperedge. For

---

[2]Overlapping community detection algorithms can also be used for the line graph. But since a link (hyperedge) is usually associated with one particular semantics, we consider only nonoverlapping community detection algorithms.

example, the i-th entry in the vector representation for a particular hyperedge e will be 0 if there is no path from node i to any of the end nodes of e; otherwise, the i-th entry will be the inverse of the product of degrees of the intermediate vertices in the shortest path from node i to any of the end nodes of e. Different entries in the vector can be easily generated by formulating this as a random walk problem which will take care of the situations like presence of multiple shortest paths. These vectors express the closeness as well as the importance of a particular node to a particular hyperedge. With such representation, standard vector-similarity metrics such as cosine similarity or Pearson correlation can be used to find similarity between hyperedges.

Considering Vertex Neighbourhoods: Similarity between the hyperedges can be measured by utilizing the neighbour set of their end vertices. We measure the similarity between only those hyperedges which are *adjacent* in the hypergraph G (i.e. which have at least one node in common). Nonadjacent hyperedges are considered to have zero similarity.

It can be noted that the adjacency of two hyperedges can be defined in two ways: (i) they have at least one node in common and (ii) they have exactly two nodes in common. Although the second definition is a special case of the first definition, the choice will have some impact on the overall performance of the algorithm. The line graph G' will be sparser in case (ii) than in case (i). So, G' in case (ii) will contain more disconnected components. Detecting communities from this sparser G' will be more difficult. Also, in case of real folksonomies, the condition of having two nodes common is too rigid. Therefore, we have considered only the first definition of adjacency.

Two popular metrics are considered for measuring the similarity among the hyperedges: (i) *matching similarity*, which is the size of the overlap between the neighbour sets of the end points, and (ii) *Jaccard similarity* which is the size of overlap normalized by the size of the union of the neighbour sets of the end points. Jaccard similarity value can range from 0 to 1.

Choice of a Similarity Metric: Vector-based similarity metrics are global metrics whose computation requires the knowledge of the entire hypergraph; hence, they are inefficient for use on the large real folksonomies. On the other hand, neighbourhood-based metrics can be efficiently computed locally for a pair of hyperedges. Also, experiments on synthetically generated hypergraphs (details in Sect. 4.3) show that Jaccard similarity gives the best performance compared to other similarity metrics. Further, a metric similar to it was found to perform well in detecting overlapping communities in unipartite graphs [1]. Hence, we selected Jaccard similarity as the similarity metric for OHC algorithm.

The Jaccard similarity between two hyperedges $e_1 = (a, b, c)$ and $e_2 = (p, q, r)$ (assuming the common node is $a = p$) is computed using Algorithm 5, where $N^X(i)$, $N^Y(i)$ and $N^Z(i)$ denote the set of neighbours of node i of type V^X, V^Y and V^Z, respectively. Note that if $i \in V^X$, then $N^X(i) = \phi$ since the nodes in the same partition are not linked.

Algorithm 5 Compute similarity between two hyperedges

Input: hyperedges $e_1 = (a, b, c)$ and $e_2 = (p, q, r)$; $a, p \in V^X$; $b, q \in V^Y$; $c, r \in V^Z$
Output: sim, Similarity between e_1 and e_2

if $a \neq p$ AND $b \neq q$ AND $c \neq r$ **then**
  $sim \leftarrow 0$ /* Hyperedges are non-adjacent */
else
  /* w.l.o.g., let $a = p$; Any of the other pairs may be common as well */

$$S_1 \leftarrow N^X(b) \bigcup N^X(c), \;\; S_2 \leftarrow N^Y(c), \;\; S_3 \leftarrow N^Z(b)$$
$$S_1^{'} \leftarrow N^X(q) \bigcup N^X(r), \;\; S_2^{'} \leftarrow N^Y(r), \;\; S_3^{'} \leftarrow N^Z(q)$$

$$sim \leftarrow \frac{|S_1 \bigcap S_1^{'}| + |S_2 \bigcap S_2^{'}| + |S_3 \bigcap S_3^{'}|}{|S_1 \bigcup S_1^{'}| + |S_2 \bigcup S_2^{'}| + |S_3 \bigcup S_3^{'}|}$$

end if
return sim

3.3 Detecting Communities in Line Graph

With the similarity measure in Algorithm 5, OHC converts the hypergraph to its corresponding line graph where any unipartite community detection algorithm can be used. We experimented with different community detection algorithms to find the best candidate.

Hierarchical Clustering: We can use single-linkage hierarchical clustering to construct a dendrogram. We start with each node in the line graph as an individual cluster, then, at each step, the two most similar clusters are merged. This procedure continues until all nodes belong to a single cluster, and cutting this dendrogram at some suitable level gives the final clusters of nodes. The optimal level for the cut can be decided based on the *partition density* metric [1].

Fast Modularity Optimization: Clauset et al. [10] proposed a fast and greedy approach to implement the modularity maximization technique proposed by Newman [29]. Starting from a set of isolated nodes in the graph, the links (which are present in the original graph) are iteratively added to produce the largest possible increase in the modularity at each step. Time complexity of this algorithm is $O(n \cdot \log^2 n)$ where n is the number of nodes in the graph.

Louvain Method: Blondel et al. [3] proposed a multistep technique (named "Louvain algorithm") where communities are first detected by locally optimizing modularity in each node's neighbourhood; in the next step, a weighted graph is formed where the communities detected are super-nodes. These two steps are iterated until modularity (which is always computed in the original graph) does not increase any further. This algorithm is of complexity $O(m)$ where m is the number of edges in the original graph.

Infomap: Rosvall et al. [33] showed that finding the best cluster structure of a graph is equivalent to optimizing *minimum description length* of a random walk taking place on the graph. This optimization can be carried out with greedy search and simulated annealing. Time complexity of this algorithm (named as "Infomap" by the authors) is also $O(m)$.

Choice of a Community Detection Algorithm: We compared the performances of all the above community detection algorithms using synthetic hypergraphs (details in Sect. 4.3). The Infomap algorithm is found to perform better than the other algorithms. Lancichinetti et al. [22] also showed that for community detection in large graphs, Infomap can identify communities more accurately as compared to several other algorithms. Further, the relatively low computational complexity of Infomap allows its use on line graphs of large real folksonomies. Therefore, Infomap is used as the community detection algorithm.

3.4 Computational Complexity of OHC Algorithm

Now that the choice of the similarity metric for hyperedges and the unipartite community detection algorithm has been made, we analyse the computational complexity of OHC algorithm. Let the number of nodes in the hypergraph be n and average node degree be d, which implies that the number of hyperedges will be $\frac{n \cdot d}{3}$. Each hyperedge will, on average, be adjacent to $3 \cdot (d-1)$ other hyperedges. So, the *line graph* will have $\frac{n \cdot d}{3}$ nodes and $\frac{n \cdot d}{3} \times 3 \cdot (d-1) = n \cdot d \cdot (d-1) = O(n \cdot d^2)$ edges. Since complexity of Infomap algorithm is linear in the size of the graph and similarity calculation in the hypergraph also takes $O(n \cdot d^2)$ time, the time complexity of OHC is $O(n \cdot d^2)$.

It is to be noted that the real-world folksonomies are known to be sparse, having small average degree d. So, essentially the complexity of OHC becomes $O(n)$ which makes this algorithm scalable to large real-world folksonomies. The performance of OHC is evaluated in the next section.

4 Experiments and Evaluation

In this section, we evaluate the performance of the presented OHC algorithm. We first compare different choices of similarity metrics as well as community detection algorithms for line graph to be used in OHC (as discussed in Sects. 3.2 and 3.3, respectively). Then, we compare the performance of OHC algorithm with the algorithms by Wang et al. [35] and Papadopoulos et al. [32], which are henceforth referred to as "CL" and "HGC", respectively.[3]

Since evaluation of clustering is difficult without the knowledge of the 'ground truth' regarding the community memberships of the nodes, we have used *synthetically generated hypergraphs* with a known community structure for evaluation of the algorithms. We now discuss the generation of synthetic hypergraphs and the metric used to evaluate the algorithms, followed by the results of experiments on synthetic hypergraphs.

4.1 Generation of Synthetic Hypergraphs

Synthetic hypergraphs are generated using a modified version of the method used in [35]. The generator algorithm takes the following as input—(i) number of nodes in a partition (all three partitions are assumed to contain equal number of nodes), (ii) number of communities C, (iii) fraction γ of all nodes which belong to multiple communities and (iv) hyperedge density β (i.e. the fraction of total number of possible hyperedges that actually exist in the hypergraph).

Initially, the nodes in each partition are evenly distributed among each community under consideration (e.g. $|V^X|/C$ nodes in the partite set V^X are assigned to each of the C communities). Subsequently, γ fraction of nodes is selected at random from each of V^X, V^Y and V^Z, and each selected node is assigned to some randomly chosen communities apart from the one it has already been assigned to. Nodes assigned to the same community are then randomly selected, one from each partition, and interconnected with hyperedges. The number of hyperedges is decided based on the specified density β.

Users in the real-world folksonomies often tag a few resources related to the topics that are different from their topics of primary interest, according to their transient interests at different times. Though such tagging is typically much fewer than those related to the primary interests of users, it can adversely affect the performance of algorithms that assign a single community to nodes. To investigate how the algorithms perform in presence of such links, a second set of synthetic hypergraphs is generated, where 1% of the generated hyperedges interconnect randomly selected nodes from *different* communities. We shall refer to these hyperedges as "scattered" hyperedges.

The above assignment of communities to nodes constitutes the "ground truth". After hypergraphs are generated, information about the communities is hidden and then communities are detected from the hypergraph by the three different community detection algorithms. The community structure detected by each algorithm is compared with the ground truth using the metric "normalized mutual information (NMI)" which is explained below.

---

[3]We acknowledge the authors of [32, 35] for providing us with the implementations of their algorithms.

4.2 Normalized Mutual Information

Normalized mutual information (NMI) is an information-theoretic measure of similarity between two partitionings of a set of elements, which can be used to compare two community structures for the same graph (as identified by different algorithms). The traditional definition of NMI does *not* consider the case of a node being present in multiple communities. Hence, Lancichinetti et al. [23] proposed an alternative definition of NMI considering overlapping communities. According to [23], given two community structures/node partitions X and Y, NMI is defined as

$$NMI(X,Y) = 1 - \frac{1}{2}\Big(H(X|Y)_{norm} + H(Y|X)_{norm}\Big)$$

where

$$H(X|Y)_{norm} = \frac{1}{N_X}\sum_i \frac{\min_{j\in\{1,2,\ldots,N_X\}} H(X_i|Y_j)}{H(X_i)}$$

$$H(Y|X)_{norm} = \frac{1}{N_Y}\sum_i \frac{\min_{j\in\{1,2,\ldots,N_Y\}} H(Y_i|X_j)}{H(Y_i)}$$

Here, $H(X)$ and $H(Y)$ are entropies of X and Y. $H(Y|X)$ and $H(X|Y)$ are conditional entropies and N_X and N_Y are number of clusters in X and Y, respectively.

The NMI is computed in two steps. First, the pairs of clusters that are closest to each other are found from two clusterings. Second, the mutual information between those pairs of clusters is averaged. The NMI value is in the range [0, 1]; the higher the NMI value, the more similar are the two community structures (refer to [23] for details).

4.3 Design Choices for OHC

To decide a similarity metric and a community detection method (as required by OHC), we generated synthetic hypergraphs having various hyperedge densities $\beta = 0.1, 0.2, \ldots, 1.0$. In each of these hypergraphs, 10% of the nodes in each partition belonged to multiple communities (i.e. $\gamma = 0.1$).

First, we compare the performances of the different similarity metrics. Infomap is used as the community detection method in line graph for all cases. The NMI values (considering the true and detected community structures) are shown in Fig. 1a. We can see that the Jaccard similarity metric consistently gives the best result.

Once Jaccard similarity has been chosen as the desired similarity metric, we compare different community detection methods which can be applied on line graph (discussed in Sect. 3.3). Figure 1b shows the comparison of NMI values—across all hyperedge densities, Infomap algorithm is found to perform better than the other algorithms.

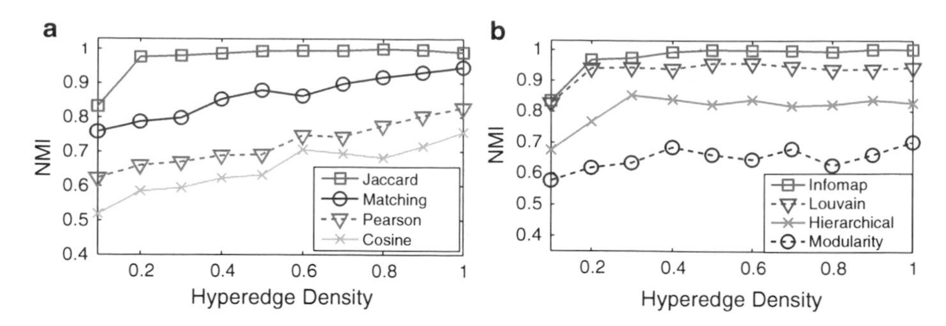

Fig. 1 Comparison of the NMI values with varying hyperedge density for (**a**) different similarity metrics and (**b**) different community detection algorithms

4.4 Comparing OHC with the Other Algorithms

The CL and HGC algorithms produce only the user and the tag communities, respectively. Hence, while calculating the NMI value for these algorithms, we have used the community memberships of only the user (respectively, the tag) nodes according to the ground truth. On the other hand, the proposed OHC algorithm gives composite communities containing all three types of nodes. Hence, to evaluate the performance of OHC, we have considered the community memberships of all three types of nodes.

For all the following experiments, $|V^X| = |V^Y| = |V^Z| = 200$ and number of communities $C = 20$. For each result, random hypergraphs were generated 50 times using the same set of parameter values, and the average performances over all those 50 runs are reported.

Performance w.r.t. Number of Hyperedges: To study how the number of hyperedges affects the performance of the clustering algorithms, we generated synthetic hypergraphs having various hyperedge densities $\beta = 0.1, 0.2, \ldots, 1.0$. In each of these hypergraphs, 10% of nodes in each partition belonged to multiple communities (i.e. $\gamma = 0.1$). The NMI values for the three algorithms are shown in Fig. 2a. Across all the hyperedge densities, OHC performs significantly better than HGC and CL algorithms. A possible explanation for this is that the proposed OHC algorithm utilizes the complete tripartite structure of the hypergraph, whereas both CL and HGC algorithms work on unweighted projections which is known to result in loss of a significant part of the information contained in the original tripartite network [18].

Note that even for very low hyperedge densities, when detecting community structures is difficult, the proposed OHC algorithm performs very well, resulting in NMI scores above 0.8. This makes OHC suitable for real-world folksonomies where the density of hyperedges is typically low.

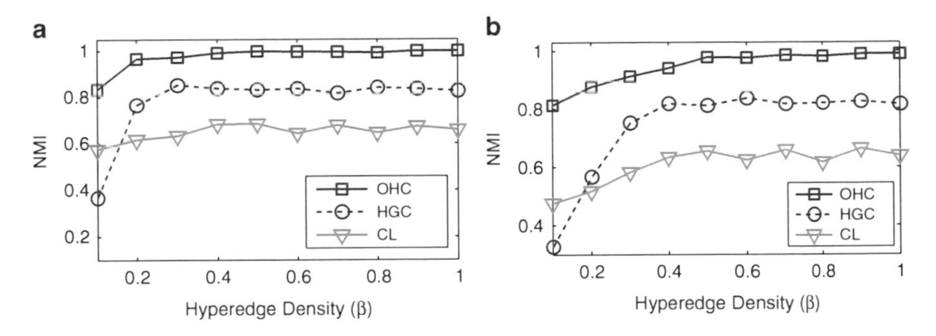

Fig. 2 Comparison among OHC, CL and HGC algorithms—variation of the NMI values with varying hyperedge density (**a**) without scattered hyperedges (**b**) in presence of 1% scattered hyperedges

Performance in Presence of Scattered Hyperedges: We have also experimented with synthetic hypergraphs having 1% of total hyperedges as "scattered". Figure 2b shows the result—as the presence of scattered hyperedges disturbs the community structure, the performance of all three algorithms degrades as expected. However, the performance of OHC is still better than HGC and CL algorithms—the NMI scores for OHC remain above 0.8 which signifies its effectiveness in detecting community structure even in the presence of noisy or scattered hyperedges.

Performance w.r.t. Fraction of Nodes in Multiple Communities: A node, which belongs to multiple communities, creates links to nodes in all those communities. Hence, from the perspective of a particular community, the links created by this member node to nodes in other communities reduce the exclusivity of this particular community. Therefore, as the number of nodes in multiple communities increases, the community structure becomes more difficult to identify. We now study how this affects the performance of the algorithms.

We generated synthetic hypergraphs by varying the fraction of nodes in multiple communities $\gamma = 0.1, 0.2, \ldots, 1.0$ while keeping hyperedge density β constant at 0.2. This low value of hyperedge density was chosen to measure the effectiveness of the algorithms in sparse environment (as in real folksonomies). Figure 3a shows that OHC performs consistently better than HGC and CL algorithms. As the community structure becomes more and more complex, information loss as a result of projections becomes increasingly more crucial. Hence, the performance of HGC and CL algorithms degrades sharply with the increase in γ, while the performance of OHC algorithm shows relatively much greater stability.

Performance w.r.t. Size of Real Community: We also observed how the performances of different algorithms are affected by the size of each real community. Hypergraphs having 200 nodes in each partition were generated while changing the number of real communities. Here, hyperedge density is fixed at 0.2 and 10% of total nodes belong to multiple communities. The results are shown in Fig. 3b. When

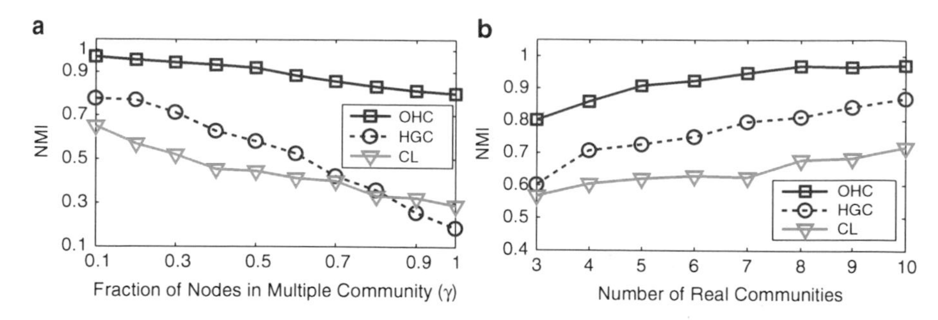

Fig. 3 Another comparison among OHC (proposed), CL and HGC algorithms—variation of the NMI values keeping hyperedge density constant at 0.2 and changing (**a**) fraction of nodes in multiple communities and (**b**) number of "real" communities

number of nodes in one community is large, random assignment of hyperedges during the generation of synthetic hypergraphs may create smaller communities inside one large community. Community detection algorithms find these smaller communities rather than the large encompassing community. For this reason, as the number of real communities increases, the size of each community decreases, enabling better NMI performance. Here also, the performance of OHC is superior than CL and HGC algorithms.

The above experiments clearly validate the motivation that considering the complete tripartite structure of hypergraphs can result in better identification of the community structure, as compared to considering projections (as done in prior studies). In the next section, we use OHC to study the community structure of real-world folksonomies.

5 Experiments on Real Folksonomies

In this section, we apply OHC algorithm to gain insights into the community structures prevalent in the real-world folksonomies. For this, we use publicly available datasets [6] consisting of snapshots of the popular folksonomies Delicious, LastFm and MovieLens. The statistics of these datasets are summarized in Table 1.

5.1 Using OHC to Detect Overlapping Communities

For all three datasets, OHC successfully groups semantically related resources as well as the tags and the users tagging these resources. As an illustration, Table 2 shows the resources and the tags placed in some example communities for each of

Table 1 Statistics of real folksonomy datasets

Dataset	# users	# resources	# tags	# hyperedges
Delicious	1,867	69,226	53,388	437,593
LastFm	1,892	17,632	11,946	186,479
MovieLens	2,113	10,197	13,222	47,957

Table 2 Examples of communities detected by proposed OHC algorithm. The algorithm successfully clusters nodes related to a common semantic theme (Column 2). Nodes related to multiple themes (boldfaced and italicized) are placed in overlapping communities

Community	Theme	Example of member nodes
LastFm artists	Hard rock	***Van Halen, Deep Purple, Aerosmith***, Alice Cooper, Guns N' Roses, Scorpions, White Lion, Bad Company, Bon Jovi, Hardline
(resources)	Heavy metal	***Van Halen, Deep Purple, Aerosmith***, Iron Maiden, Motorhead, Black Sabbath, Metallica, Twisted Sister, Crazy Lixx
LastFm tags	Metal	***Blues rock, psychedelic rock, rap metal, nu metal***, metal, progressive metal, speed metal, metalcore, viking metal, power metal
	Rock	***Blues rock, psychedelic rock, rap metal, nu metal***, progressive rock, soft rock, gothic rock, punk rock, hard rock, pop rock
MovieLens movies	Superhero	***The Incredibles, Shrek, Shrek 2, The Incredible Hulk***, Batman Begins, Batman Returns, Spider-Man, Superman, X-Men
(resources)	Animation	***The Incredibles, Shrek, Shrek 2, The Incredible Hulk***, Kung fu Panda, Beowulf, Ratatouille, Finding Nemo, Toy Story
MovieLens tags	Criticism	***Violent, brutal***, too violent, waste of celluloid, disturbing, junk, tragically stupid, lousy script, waste of money, confusing plot
	Violence	***Violent, brutal***, violence, murder, fatality, civil war, great villain, dark, Spanish civil war, serial killer, Vietnam war, world war II
Delicious tags	Web 2.0	Social networking, social web, php, drupal, xml, cms, webdesign, css3, Twitter, Skype, Ruby, Facebook, Snippets, Wikipedia, blog

the three datasets. It is evident that the resources and the tags that are placed in the same community are often related to a common semantic theme.

A closer look at Table 2 reveals that the algorithm also correctly identifies nodes that are related to multiple overlapping communities (themes). For instance, the band *Van Halen* is placed in two different communities detected from LastFm. The

Wikipedia article about Van Halen[4] justifies this placement pointing their genre as both "hard rock" and "heavy metal".

A *nonoverlapping* community detection algorithm would have placed this node in either of the two communities (assume "hard rock"). Thus, community-based recommendation systems (which recommend resources to users based on common memberships in communities) would have overlooked the fact that this resource is a candidate for recommendation to users who are interested in "heavy metal" as well. OHC algorithm places the resource in both communities, thus raising the chances of this resource being recommended to users interested in "hard rock" as well as "heavy metal".

5.2 Evaluation of Communities Detected

The principal difficulty in evaluating the communities detected in case of real folksonomies is the absence of the "ground truth" regarding the community memberships of nodes in folksonomies, since their huge size makes it impossible for human experts to evaluate the quality of identified communities. Hence, we use the following two methods for evaluation:

(1) We use the graph-based metric *conductance*, which has been shown to correctly conform with the intuitive notion of communities and is extensively used for evaluating the quality of communities in online social networks [24].
(2) In case of the folksonomies which allow users to form a social network among themselves, we can assume that users having similar interests are likely to be linked in the social network or at least to have a common social neighbourhood (a property known as *homophily* [26]). We utilize this notion to evaluate the user communities detected by CL algorithm and the user nodes in the communities identified by OHC algorithm.

Comparison of Conductance Value
As conductance is defined only for unipartite networks, we compare the tag communities detected by HGC with the tag nodes in the communities identified by OHC algorithm. Conductance values range from 0 to 1 where a *lower* value signifies *better* community structure [24].

Figure 4 shows the cumulative distribution of the conductance values of the detected tag communities by the two algorithms. Across all three datasets, OHC produces more communities having lower conductance values, which implies that OHC can find communities of better quality than obtained by HGC algorithm. The reason for this superior performance is that OHC groups semantically related nodes

[4] http://en.wikipedia.org/wiki/Van_Halen

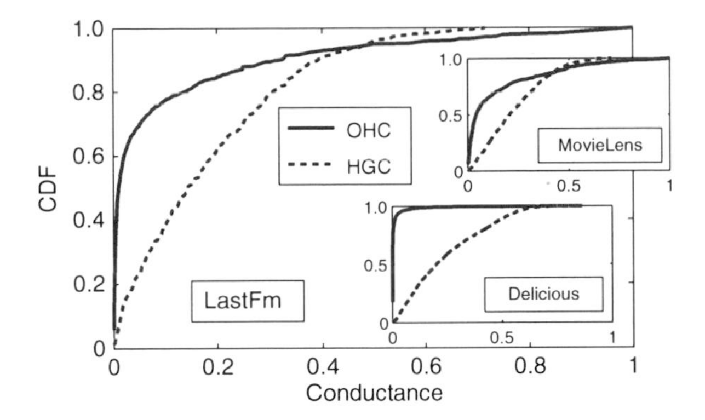

Fig. 4 Comparison between OHC (proposed) and HGC algorithms—cumulative distribution of conductance values of tag communities obtained from LastFm (main plot), Delicious and MovieLens (both inset)

into relatively smaller cohesive communities instead of creating a few number of generalized large communities. For examples of semantically related communities identified by OHC, refer to Table 2.

Comparing Detected User Communities with Social Network:
In case of the folksonomies, which allow users to form a social network,[5] there are two types of relationships among users—explicit social connections (in the social network) and implicit connections through their tagging behaviour (e.g. tagging the same resource). A community detection algorithm for hypergraphs utilizes the implicit relationships to identify the community structure, and we propose to evaluate the detected community structure using the explicit connections that the users themselves create in the social network. For instance, if a large fraction of the users who are socially linked (or share a common social neighbourhood in the social network) are placed in the same community (by an algorithm), the detected community structure can be said to group together users having common interests.

To compare the community structure identified by two algorithms, we consider the user pairs who are within a certain distance from each other in the social network (where distance 1 implies directly connected friends) and compute the fraction of such user pairs who has been placed in a common community by each algorithm. Figure 5a shows the result for the proposed OHC algorithm and the CL algorithm, for the LastFm dataset. Across all distances, OHC places a larger number of user pairs who share a common social neighbourhood, in a common community, as compared to the CL algorithm. Also, as the distance between two users in the social network increases, both algorithms put a smaller fraction of such user pairs in the same community.

---

[5]LastFm has an undirected social network, while in Delicious, a user can be a "fan" of another user, but this fan relationship may or may not be reciprocated. We assumed two users are linked

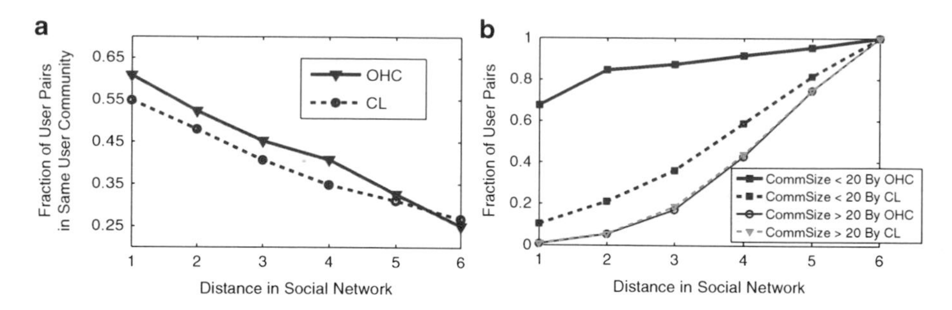

Fig. 5 Comparing the community structures detected by OHC (proposed) and CL algorithms with the social networks in (**a**) LastFm and (**b**) Delicious

We also investigate the reverse question—among the users who are placed in a common community (by a community detection algorithm), what fraction of these users is actually connected in the social network (or share a common social neighbourhood)? While investigating this question, it is to be noted that "quality" of large communities detected by community detection algorithms is known to be lower than smaller communities [24]. Hence, it is meaningful to answer this question for detected communities taking their size into consideration. Figure 5b shows the fraction of user pairs that is placed in a common community by the OHC and CL algorithms, which are within a certain distance in the social network, for the Delicious dataset. For detected user communities of size lesser than 20, about 70% of the user pairs that are placed in a common community by OHC are actually friends in the social network, whereas the corresponding value for the CL algorithm is much lesser. However, for larger detected communities (having more than 20 users), the fraction of user pairs who share a common social neighbourhood is much lower and almost identical for both algorithms.

The above results clearly show that the presented OHC algorithm can detect much better community structure in real folksonomies, as compared to the existing CL and HGC algorithms. The fact that a very large fraction of the user pairs that are placed in a common community by OHC are actually friends shows that OHC can be used to identify potential friends directly from the hypergraph structure.

6 Conclusion

In this chapter, we presented the "Overlapping Hypergraph Clustering" (OHC) algorithm which detects overlapping communities considering the full tripartite hypergraph structure of folksonomies. Through extensive experiments on synthetic

if they belong to a *mutual* fan relationship. In the LastFm and Delicious datasets, there are 12,717 and 7,668 bidirectional links, respectively, in the social network among users.

as well as real folksonomy networks, we showed that OHC algorithm outperforms existing algorithms that consider projections of hypergraphs.

In large folksonomies, it is difficult for an individual user to find other like-minded users as well as resources of her interest. OHC algorithm successfully groups nodes into multiple communities where each community represents a topic of interest. Based on these interests, like-minded users as well as resources can be identified. Thus, this algorithm can be effectively used in recommending interesting resources and friends to users in folksonomies. Our future work will be to build such a recommendation system taking advantage of the effectiveness of the algorithm.

References

[1] Y.-Y. Ahn, J.P. Bagrow, S. Lehmann, Link communities reveal multiscale complexity in networks. Nature **466**(7307), 761–764 (2010)

[2] J. Baumes, M.K. Goldberg, M.S. Krishnamoorthy, M.M. Ismail, N. Preston, Finding communities by clustering a graph into overlapping subgraphs, in *Proc. IADIS Conference on Applied Computing*, pp. 97–104, 2005

[3] V.D. Blondel, J.-L. Guillaume, R. Lambiotte, E. Lefebvre, Fast unfolding of communities in large networks. J. Stat. Mech. Theor Exp. **2008**(10), (2008)

[4] M. Brinkmeier, J. Werner, S. Recknagel, Communities in graphs and hypergraphs, in *Proc. ACM Conference on Information and Knowledge Management (CIKM)*, 2007

[5] S.R. Bulo, M. Pelillo, A game-theoretic approach to hypergraph clustering. Adv. Neural Inform. Process. Syst. **22**, 1571–1579 (2009)

[6] I. Cantador, P. Brusilovsky, T. Kuflik, 2nd Workshop on Information Heterogeneity and Fusion in Recommender Systems (HetRec 2011), in *Proc. ACM Conference on Recommender Systems (RecSys)*, 2011

[7] C. Cattuto, C. Schmitz, A. Baldassarri, V.D.P. Servedio, V. Loreto, A. Hotho, M. Grahl, G. Stumme, Network properties of folksonomies. AI Comm. **20**(4), 245–262 (2007)

[8] A. Chakraborty, Credibility measurement of users in e-learning forums, in *Proc. National Convention of Computer Engineers*, February 2012

[9] A. Chakraborty, S. Ghosh, N. Ganguly, Detecting overlapping communities in folksonomies, in *Proc. ACM Hypertext Conference*, June 2012

[10] A. Clauset, M.E.J. Newman, C. Moore, Finding community structure in very large networks. Phys. Rev. E **70**, 066111 (2004)

[11] T.S. Evans, R. Lambiotte, Line graphs, link partitions, and overlapping communities. Phys. Rev. E **80**, 016105 (2009)

[12] S. Fortunato, Community detection in graphs. Phys. Rep. **486**(3–5), 75–174 (2010)

[13] S. Ghosh, P. Kane, N. Ganguly, Identifying overlapping communities in folksonomies or tripartite hypergraphs, in *Proc. ACM Conference on World Wide Web (WWW) companion volume*, pp. 39–40, Mar 2011

[14] M. Girvan, M.E.J. Newman, Community structure in social and biological networks. Proc. Natl. Acad. Sci. **99**(12), 7821–7826 (2002)

[15] M. Girvan, M.E.J. Newman, Finding and evaluating community structure in networks. Phys. Rev. E 69 (2004)

[16] S. Gregory, Finding overlapping communities using disjoint community detection algorithms, in *Complex Networks*, vol. 207 of *Studies in Computational Intelligence* (Springer, Berlin, 2009), pp. 47–61

[17] S. Guha, R. Rastogi, K. Shim, ROCK: a robust clustering algorithm for categorical attributes. Inform. Syst. **25**(5), 345–366 (2000)

[18] R. Guimera, M. Sales-Pardo, L.A.N. Amaral, Module identification in bipartite and directed networks. Phys. Rev. E **76**, 036102 (2007)
[19] T.G. Kolda, B.W. Bader, Tensor decompositions and applications. SIAM Rev. **51**(3), 455–500 (2009)
[20] I. Konstas, V. Stathopoulos, J.M. Jose, On social networks and collaborative recommendation, in *Proc. ACM SIGIR Conference*, pp. 195–202, 2009
[21] A. Lancichinetti, S. Fortunato, Benchmarks for testing community detection algorithms on directed and weighted graphs with overlapping communities. Phys. Rev. E **80**(1), 9 (2009)
[22] A. Lancichinetti, S. Fortunato, Community detection algorithms: a comparative analysis. Phys. Rev. E **80**, 056117 (2009)
[23] A. Lancichinetti, S. Fortunato, J. Kertesz, Detecting the overlapping and hierarchical community structure in complex networks. New J. Phys. **11**, 033015 (2009)
[24] J. Leskovec, K.J. Lang, A. Dasgupta, M.W. Mahoney, Statistical properties of community structure in large social and information networks, in *Proc. ACM Conference on World Wide Web (WWW)*, 2008
[25] Y.-R. Lin, J. Sun, P. Castro, R. Konuru, H. Sundaram, A. Kelliher, Metafac: community discovery via relational hypergraph factorization, in *Proc. ACM SIGKDD Conference*, pp. 527–536, 2009
[26] M. McPherson, L. Smith-Lovin, J. Cook, Birds of a feather: Homophily in social networks. Ann. Rev. Sociol. **27**, 415–444 (2001)
[27] T. Murata, Detecting communities from social tagging networks based on tripartite modularity, in *Proc. Workshop on Link Analysis in Heterogeneous Information Networks*, July 2011
[28] N. Neubauer, K. Obermayer, Towards community detection in k-partite k-uniform hypergraphs, in *Proc. Workshop on Analyzing Networks and Learning with Graphs*, pp. 1–9, 2009
[29] M.E.J. Newman, Fast algorithm for detecting community structure in networks. Phys. Rev. E **69**, 066133 (2004)
[30] V. Nicosia, G. Mangioni, V. Carchiolo, M. Malgeri, Extending the definition of modularity to directed graphs with overlapping communities. J. Stat. Mech. Theor Exp. **3**, 03024 (2008)
[31] G. Palla, I. Derenyi, I. Farkas, T. Vicsek, Uncovering the overlapping community structure of complex networks in nature and society. Nature **435**, 814–818 (2005)
[32] S. Papadopoulos, Y. Kompatsiaris, A. Vakali, A graph-based clustering scheme for identifying related tags in folksonomies, in *Proc. Conference on Data Warehousing and Knowledge Discovery (DaWaK)*, pp. 65–76, 2010
[33] M. Rosvall, C.T. Bergstrom, Maps of random walks on complex networks reveal community structure. Proc. Natl. Acad. Sci. **105**, 1118–1123 (2008)
[34] A. Vazquez, Finding hypergraph communities: a Bayesian approach and variational solution. J. Stat. Mech. Theor Exp. **2009**, P07006 (2009)
[35] X. Wang, L. Tang, H. Gao, H. Liu, Discovering overlapping groups in social media, in *Proc. IEEE Conference on Data Mining (ICDM)*, pp. 569–578, 2010
[36] S. Xu, S. Bao, B. Fei, Z. Su, Y. Yu, Exploring folksonomy for personalized search, in *Proc. ACM SIGIR Conference*, pp. 155–162, 2008
[37] D. Zhou, J. Huang, B. Scholkopf, Learning with hypergraphs: clustering, classification, and embedding, in *Proc. Advances in Neural Information Processing Systems*, 2006

The Stability of a Graph Partition: A Dynamics-Based Framework for Community Detection

Jean-Charles Delvenne, Michael T. Schaub, Sophia N. Yaliraki, and Mauricio Barahona

1 Introduction

Recent years have seen a surge of interest in the analysis of complex systems. This trend has been facilitated by the availability of relational data and the increasingly powerful computational resources that can be employed for their analysis. A unifying concept in the study of complex systems is their formalisation as *networks* comprising a large number of non-trivially interacting agents. By considering a network perspective, it is hoped to gain a deepened understanding of system-level properties beyond what could be achieved by focussing solely on the constituent units. Naturally, the study of real-world systems leads to highly complex networks

J.-C. Delvenne (✉)
Institute of Information and Communication Technologies, Electronics and Applied Mathematics (ICTEAM) and Center for Operations Research and Optimisation (CORE), Université catholique de Louvain, Louvain-la-Neuve, Belgium

Namur Center for Complex Systems (naXys), Facultés Universitaires Notre-Dame de la Paix, Namur, Belgium
e-mail: jean-charles.delvenne@uclouvain.be

M.T. Schaub
Department of Chemistry, Imperial College London, London, UK

Department of Mathematics, Imperial College London, London, UK
e-mail: michael.schaub09@imperial.ac.uk

S.N. Yaliraki
Department of Chemistry, Imperial College London, London, UK
e-mail: s.yaliraki@imperial.ac.uk

M. Barahona
Department of Mathematics, Imperial College London, London, UK
e-mail: m.barahona@imperial.ac.uk

A. Mukherjee et al. (eds.), *Dynamics On and Of Complex Networks, Volume 2*, Modeling and Simulation in Science, Engineering and Technology,
DOI 10.1007/978-1-4614-6729-8_11,

and a current challenge is to extract intelligible, simplified descriptions from the network in terms of relevant subgraphs (or *communities*), which can provide insight into the structure and function of the overall system.

Sparked by the seminal work by Newman and Girvan [1, 2], an interesting line of research has been devoted to investigating modular community structure in networks, revitalising the classic problem of graph partitioning. In a prescient piece, Simon [3] hypothesised that a modular structure exhibits evolutionary advantages. In particular, the ability to reuse components or to facilitate efficient local processing has been extensively studied thereafter. In recent years, a myriad of studies have gone on to detect communities and hierarchies in real-world systems, ranging from social systems to technological and biochemical systems (for a recent review, see [4]). Apart from the interest in the structure of these networks, the hope is that by finding a meaningful decomposition of a network, we will gain understanding of the relationship between the structural and the functional (dynamical) behaviour of the system. As Francis Crick put it: "If you want to understand function, study structure."

However, modular or community structure in networks has notoriously evaded rigorous definition. The most accepted notion of community is perhaps that of a group of elements which exhibit a stronger level of interaction within themselves than with the elements outside the community. This concept has resulted in a plethora of computational methods and heuristics for community detection. Nevertheless, a firmly grounded theoretical understanding of most of these methods, in terms of how they operate and what they are supposed to detect, is still lacking to date. For a meaningful application of community detection methods, such an understanding is however essential. In the following, we will develop a dynamical perspective towards community detection enabling us to define a measure named the *stability of a graph partition* [5,6]. It will be shown that a number of previously ad hoc defined heuristics for community detection can be seen as particular cases of our method providing us with a dynamic reinterpretation of those measures. Our dynamics-based approach thus serves as a unifying framework to gain a deeper understanding of different aspects and problems associated with community detection and allows us to propose new dynamically inspired criteria for community structure. Further discussion on dynamics-based community detection, along with a description of some earlier methods, is proposed in another chapter in this book [7].

The chapter is organised as follows. The stability of a graph partition is defined and analysed in Sect. 2. In Sect. 3, we show how it encompasses several other community detection methods. Section 4 discusses how to find the optimal partition with respect to stability and how to evaluate the meaningfulness of the results. Section 5 illustrates different capabilities of the method for multiscale community detection through applications to diverse examples.

1.1 Notation

For simplicity, in the following we consider only undirected, connected and non-bipartite graphs with N nodes and E edges. It is important to remark that the methodology extends seamlessly to directed graphs as discussed in [6]. We use the following notation hereafter. The topology of a graph is encoded in the weighted adjacency matrix $A \in \mathbb{R}^{N \times N}$, where the weight of the link between node i and node j is given by A_{ij} and $A = A^T$ due to the undirected nature of the graph. The weighted degrees (or strengths) of the nodes are given by the vector $d = A\mathbf{1}$, where $\mathbf{1}$ is the $N \times 1$ vector of ones. We further define the diagonal matrix $D = \text{diag}(d)$ and denote the total weight of the edges by $m = d^T\mathbf{1}/2$. Finally, the *combinatorial* graph Laplacian is defined as $L = D - A$, while the *normalised* graph Laplacian is defined as $\mathcal{L} = D^{-1/2}LD^{-1/2}$. Both Laplacians are symmetric non-negative definite, with a simple zero eigenvalue when the graph is connected [8].

2 Dynamics on Graphs and Community Detection: The Stability of a Graph Partition

As outlined above, a desirable outcome of community detection is to find a meaningful, simplified structural decomposition of a network that can shed light on the functional (dynamical) behaviour of the network and its components. Conversely, when a dynamics takes place on a network, it will be constrained by the graph structure and could potentially reveal features of the structural organisation of the network. This notion, i.e. the structure and the dynamical behaviour are inextricably linked and influence each other, is the seed from which our perspective on community has emerged.

The central idea underpinning our dynamical approach is the following: Considering a diffusion process on a network, we ask ourselves what this simplest form of dynamics can reveal about the underlying community structure of the graph. This idea is readily illustrated by the example of a vessel filled with water in which one were to put a small droplet of ink and observe how it diffuses. If the container has no structure, the dye would diffuse isotropically. If the container is structured (e.g. compartmentalised or comprising a set of smaller vessels connected via some tubing), the dye would not spread isotropically but would rather get transiently trapped in certain regions for longer times until it eventually becomes evenly distributed throughout the whole vessel. Therefore, by observing the time dynamics of this diffusion process we can gain valuable information about the structural organisation of the container.

2.1 Defining the Measure: Stability as an Autocovariance

Let us now make more precise some of these ideas and consider a Markovian diffusion process defined on the graph. One of the advantages of the method is that it allows us to choose the dynamics used to reveal the structure of the network from a variety of discrete or continuous-time processes (see an extended discussion of these issues in [6, 9]). The dynamics chosen will have distinct dominant diffusion paths and thus will lead to the the emergence of potentially different community structure. The flexibility in the choice of dynamics allows us to incorporate any *a priori* information we might have about the intrinsic dynamics in the system under study or to base our detection on a particular heuristic for the defining characteristics of the communities. In this work, we discuss only the generic case of an *unbiased* random walk governed by the following dynamics:

$$\dot{\mathbf{p}} = -\mathbf{p}\,[D^{-1}L] \tag{1}$$

in continuous time or, alternatively, by

$$\mathbf{p}_{t+1} = \mathbf{p}_t D^{-1} A \equiv \mathbf{p}_t M \tag{2}$$

for the case of discrete time. Here $\mathbf{p}$ denotes the $1 \times N$ dimensional probability vector and under the assumptions made above (connected, undirected and non-bipartite graph), both dynamics converge to a unique stationary distribution given by $\pi = d^T/2m$. While the discrete-time random walker jumps from one node to the next at unit-time intervals, the continuous-time random walker waiting time at a node before the next jump is a continuous memoryless random variable distributed exponentially with unit expectation. Whenever a jump occurs, the jump probabilities between nodes are identical for both processes. Hence, the discrete-time random walk can also be interpreted as an approximation of its continuous-time counterpart (see also Sect. 3.3), and our numerics show that the two processes lead to similar community structures in most graphs of interest.

Consider also a given (hard) partition of this graph into c communities. This partition is encoded in a $N \times c$ indicator matrix H with $H_{ij} \in \{0, 1\}$, where a 1 denotes that node i belongs to community j. Observe that, for any $N \times N$ matrix B, $[H^T BH]$ is the $c \times c$ matrix whose $k\,l$–th entry is the sum of all entries B_{ij} such that node i is in community k and node j is in community l. It is also easy to see that $H^T H = \text{diag}(n_k)$, where n_k is the number of nodes in community k.

Now we define the *clustered autocovariance* matrix of the diffusion process at time t as

$$R(t) = H^T \left[\Pi P(t) - \pi^T \pi\right] H, \tag{3}$$

where $\Pi = \text{diag}(\pi)$ and $P(t)$ is the t-step transition matrix of the process: $P(t) = \exp(-tD^{-1}L)$ for continuous time and $P(t) = M^t$ for discrete time.

To see that $R(t)$ is indeed an autocovariance matrix associated with the c communities of a partition, consider the N-dimensional vector $X(t)$, a random indicator vector describing the presence of a particle diffusing under the above dynamics, i.e. $X_k(t) = 1$ if the particle is in node k at time t and zero otherwise. Given a partition H, $Y(t) = H^T X(t)$ is the community indicator vector. Using the definition of the transition matrix, it follows that

$$R(t) = H^T \operatorname{cov}[X(\tau), X(\tau + t)]\, H = \operatorname{cov}[Y(\tau), Y(\tau + t)], \quad (4)$$

where we have made use of the linearity of the covariance.

If the graph has well-defined communities *over a given time scale*, we expect that the state is more likely to remain within the starting community with a significant probability over that time scale. Therefore, the value of $Y_i(0)$ is positively correlated with the value of $Y_i(t)$ for t in that time scale leading to a large diagonal element of $R(t)$ and hence a large trace of $R(t)$. In other words, if we observe the Markov process only in terms of a colouring assigned to each community, a good partition is indicated by a slow decay of the autocovariance of the observed signal. This statement presents a dual (but equivalent) view to the description of a community in terms of a subgraph where the diffusion probability gets transiently trapped.

We now define the *stability of a partition* encoded by H at time t as

$$r(t, H) = \text{trace } R(t), \quad (5)$$

which is a global quality function for a given graph and partition that changes as a function of time (according to the chosen dynamics). In view of the formalism outlined above, stability serves as a quality function to evaluate the goodness of a partition in terms of the persistence of the probability flows in the graph at time t.

A brief technical aside: The definition of stability in the discrete-time case

$$r_t(H) = \min_{0 \le s \le t} \text{trace } \left[H^T \left[\Pi M^s - \pi^T \pi \right] H \right] \quad (6)$$

includes a minimisation over the time interval $[0, t]$, which is not necessary in the continuous-time case.[1] This technicality ensures maximum generality of the definition, but in most cases of interest it is not necessary.[2] Indeed, numerical experiments [10] have shown that

$$r_t(H) = \min_{0 \le s \le t} \text{trace } R_s \approx \text{trace } R_t$$

for the discrete-time case too. In the rest of this chapter, we assume this to be valid. The stability $r(t, H)$ can be used to rank partitions of a given graph at different time scales, with higher stability indicating a better partition into communities at

---

[1]In the continuous case, trace $R(t)$ is monotonically decreasing with time. To prove this, note that $P(t) = D^{-1/2} \exp(-t\mathcal{L}) D^{1/2}$ and $d(\text{trace } R(t))/dt = -\text{trace } H^T \Pi D^{-1/2} \mathcal{L}^{1/2} \exp(-t\mathcal{L}) \mathcal{L}^{1/2} D^{1/2} H = -\text{trace } H^T D^{1/2} \mathcal{L}^{1/2} \exp(-t\mathcal{L}) \mathcal{L}^{1/2} D^{1/2} H / 2m$. This is obviously strictly negative since the matrix $\exp(-\mathcal{L})$ is symmetric positive definite.

[2]In particular cases, such as a bipartite graph, trace R_s can oscillate in the discrete-time case, indicating poor communities or even "anti-communities" with a rapid alternance of random walkers between communities. We therefore take the lowest point of the R_s over the interval as the quality function.

a specific time scale. Alternatively, $r(t, H)$ can be maximised for every time t in the space of all possible partitions—resulting in a sequence of partitions, each of which is optimal over some time interval. Although, as is the case for many graph theoretical problems, this optimisation is NP-hard, a variety of optimisation heuristics for graph clustering can be used (see Sect. 4). The effect of time in the stability measure is intuitive: as time increases, the Markov process explores larger regions of the graph, such that the Markov time acts as a resolution parameter that enables us to identify community structure at different scales. In general, the relevant partitions become coarser as the Markov time increases. Importantly, stability does not aim to find *the* best partition for the graph but rather tries to reveal relevant clusterings at different scales through the systematic zooming process induced naturally by the dynamics. A relevant partition should be both persistent over a comparably long time scale and robust with respect to slight variations in the graph structure and/or the optimisation [11, 12] (see also Sect. 4). We remark that the systematic sweeping across scales provided by the Markov dynamics is a fundamental ingredient in our approach leading to a multiscale analysis of the community structure.

2.2 Stability from a Random Walk Perspective

Stability can be readily interpreted in terms of a random walk by noting that each entry $[R(t)]_{ij}$ of the clustered autocovariance (see Eq. (3)) is the difference between two quantities. The first is the probability for a random walker to start in community i at stationarity and end up in community j after time t, while the second is the probability of two independent walkers to be in i and j at stationarity. In this view, communities correspond to groups of nodes within which the probability distribution of the Markov process is more contained after a time t than otherwise expected at stationarity. To make this perspective more explicit, one can rewrite the stability of a partition H with communities $\mathcal{C}$ as:

$$r(t, H) = \sum_{\mathcal{C}=1}^{c} P_{\text{AtHome}}(\mathcal{C}, t) - P_{\text{AtHome}}(\mathcal{C}, \infty), \tag{7}$$

where $P_{\text{AtHome}}(\mathcal{C}, t)$ is the probability for a random walker to be in the same community $\mathcal{C}$ at time zero and at time t (possibly leaving and coming back a number of times in between). By ergodicity, the walker at infinite time holds no memory of its initial position, and $P_{\text{AtHome}}(\mathcal{C}, \infty)$ coincides with the probability of two independent walkers to be in the same community.

2.3 A Free Energy Perspective: Stability as a Trade-Off Between Entropy and Generalised Cut

Stability may also be interpreted as a trade-off between an entropy and a generalised cut measure, thus providing a link to the concept of free energy. To consider this perspective, we first observe that stability at time $t = 0$ is essentially equivalent to the so called Gini-Simpson *diversity index* [13]:

$$r(0, H) = \text{trace}\left[H^T \left(\Pi - \pi^T \pi\right) H\right] = 1 - \sum_{C=1}^{c} (\pi \mathbf{h}_C)^2 \equiv \text{Diversity}, \quad (8)$$

where $\mathbf{h}_C$ is the C-th column of the indicator matrix H. The diversity index (or equivalent quantities up to some simple transformation) has appeared under numerous names in various fields: the Hirschman-Herfindahl index in economics [14], the Rényi entropy of order 2 in information theory [15], or the Tsallis entropy of parameter 2 in non-extensive thermodynamics [16]. This quantity may be interpreted as a measure of entropy and favours partitions into many communities of equal sizes (in terms of degree weights for the unbiased dynamics) over partitions where one community is much bigger (in terms of degree weights) than the others. For instance, the diversity index of a k way equal-size partition is $1 - (1/k)$, while a partition with a very dominant community and $k - 1$ small communities has a diversity index of almost zero. The diversity index is minimal for the all-in-one partition and maximal for the partition into one-node communities.

Secondly, consider the variation of stability between time zero and time t:

$$r(0, H) - r(t, H) = 1 - \text{trace}\left[H^T \Pi P(t) H\right] \equiv \text{Generalised Cut}(t). \quad (9)$$

This quantity is the fraction of walks of length t that start and end in two different communities and may be thought of as a generalised cut size, i.e. the fraction of edge weights hanging between communities for a graph with adjacency matrix $\tilde{A} = \Pi P(t)$ induced by the dynamics of the diffusion process at time t [6,9].

It then follows that stability can be written as

$$r(t, H) = \text{Diversity} - \text{Generalised Cut}(t), \quad (10)$$

which provides us with an interpretation of stability as a trade-off between an entropy and a cut measure moderated by the Markov time. We will see below that when considering the linearised version of the stability $r(t, H)$, this interpretation allows us to write stability as a kind of free energy that balances an entropy (diversity) and an energy (cut) with the time t playing the role of an inverse temperature.

3 Stability as a Unifying Framework for Other Community Detection Methods

Stability provides a unifying framework for a number of different graph partitioning and community detection techniques and heuristics that have been postulated in the literature under different premises. We now highlight some of these relations to provide further insight into the stability framework while at the same time establishing a dynamical reinterpretation of some key measures widely used in community detection. Most of the results shown below can be found in [5, 6].

In the following, we will use shorthand to distinguish between the discrete-time version of stability (denoted as $r_t(H)$) and the continuous-time version (denoted by $r(t, H)$).

3.1 Discrete-Time Stability at Time One Is Modularity

Modularity [1] is usually defined as

$$\text{Modularity} = \frac{1}{2m} \sum_{C} \sum_{i,j \in C} A_{ij} - \frac{k_i k_j}{2m} \tag{11}$$

where the sum is taken over nodes i, j that are in the same community, k_i is the degree of node i and m is the number of edges. Modularity is a popular cost function which is maximised to find the best partition of a graph. Noting that the stationary probability $\pi_i = k_i/2m$, it follows easily from our formalism that modularity is equivalent to discrete-time stability at time 1:

$$r_1(H) = \text{trace}\left[H^T \left(\frac{A}{2m} - \pi^T \pi\right) H\right] = \text{Modularity}. \tag{12}$$

Furthermore, from Sect. 2.3, it is easy to see that Generalised Cut at time $t = 1$ in the discrete-time case is simply the fraction of edges between communities, i.e. the (standard) Cut:

$$\text{Generalised Cut}(1) = 1 - \text{trace}\left[H^T \frac{A}{2m} H\right] = \text{trace}\left[H^T \frac{L}{2m} H\right] = \text{Cut}, \tag{13}$$

whence it follows that

$$r_1(H) = \text{Diversity} - \text{Cut} = \text{Modularity}. \tag{14}$$

Therefore, modularity can be seen as a compound quality function with two competing objectives: minimise the Cut size while at the same time try to maximise the diversity index (which favours a large number of equally sized communities) thus resulting in more balanced partitions.

3.2 Stability at Large Time Is Optimised by the Normalised Fiedler Partition

The asymptotic behaviour of stability at large t is determined by the spectral properties of the graph and leads to the dominance of the Fiedler bipartition (a classic heuristic in graph partitioning) as $t \to \infty$. Consider first the discrete-time case. Elementary spectral theory implies that the eigenvalues λ_i of $M \equiv D^{-1}A$ as well as the left and right eigenvectors of M are real and obey the relations $M\mathbf{v}_i = \lambda_i \mathbf{v}_i$, $\mathbf{u}_i M = \lambda_i \mathbf{u}_i$ and $\mathbf{u}_i^T = \Pi \mathbf{v}_i$. Consider the spectral decomposition of M:

$$M = \mathbf{1}\pi + \lambda_2 \mathbf{v}_2 \mathbf{u}_2 + \lambda_3 \mathbf{v}_3 \mathbf{u}_3 + \ldots \tag{15}$$

and assume that the second largest eigenvalue λ_2 is not dominated by the smallest (i.e. $\lambda_2 > |\lambda_N|$), which is the case if the graph is not "almost bipartite". Then asymptotically we have

$$\lim_{t\to\infty} r_t(H) \sim \lambda_2^t \operatorname{trace} H^T \mathbf{v}_2 \mathbf{u}_2 H.$$

It is not difficult to see [5] that this leading term is maximised for the partition H that classifies the nodes into two communities according to the sign of the entries of $\mathbf{v}_2$ (or, equivalently, $\mathbf{u}_2$). This partition is called the normalised Fiedler partition [17], a variant of the classic Fiedler partition proposed as a heuristic in spectral clustering [18, 19].

Therefore, spectral clustering, usually presented as a heuristic for the solution of combinatorial optimisation problems (see, e.g. [17]) emerges here as the exact solution of the large time scale stability optimisation. It is worth noting that for the continuous-time case we can obtain similar results—here $P(t)$ behaves asymptotically as $e^{(\lambda_2-1)t}\mathbf{v}_2\mathbf{u}_2$ leading to the same optimal partition.

3.3 The Meaning of the Linearisation of Stability at Short Times

We now show that several heuristics for community detection that have been introduced from a statistical physics perspective appear from the linearisation of stability at small times.

Consider the continuous version of stability in the limit of small times $t \to 0$ and expand to first order. This leads to the *linearised stability* [5, 6]:

$$r_{\text{lin}}(t, H) = r(0, H) + t\, \frac{dr(t, H)}{dt}\bigg|_{t=0} = r(0, H) - t \operatorname{trace}\left[H^T \frac{L}{2m} H\right]. \tag{16}$$

Linearised Stability as a Type of Free Energy

It then follows from (13) that

$$r_{\text{lin}}(t, H) = \text{Diversity} - t\,\text{Cut}. \tag{17}$$

This linearised stability can also be seen as a linear interpolation for the discrete-time stability between time zero and time one: $r_{\text{lin}}(t, H) = (1-t)r_0(H) + t\,r_1(H)$. Incidently, this highlights a further connection between the discrete-time and the continuous-time stabilities, since the former can be seen as a linear approximation of the latter. We emphasise however that linearised stability is meaningful for all times, whether smaller or larger than one. Indeed, renormalising this expression as [5]

$$-\frac{r_{\text{lin}}(t, H)}{t} = \text{Cut} - \frac{1}{t}\,\text{Diversity} \tag{18}$$

makes it clear that the linearised stability can be viewed as a free energy to be minimised, the diversity index acts as a type of entropy measure, the Cut acts as an energy cost (as an analogy, think of the edges as bonds to be broken, whose energy is the weight of the edge), and the inverse Markov time plays the role of temperature.

Linearised Stability Is Equivalent to the Reichardt and Bornholdt Potts Model Heuristic

Reichardt and Bornholdt (RB) proposed a Potts model heuristic for multiscale community detection based on the minimisation of the Potts Hamiltonian [20, 21].

$$\mathcal{H}_{\gamma}^{\text{RB}} = -2m \text{ trace}\left[H^T\left(\frac{A}{2m} - \gamma\pi^T\pi\right)H\right], \tag{19}$$

where γ is a tunable resolution parameter. From above, we observe that

$$r_{\text{lin}}(t, H) = (1-t) - t\left(\frac{1}{2m}\mathcal{H}_{1/t}^{\text{RB}}\right). \tag{20}$$

Hence, maximising the linearised stability is equivalent to minimising the RB Potts Hamiltonian $\mathcal{H}_{\gamma}^{\text{RB}}$ with $\gamma = 1/t$. Having established this connection enables us to provide an explicit interpretation of the "resolution parameter" γ in terms of the Markov time t and also to interpret the RB Potts model heuristic as a free energy, corresponding to a trade-off between entropy and cut [5]. Note that this view differs slightly from RB's initial interpretation in terms of energy only.

Linearised Stability with Combinatorial Laplacian Dynamics Is Related to the Constant Potts Model

It is worth mentioning briefly that several other recent heuristics also correspond to the linearised stability obtained under a different dynamical evolution, namely when the dynamics is governed by the *combinatorial* Laplacian [6]

$$\dot{\mathbf{p}} = -\mathbf{p}\frac{L}{\langle d \rangle}, \tag{21}$$

where $\langle d \rangle$ is the average strength of the nodes in the graph.

It is then easy to show that maximising the linearised stability $r_{\text{lin}}^{\text{CL}}(t, H)$, which follows seamlessly from this dynamics, is equivalent to minimising the Hamiltonian of the so-called constant Potts model (CPM) introduced by Traag et al. [22]:

$$\mathcal{H}_{\mu}^{\text{CPM}} = -2m \text{ trace } \left[H^T \left(\frac{A}{2m} - \frac{\mu}{2m} \right) H \right], \tag{22}$$

where μ is a tunable "resolution parameter". In fact,

$$r_{\text{lin}}^{\text{CL}}(t, H) = (1 - t) - t \left(\frac{1}{2m} \mathcal{H}_{2m/tN^2}^{\text{CPM}} \right), \tag{23}$$

and the optimization of the combinatorial linearised stability is equivalent to minimising the CPM with resolution parameter $\mu = 2m/N^2 t$.

A similar reasoning can be applied to the Potts heuristic proposed by Nussinov and co-workers [23, 24].

4 Computational Aspects of the Stability Framework

4.1 *Optimisation of the Stability Measure*

Stability defines a quality measure on partitions and one can use in principle any graph clustering algorithm (even if based on different principles) and assess and rank the quality of the found partitions *a posteriori* with our stability measure. Some well known algorithms that we have used in this way include the methods proposed by Shi and Malik [17] and Kannan, Vempala and Vetta [25], which operate divisively to obtain finer and finer partitions via recursive spectral bipartitioning. The partitions thus obtained can then be evaluated at different time scales according to stability: as outlined above, for short time scales we expect the finer partitions to be more relevant, whereas for larger time scales coarser partitions will dominate. Using the spectral clustering methods in such a way turns them effectively into a heuristic for optimising stability, which performs well in many cases [5]. See Sect. 5.1 for an example of this type of analysis.

Stability, whether linearised, continuous time or discrete time, can also be considered as an objective function to be optimised, i.e. for each time we have to find the partition $\hat{H}$ with maximal stability

$$\hat{H} = \arg\max_{H} r(t, H). \tag{24}$$

However, as is the case for most non-trivial problems in clustering and graph partitioning, including modularity optimisation [26], the optimisation of stability over the space of all possible partitions is an NP-hard problem. Therefore, except for asymptotic times $t \to 0$ and $t \to \infty$, we must employ heuristics to optimise stability and these come with no guarantee on the obtained results.

A helpful feature of stability is the fact that it can be written as the modularity of a time-dependent network evolving under the Markov process [6]. Therefore, all the computational heuristics that have been developed for modularity optimisation can be naturally applied to stability optimisation too. These include not only divisive spectral methods [27] but also agglomerative methods, such as the efficient Louvain method [28] based on a fast greedy optimisation that can deal with very large graphs. Examples of the application of the Louvain algorithm to stability optimisation are shown in Sects. 5.2 and 5.3. Other methods have been devised specifically for stability optimisation over a time interval [10].

One may as well combine different methods to optimise stability, at the expense of greater computational cost. For instance, one can improve a Shi-Malik spectral partition with a local Kernighan-Lin scheme [29]. A wide range of heuristic stability optimisation procedures can therefore be devised.

As is the case with hard non-convex optimisation problems, it is important to assess whether these different optimisation methods yield consistent results, i.e. to check that the value of stability obtained by the use of different optimisation algorithms is robust. Our numerical experiments have shown that this is indeed the case specifically when the partitions found are most relevant [5, 12, 30].

4.2 Assessing the Robustness of a Partition

Related to the question of the robustness of the optimisation is the question of the robustness of the partitions found, which we use as an indicator of the relevance of the partitions found. One simple mechanism to detect the robustness of a partition is already built into the stability measure via the Markov time: a robust partition should ideally be persistent over an extended range of time scales. When the graph has a strong community structure, this can be observed as plateaux extending over Markov time intervals during which each corresponding partition is found.

As a second indicator for the robustness of the partitions, we evaluate the dispersion induced in the optimised partitions by a randomisation of the optimisation algorithm. In particular, for each Markov time we compute an ensemble of randomised Louvain optimisations of stability (started from a random initial node ordering) and check how different the optimised partitions are, as established by the variation of information [31]. Let $p(\mathcal{C}_\alpha)$ be the relative frequency of finding a node in community $\mathcal{C}$ in partition $\mathcal{P}^\alpha$, i.e. $p(\mathcal{C}_\alpha) = n_{\mathcal{C}_\alpha}/N$, where $n_{\mathcal{C}_\alpha}$ is the number of nodes in community $\mathcal{C}_\alpha$. The variation of information between two partitions $\mathcal{P}^\alpha$ and $\mathcal{P}^\beta$ is defined as [31]

$$\mathrm{VI}(\mathcal{P}^\alpha, \mathcal{P}^\beta) = 2H(\mathcal{P}^\alpha, \mathcal{P}^\beta) - H(\mathcal{P}^\alpha) - H(\mathcal{P}^\beta), \tag{25}$$

where $H(\mathcal{P}) = -\sum_{\mathcal{C}} p(\mathcal{C}) \log p(\mathcal{C})$ is the Shannon entropy and $H(\mathcal{P}^\alpha, \mathcal{P}^\beta)$ is the Shannon entropy of the joint probability. Here we use a normalised variant of the variation of information obtained by dividing the quantity in Eq. (25) by its maximum, $\log N$. A low variation of information indicates optimised partitions that are very similar to each other, indicating that the partition is robust to changes in the starting point of the optimisation. Borrowing freely some terminology from dynamical systems, robust partitions can thus be considered to possess an attractor with a large basin of attraction for the optimisation process.

Finally, one can also consider the robustness of a partition to small changes in the underlying graph either through random perturbations [11] or through the creation of surrogates for specific applications [12]. All these measures of robustness are used to select the relevant partitions across Markov times.

5 Examples of Community Detection with the Stability Framework

In this section, we showcase a few applications of the stability framework. Apart from the examples discussed below, this approach has already been applied to a variety of fields, such as image segmentation, social networks and computational biology, among others [5, 12, 30, 32].

5.1 A First Example: Community Structure in a Collaboration Network

Community detection methods are commonly used for social networks, with the aim to identify sets of strongly interdependent people so as to shed light on the global structure of the social group. As an example of this type of application, we analyse in Fig. 1a graph of co-authorships between 379 researchers in network science (see [27]). Figure 1b shows the hierarchy of partitions associated with the maximisation of discrete-time stability over all partitions obtained by using the Kannan, Vempala and Vetta (KVV) algorithm [25] across all Markov times. The KVV method is a conductance-based spectral divisive algorithm that splits the network in two parts repeatedly (in a kind of bisection scheme) producing a set of

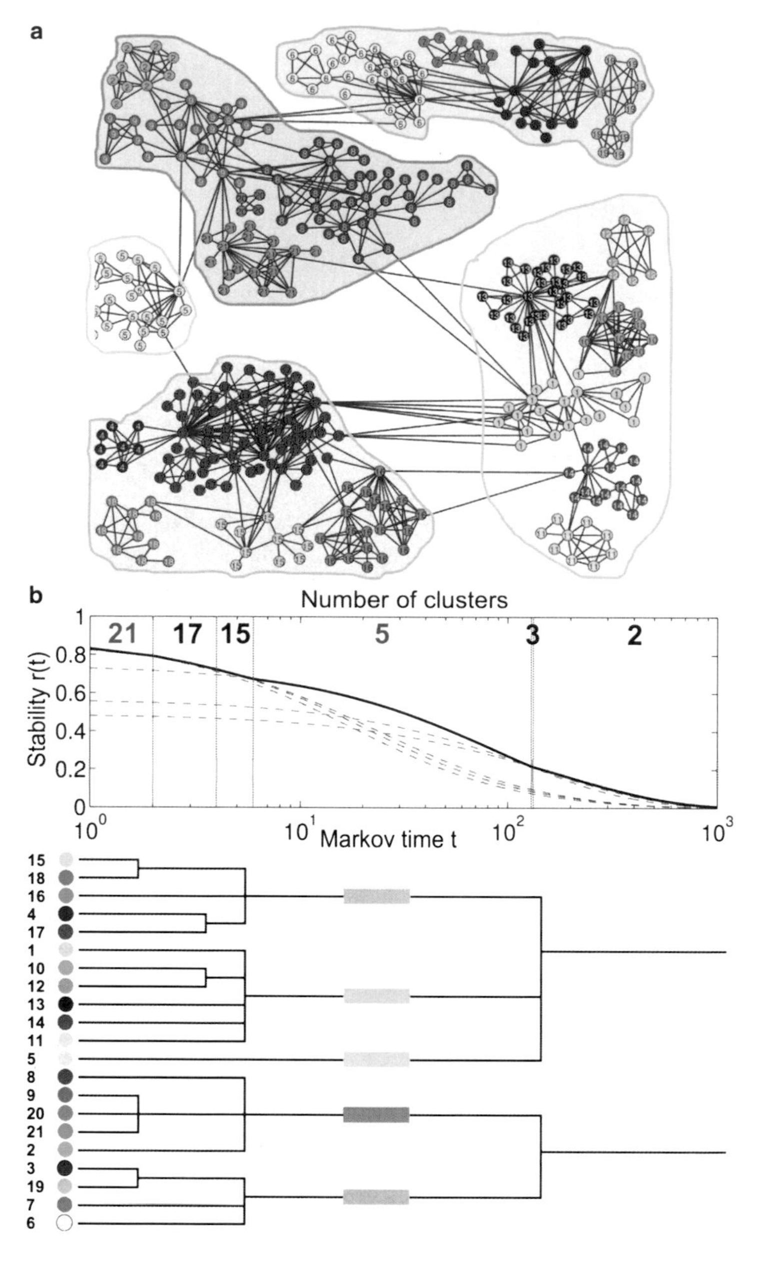

Fig. 1 *A network of scientific co-authorship [33].* (**a**) The vertices correspond to 379 researchers in network science, indexed by the 21-community partition obtained by maximising the time-one discrete-time stability (i.e. modularity). (**b**) We estimate the optimal partition for every time with a divisive KVV algorithm and exhibit the optimal stability curve as a function of time (*top*) and the corresponding dendrogram of the hierarchy of partitions (*bottom*). The dendrogram is not a binary tree (as compared with most of the hierarchical clustering methods) and has relatively few levels. Except for the two-way partition, the decomposition into five communities (represented by the coloured shaded areas in **a**) stands out by its longevity [5]

partitions that are then ranked according to stability at each time. This procedure results in a streamlined *nonbinary* dendrogram of optimal partitions, which differs from the notably more complicated binary trees produced by most hierarchical methods using different local heuristics. The communities shown have been selected according to their long persistence in time. Note in particular the very long-lived partition into 5 communities, which constitute well-formed thematic and national groups of authors, as a close look at the identity of the nodes reveals [5].

By construction, the particular optimisation algorithm used here as a heuristic for this NP-hard problem outputs a perfect hierarchy, where partitions are embedded into one another. However, there are many instances where the actual partitions at different time scales are not hierarchical, i.e. they are not necessarily exact progressive refinements of each other. As shown below, optimisation of stability with other algorithms can compute a nonhierarchical, yet multiscale, community structure across time scales. One can then find if the sequence of partitions is close to being hierarchical or not [6].

5.2 *Non-clique-Like Communities and the Field-of-View Limit*

Stability as a Multistep Method *vs.* Single-Step Methods

In contrast to stability, most commonly used community detection algorithms can be reformulated as single-step methods from our dynamical viewpoint, i.e. these methods only take into account nearest neighbours in the graph with no paths of length greater than one considered. For modularity and derived measures (like the RB Potts model), this is obvious from our discussion in Sect. 3. It can also be shown that the popular Map equation method [34] is effectively based on single-step, block-averaged transitions of a random walker [30, 35]. However, by using one-step measures, such methods effectively introduce an upper scale into the detection of communities. Metaphorically, these algorithms suffer from a limited "field-of-view" (see [30, 35] for an in-depth discussion) that precludes them from detecting communities with large internal distances. In contrast, the stability framework has no preferred scale (or number of steps) *a priori*, i.e. the scanning across all scales imposed by the dynamics is intrinsic and an essential feature of the algorithm. Therefore, if there are one or more scales, they can be found through the dynamic sweeping.

Advantages of a Multistep Approach: Detection of Non-clique-Like Communities

If one were to consider the simplest notion of community structure, one would probably think of a graph in which clique-like graphs with strong connections within them are weakly connected to each other in an "all-to-all" manner. The community structure thus emerges from the different intra- and intercommunity weights leading to a stochastic "clique-of-cliques". The corresponding notion of community as a homogeneously connected dense substructure homogeneously embedded in the whole graph underlies most of the popular benchmarks for community detection [1, 36–39]. Although such notion of community structure is a good description in several application areas (e.g. networks constructed from correlation measurements or in some instances of social groupings), there exists a wide range of real networks that do not display such an "all-to-all" connection pattern.

Prominent examples of networks without clique-like connection patterns are geographically embedded networks, such as sensor-networks, power grids, river-networks, road-networks and train-networks, as well as other transport, supply or distribution networks. More generally, networks underpinned by intrinsic constraints dictated by geometry or other cost functions (e.g. higher-dimensional grid- or lattice-like structures originating from physical and biological systems) will not in general display homogeneous, block-like structures in their connectivity patterns. Yet such networks may still have a pronounced community structure in its general sense: a *non-clique-like* community may in this case be thought of as a group of nodes which possess a stronger direct or *indirect* influence on each other than on nodes outside their community, e.g. if entities are coupled via a chain of local interactions [30]. In these instances, one-step methods, such as modularity and the Map method, will fail to identify such non-clique-like communities due to their large effective diameter, which puts them beyond the field-of-view of standard single-step methods. It is important to remark that in the case of clique-like community structure, stability still recovers the results of standard methods, which can be seen as a particular case of our framework.

Analysis of the Graph of the European Power Grid

The analysis of the European power grid, a network in which non clique-like community structure plays a significant role, is used here to exemplify the advantage of using an intrinsic multistep method such as stability [30]. This network[3] is based on data from the Union for the Coordination of Transmission Energy (UCTE) and has been analysed previously for robustness to targeted attacks [40, 41]. It consists of 2,783 nodes corresponding to generators and substations with 3,762 (unweighted) links corresponding to high voltage (110–400 kV) transmission lines (see also [40, 41]). Geography and engineering costs impose some extrinsic local constraints on the power grid and hence its connection patterns are far from a homogeneous all-to-all type. Hence, we expect the relevant community structure to be of a non-clique-like type. The dynamic sweeping of the stability framework over different Markov times is shown in Fig. 2. In this example, the analysis has been performed with the continuous-time-variant of stability, optimised via the

---

[3]Dataset taken from http://www.termoenergetica.upc.edu/marti/index.htm.

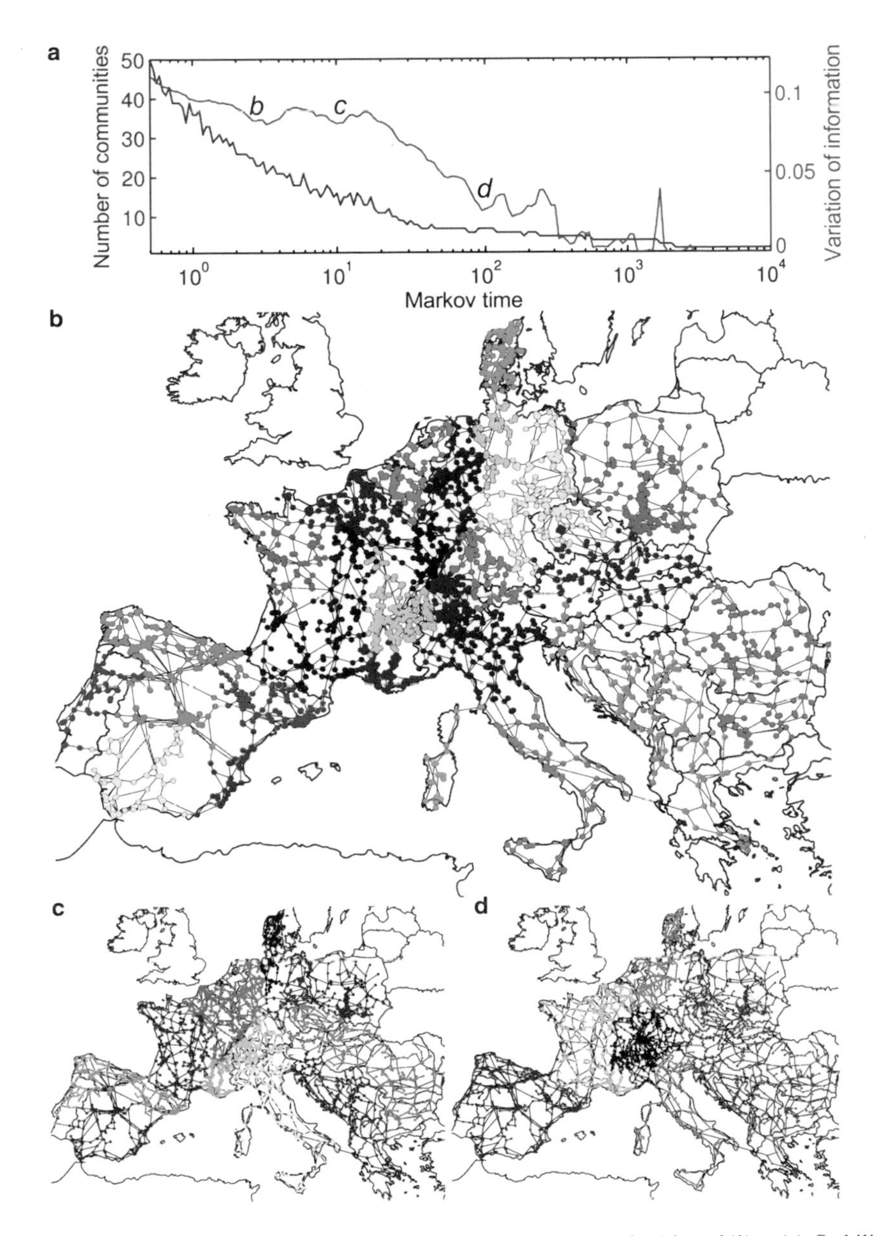

Fig. 2 *Community structure analysis of the European power grid with stability.* (**a**) Stability analysis on the European power grid: number of communities found (*blue*) *vs.* Markov time. A relatively low average variation of information distance [31] (*green*) for a given Markov time indicates a robust partition. (**b–d**) The robust partitions for different Markov times seem consistent with known structures in the power grid. Those partitions correspond to Markov times: (**b**) $t = 2.63$ (25 communities), (**c**) $t = 11.76$ (15 communities) and (**d**) $t = 94.79$ (7 communities). Analysis of the same graph with popular methods, such as modularity and the Map equation, lead to severe over-partitioning, due to the intrinsic one-step nature of these methods [30]

Louvain algorithm adapted for stability. The Markov times highlighted in Fig. 2 have been selected according to the relative decrease in variation of information, which indicates a more robust partitioning.

The analysis reveals a multiscale community structure within the power grid, with distinct relevant partitions at various Markov times corresponding to subregions which, interestingly, reflect meaningful historical and commercial features of the power grid network. For long Markov times (Fig. 2d), the coarse-grained subregions of the network can be identified mostly with nations or supranational structures, which may be associated with big historical monopolies, while concurrently the German power grid remains split between four large companies (with one covering the eastern part of Germany).[4] For shorter times (Fig. 2b,c), communities on a subnational scale are obtained. These are associated with regional operators, e.g. the French power grid is divided into several regions which overlap well with the communities found in our analysis.[5] Similar results hold for Spain, Italy and Switzerland. For more details see [30].

5.3 *Detecting Communities of Links in the Internet Autonomous Systems Network*

As a final example, we look at another classical network system, namely the Internet at the level of autonomous systems (AS). An autonomous system is a large group of connected routing prefixes implementing a clearly defined, consistent routing policy.[6] As this network is only known approximately, we use measurements obtained by the Cooperative Association for Internet Data Analysis (CAIDA) [42]. To construct the network, we select only direct links between two different and unique AS and we ignore all multiple origin AS (MOAS) edges (i.e. links where a prefix originates from multiple AS). Furthermore, we limit ourselves to AS of large Internet service providers (ISPs), small ISPs and universities corresponding to the categories "t1", "t2" and "edu" from the Autonomous System Taxonomy Repository.[7] Note that, in general, there is no equivalence between an ISP and a AS: although large enough ISPs tend to form their own AS, an AS can also consist of several smaller companies or correspond to the network of a university. The analysed network consists of 3,951 nodes and 9,870 unweighted, undirected links.

To exemplify the use of our framework with a different objective, we performed the community analysis on the space of edges. This can be achieved by analysing

---

[4] See http://de.wikipedia.org/wiki/Stromnetz#Netzbetreiber.

[5] For a map of the French regional electrical companies see http://www.rte-france.com/fr/nous-connaitre/qui-sommes-nous/organisation-et-gouvernance/le-siege-et-les-unites-regionales.

[6] see http://en.wikipedia.org/wiki/Autonomous_system_(Internet).

[7] http://www.caida.org/data/active/as_taxonomy/.

the *line graph* of the network. The line graph of a network G is a graph whose nodes are the edges of G, with a link between two nodes if the corresponding edges in G are adjacent. The study of the line graph allows us to focus on the links in the network, which carry the actual information between the AS, to find whether there is a structural organisation in the edges, corresponding to topological "traffic communities" in the global network structure. Since community labels are assigned to edges, rather than to the AS nodes, this allows for multiple community memberships of an AS node [43]. The results of the community analysis of the line graph using the Louvain algorithm to optimise the continuous-time stability are shown in Fig. 3.

Our analysis reveals interesting structural features. The links to/from each of the large ISPs belong predominantly to one cluster. For smaller ISPs, a different picture emerges. Some small ISPs appear to be *stub AS*, in that they display only a small number of connections (usually one or two) linking them to one large t1 AS. The links from the large t1 AS to its stubs typically all belong to the same community (Fig. 3b). On the other hand, we find a second class of small ISPs which one may call *diversified AS*: they have edges belonging to different communities, connecting them to multiple large ISPs, suggesting that they "diversify" their connectivity patterns so as to enhance the robustness of their connectivity to the Internet. An example of such a small ISP with a diversified connection pattern is the AS of "yahoo!" with nine connections to different large ISPs (see Fig. 3a) all belonging to different communities.

Clearly, the significance of these results is limited by the data and the techniques used to infer the connections between the AS, as well as by the fact that we do not consider the actual data volume transferred over the links. However, the fact that this simplified analysis reveals interesting groupings within the AS network indicates the potential for future work in this direction. In particular, it may be interesting to test to what extent community detection might be used for a classification of the AS, possibly in comparison to what has been inferred by more classical measures. This will be the objective of future work.

6 Conclusions and Future Work

In this chapter we have shown that adopting a dynamical perspective towards community detection through the introduction of the stability measure provides us with a unifying framework that enables us to gain deeper understanding of community detection. In particular, it was shown how different community detection methods can be seen as special cases of the introduced stability and thus be interpreted from a dynamical viewpoint as well.

It has been further illustrated with a few examples how the stability framework can be used in different application areas. The framework provides a surprisingly versatile set of tools and can be easily extended and adopted for different application scenarios. It is important to mention that the framework naturally extends to directed

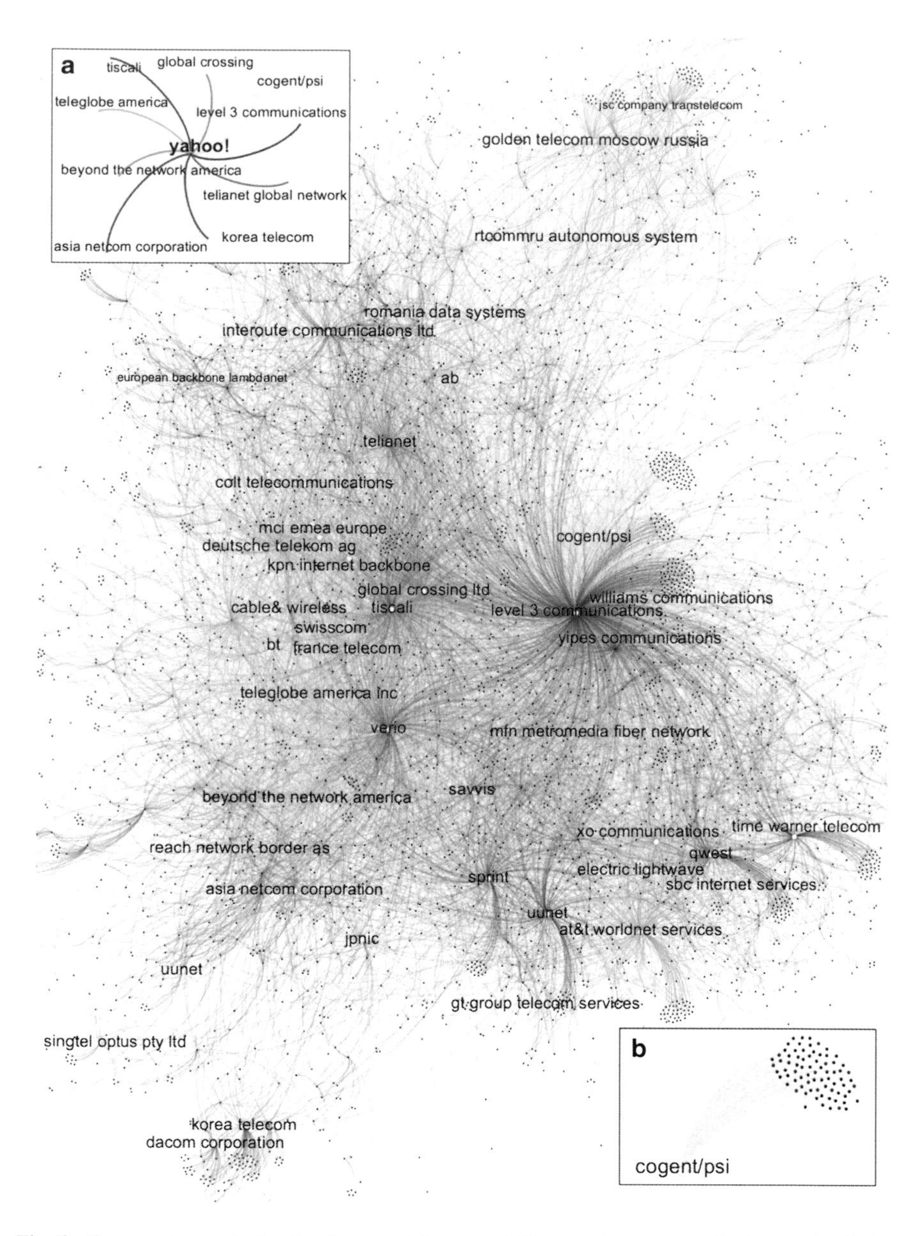

Fig. 3 *Community analysis of an Internet AS network.* Community structure in the graph of edges of an AS network with 3,951 nodes and 9,870 (unweighted) links. The links are grouped into different communities by optimising the continuous-time stability of the *line graph* of this network. For Markov time $t = 0.52$, we find 31 communities for the edges (indicated by different colours). The nodes of the graph are coloured according to their type: t1 nodes in white, t2 in black, edu in blue. All 43 large ISPs (t1 nodes) in our dataset are labelled explicitly. Insets: (**a**) A small ISP node (yahoo!) is connected to several large ISP neighbours (t1) through edges that belong to different communities. (**b**) Example of a stub AS connected to the t1 node "cogent/psi" where all edges belong to the same community

networks [6], another important class of networks which arise in wide spectrum of applications, especially when there is a flow between different entities in the network.

A whole set of new tools can be derived by considering different types of dynamics, including different types of random walks taking place on the network [6]. This feature of the stability is especially interesting if one has some pre-knowledge about the flow pattern or the dynamics taking place on the network under consideration. By adapting the dynamics to prior knowledge, one can potentially tailor the stability measure to the specific problem at hand and establish the most relevant community structure as it pertains to the dynamical functionality of the network. In particular, our measure has been adopted for the study of time-evolving networks [44], an area of current focus in network research.

Acknowledgements J.-C. D. acknowledges support from the grant "Actions de recherche concertées—Large Graphs and Networks" of the Communauté Française de Belgique, the EULER project (Grant No.258307) part of the Future Internet Research and Experimentation (FIRE) objective of the Seventh Framework Programme (FP7) and from the Belgian Network DYSCO (Dynamical Systems, Control, and Optimization) funded by the Interuniversity Attraction Poles Programme initiated by the Belgian State Science Policy Office. S.N.Y. and M.B. acknowledge funding from grant EP/I017267/1 from the EPSRC (Engineering and Physical Sciences Research Council) of the UK under the *Mathematics Underpinning the Digital Economy* programme and from the Office of Naval Research (ONR) of the USA.

References

[1] M.E.J. Newman, M. Girvan, Phys. Rev. E **69**(2), 026113 (2004)
[2] M.E.J. Newman, Proc. Natl. Acad. Sci. **103**(23), 8577 (2006). DOI 10.1073/pnas.0601602103
[3] H.A. Simon, Proc. Am. Phil. Soc. **106**(6), 467 (1962)
[4] S. Fortunato, Phys. Rep. **486**(3–5), 75 (2010). DOI 10.1016/j.physrep.2009.11.002
[5] J.C. Delvenne, S.N. Yaliraki, M. Barahona, Proc. Natl. Acad. Sci. **107**(29), 12755 (2010). DOI 10.1073/pnas.0903215107
[6] R. Lambiotte, J.C. Delvenne, M. Barahona, Laplacian Dynamics and Multiscale Modular Structure in Networks (2009). ArXiv:0812.1770
[7] R. Lambiotte, Multi-scale modularity and dynamics in complex networks, in *Dynamics on and of Complex Networks, vol. 2: Applications to Time-Varying Dynamical Systems*, ed. by A. Mukherjee, M. Choudhury, F. Peruani, N. Ganguly, B. Mitra (Springer, New York, 2013)
[8] F. Chung, *Spectral Graph Theory*. No. 92 in Regional Conference Series in Mathematics (American Mathematical Society, 1997)
[9] R. Lambiotte, R. Sinatra, J.C. Delvenne, T.S. Evans, M. Barahona, V. Latora, Phys. Rev. E **84**(1), 017102 (2011)
[10] E. Le Martelot, C. Hankin, Multi-scale Community Detection using Stability Optimisation, The International Journal of Web Based Communities (IJWBC) Special Issue on Community Structure in Complex Networks, **9**, to appear. (2013)
[11] R. Lambiotte, in *Proceedings of the 8th International Symposium on Modeling and Optimization in Mobile, Ad-Hoc and Wireless Networks (WiOpt 2010)*, University of Avignon, Avignon, 31 May–4 June 2010 (IEEE, 2010), pp. 546–553
[12] A. Delmotte, E.W. Tate, S.N. Yaliraki, M. Barahona, Phys. Biol. **8**(5), 055010 (2011)

[13] E. Simpson, Nature **163**(4148), 688 (1949)
[14] A. Hirschman, Am. Econ. Rev. **54**(5), 761 (1964)
[15] A. Renyi, in *Fourth Berkeley Symposium on Mathematical Statistics and Probability*, pp. 547–561, 1961
[16] C. Tsallis, J. Stat. Phys. **52**(1), 479 (1988)
[17] J. Shi, J. Malik, IEEE Trans. Pattern Anal. Mach. Intell. **22**(8), 888 (2000). DOI 10.1109/34.868688
[18] M. Fiedler, Czech. Math. J. **23**(2), 298 (1973)
[19] M. Fiedler, Czech. Math. J. **25**(4), 619 (1975)
[20] J. Reichardt, S. Bornholdt, Phys. Rev. Lett. **93**(21), 218701 (2004). DOI 10.1103/PhysRevLett.93.218701
[21] J. Reichardt, S. Bornholdt, Phys. Rev. E **74**(1), 016110 (2006). DOI 10.1103/PhysRevE.74.016110
[22] V.A. Traag, P. Van Dooren, Y. Nesterov, Phys. Rev. E **84**(1), 016114 (2011)
[23] P. Ronhovde, Z. Nussinov, Phys. Rev. E **80**, 016109 (2009). DOI 10.1103/PhysRevE.80.016109
[24] P. Ronhovde, Z. Nussinov, Phys. Rev. E **81**(4), 046114 (2010). DOI 10.1103/PhysRevE.81.046114
[25] R. Kannan, S. Vempala, A. Veta, in *Proceedings. 41st Annual Symposium on Foundations of Computer Science, 2000*, pp. 367–377, 2000
[26] U. Brandes, D. Delling, M. Gaertler, R. Gorke, M. Hoefer, Z. Nikoloski, D. Wagner, IEEE Trans. Knowl. Data Eng. **20**(2), 172 (2008)
[27] M.E.J. Newman, Phys. Rev. E **74**(3), 036104 (2006). DOI 10.1103/PhysRevE.74.036104
[28] V.D. Blondel, J.L. Guillaume, R. Lambiotte, E. Lefebvre, J. Stat. Mech. Theor Exp. **2008**(10), P10008 (2008)
[29] B. Kernighan, S. Lin, Bell Syst. Tech. J. **49**(2), 291 (1970)
[30] M.T. Schaub, J.C. Delvenne, S.N. Yaliraki, M. Barahona, PLoS ONE **7**(2), e32210 (2012). DOI 10.1371/journal.pone.0032210
[31] M. Meila, J. Multivariate Anal. **98**(5), 873 (2007). DOI 10.1016/j.jmva.2006.11.013
[32] D. Meunier, R. Lambiotte, E.T. Bullmore, Modular and hierarchically modular organization of brain networks. Front. Neurosci. 4:200. doi: 10.3389/fnins.2010.00200 (2010)
[33] M.E.J. Newman, Phys. Rev. E **74**(3) (2006)
[34] M. Rosvall, C.T. Bergstrom, Proc. Natl. Acad. Sci. **105**(4), 1118 (2008). DOI 10.1073/pnas.0706851105
[35] M.T. Schaub, R. Lambiotte, M. Barahona, Encoding dynamics for multiscale community detection: Markov time sweeping for the map equation. Phys. Rev. E. **86**, pp. 026112. American Physical Society. DOI 10.1103/PhysRevE.86.026112 (2012)
[36] A. Lancichinetti, S. Fortunato, F. Radicchi, Phys. Rev. E **78**(4), 046110 (2008). DOI 10.1103/PhysRevE.78.046110
[37] A. Lancichinetti, S. Fortunato, Phys. Rev. E **80**(1), 016118 (2009). DOI 10.1103/PhysRevE.80.016118
[38] L. Danon, A. Díaz-Guilera, A. Arenas, J. Stat. Mech. Theor Exp. **2006**(11), P11010 (2006)
[39] B. Karrer, M.E.J. Newman, Phys. Rev. E **83**(1), 016107 (2011). DOI 10.1103/PhysRevE.83.016107
[40] M. Rosas-Casals, S. Valverde, R.V. Solé, Int. J. Bifurcat. Chaos **17**(7), 2465 (2007)
[41] R.V. Solé, M. Rosas-Casals, B. Corominas-Murtra, S. Valverde, Phys. Rev. E **77**(2), 026102 (2008). DOI 10.1103/PhysRevE.77.026102
[42] Y. Hyun, B. Huffaker, D. Andersen, E. Aben, M. Luckie, K. Claffy, C. Shannon, The IPv4 Routed /24 AS Links Dataset - Data from 3.1.2012. URL http://www.caida.org/data/active/ipv4_routed_topology_aslinks_dataset.xml
[43] T.S. Evans, R. Lambiotte, Phys. Rev. E **80**(1), 016105 (2009)
[44] P.J. Mucha, T. Richardson, K. Macon, M.A. Porter, J.P. Onnela, Science **328**(5980), 876 (2010)

Algorithms for Finding Motifs in Large Labeled Networks

Maleq Khan, V.S. Anil Kumar, Madhav V. Marathe, and Zhao Zhao

1 Introduction

Large labeled networks/graphs form a common abstraction for capturing abstract relationships between entities in diverse applications, ranging from public health, sociology and economics, online networks formed by Facebook and Twitter, and infrastructure systems, including the transportation network, power grid, and the Internet. Formally, such a graph, denoted by $G = (V, E)$, is induced by a set V of entities or agents (which could be people, computers, etc., depending on the application) with edges $e = (u, v) \in E$ if agents u, v are related (e.g., individuals u, v visit the same location, in the case of a social contact network). Labeled graphs provide a powerful abstraction for understanding complex, nonlocal, and hidden relationships between different entities.

One approach for capturing such relationships is through motifs or commonly occurring subgraphs. Informally, we say that a subgraph H of G is isomorphic to a pattern or template T, if H is a "copy" of T, possibly by renumbering the nodes (a formal definition is given in Sect. 2). The general problems of subgraph analysis involve determining if and how many copies of template T occur in G. Such problems are commonly used in a number of applications, such as social network analysis [17], data mining [10, 34, 55, 56], fraud detection [7], chemical informatics [9], Web information management [46], and bioinformatics [23, 40], where analysts are interested in finding subsets of nodes with specific labels or attributes and mutual relationships that match a specific template.

Different variants of subgraph analysis problems arise in specific examples, making it a very rich research area that overlaps with a number of application

M. Khan (✉) • V.S. Anil Kumar • M.V. Marathe • Z. Zhao
Virginia Bioinformatics Institute, Virginia Tech, Blacksburg, VA 24060, USA
e-mail: maleq@vbi.vt.edu; akumar@vbi.vt.edu; mmarathe@vbi.vt.edu; zhaozhao@vbi.vt.edu

A. Mukherjee et al. (eds.), *Dynamics On and Of Complex Networks, Volume 2*, Modeling and Simulation in Science, Engineering and Technology, DOI 10.1007/978-1-4614-6729-8_12,

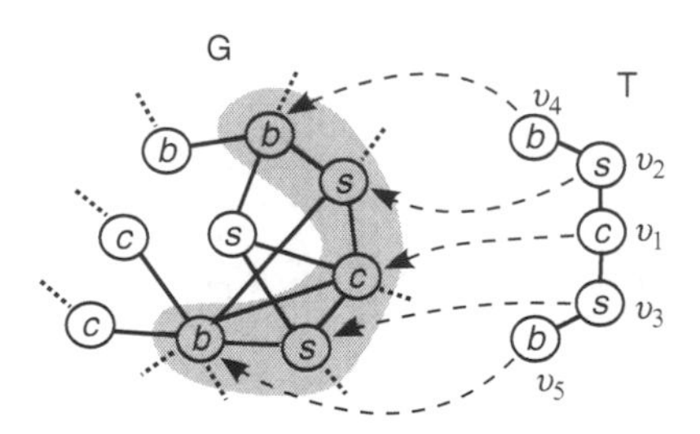

Fig. 1 An example application of detecting subgraphs matching a given template in financial networks (as in [7, 10]): G is an underlying graph whose nodes represent legitimate customers (c), suspicious entities, (s) or banks (b). The template T is an abnormal transaction pattern, in which a customer v_1 uses suspicious intermediaries v_2 and v_3 to reach banks v_4 and v_5. The shaded subgraph shows a matching of the abnormal pattern

areas (see, e.g., [17]), including frequent subgraph counting [3, 25] and graphlet distribution computing [44] in bioinformatic networks, relational subgraph querying in semantic Webs [7, 10] and RDF stores [24, 47], graph grammar analysis to memory diagnose [39], etc. We discuss some of the representative problems below.

- Counting the number of subgraphs that are isomorphic to a given template is fundamental to many applications. For instance, Milo et al. find in [40] that subgraphs with abnormal high frequencies or motifs form the building blocks of many real-world networks from biochemistry, neurobiology, ecology, and engineering. Motif is mostly used in categorizing biological networks [3, 32]. A variation to subgraph counting is the graphlet frequency distribution computing, which does not compute the overall counts of a subgraph but the number of subgraphs "touched" on each node [44, 45].
- Other than subgraph counting, subgraph enumeration/querying is also very popular in many areas. The subgraph queries can be used for detecting anomalous patterns in applications such as financial networks [7, 10], where nodes are banks and individuals and edges represent financial transactions. An investigator might be interested in specific transaction patterns from an individual to banks, e.g., through suspicious intermediaries to deflect attention, which can be formulated as specific labeled subgraphs (see Fig. 1). Subgraph enumeration and analysis is used as a subroutine in different systems, networking, and software engineering applications. For instance, enumeration of subgraphs with specific constraints is used for automatic custom instruction set generation [8]. They are also used in localizing software bugs and memory leaks [15, 39].

These problems are computationally very challenging, and there has been a lot of interesting theoretical and practical research in sequential and parallel algorithms. Efficient parallel algorithms are the only choice for scaling to very large graphs. The goal of this chapter is to introduce the different kinds of subgraph analysis problems and discuss some of the important parallel algorithmic techniques that have been developed for them. Because of limited space, we only briefly discuss the broad literature in this area and focus primarily on the problem of counting the number of occurrences of a given subgraph. We consider some special classes of subgraphs including trees, triangles, and cliques.

The rest of the chapter is organized as follows. Section 2 describes the definitions and notations used in this chapter. In Sect. 3, we discuss the current state-of-the-art research and results on the motif counting problem, the challenges in devising parallel algorithms for this problem, and some general techniques to cope with these challenges. Then we discuss a technique called color coding used in approximation algorithms for counting motifs and two parallel algorithms using this color coding technique in Sect. 4. Several exact and approximation parallel algorithms for counting triangles in a network using MapReduce framework are given in Sect. 5. Section 6 presents a sequential algorithm to count the maximal cliques and its parallelization. We conclude in Sect. 7 with a discussion on open problems.

2 Preliminaries and Notation

We consider labeled graphs $G = (V_G, E_G, L, \ell_G)$, where V_G and E_G are the sets of nodes and edges, L is a set of labels, and $\ell_G : V \to L$ is a labeling on the nodes. A graph $H = (V_H, E_H, L, \ell_H)$ is a *non-induced subgraph* of G if we have $V_H \subseteq V_G$ and $E_H \subseteq E_G$. We say that a template graph $T = (V_T, E_T, L, \ell_T)$ is isomorphic to a non-induced subgraph $H = (V_H, E_H, L, \ell_H)$ of G if there exists a bijection $f : V_T \to V_H$ such that (i) for each $(u, v) \in E_T$, we have $(f(u), f(v)) \in E_H$ and (ii) for each $v \in V_T$, we have $\ell_T(v) = \ell_H(f(v))$. We also call H a non-induced embedding of T, e.g., Fig. 1 shows a non-induced embedding of template T. If condition (i) is changed so that we have $(u, v) \in E_T$ *if and only if* $(f(u), f(v)) \in E_H$, we say that this is an induced embedding. Let $emb(T, G)$ denote the number of non-induced occurrences of template T in the graph G. In most of this chapter, we focus on the problem of estimating $emb(T, G)$ for given T and G.

Most often the terms "subgraph counting" and "subgraph enumeration" are used interchangeably. In an absolute sense, *subgraph counting* refers to the problem of finding the number of occurrences of T in a graph G, whereas *subgraph enumeration* refers to the problem of listing all subgraphs matching T. However, in most cases, these two problems are almost equivalent: an algorithm for enumerating subgraphs can easily be used for counting them, and on the other hand, the algorithms for counting subgraphs also identify the occurrences of the subgraphs in the process of counting them, and thus such algorithms can be used for enumerating subgraphs; however, note that any approximation algorithms for counting subgraphs identifies only a subset of their occurrences. As a result, these two terms are typically used interchangeably.

Further, the above subgraph counting problem has been studied in many different context such as finding network motifs, frequent subgraphs and graphlet distribution. *Network motifs* are subgraph patterns occurring in complex networks at numbers that are significantly higher than those in random networks. Such motifs can be found in networks from biochemistry, neurobiology, ecology, and engineering [40]. Finding *frequent subgraphs* in a given collection of graphs is an important problem in data mining area[14, 31]. In this setting, we are given a collection $\mathcal{G}$ of graphs and a parameter σ with $0 \leq \sigma \leq 1$. A subgraph is called frequent if it appears in at least $\sigma|\mathcal{G}|$ of the graphs in $\mathcal{G}$. *Graphlet distribution* is based on subgraph counts, that generalizes the degree distribution. A *graphlet* on node v is a subgraph containing $v \in V_G$ of G and isomorphic to the given template T. The *graphlet distribution* (relative to T) is the distribution of the number of graphlets on each node v [44, 45]. Note that when T is a single node, graphlet distribution is exactly the degree distribution. An underlying fundamental operation in all of these problems is to count the number of occurrences (embeddings) of a given subgraph. In this chapter, we discuss various algorithms to find the number of embeddings of given template T in a single graph G.

3 State of the Art

Many variants of subgraph isomorphism problems have been studied in different applications, both from a theoretical and practical perspective. We discuss some of the main results here and focus primarily on challenges and techniques for parallel algorithms that have been used for these problems.

3.1 Algorithmic Complexity and Heuristics

In general, subgraph isomorphism and enumeration problems are computationally very challenging problems. Aravind et al. [6] show that counting cliques and independent sets of size k in a given graph G are $W[1]$-hard; informally this means that these problems are unexpected to be solvable in much better than $O(n^k)$ time. Given an arbitrary template of size k, the best known rigorous result for the subgraph isomorphism problem is obtained by Eisenbrand et al. [16] with a running time of roughly $O(n^{\omega k/3})$ (which improves on the naive $O(n^k)$ time), where ω denotes the exponent of the best possible matrix multiplication algorithm.Vassilevska et al. [54] give an algorithm with an improved running time of $O(2^s n^{k-s+3} k^{O(1)})$, where the template has an independent set of size s; this algorithm is improved slightly by Kowaluk et al. [30]. Most of these algorithms are of theoretical interest and are quite challenging to use in practice.

The above results are for exact count of subgraphs, and because of their hardness, there has also been a lot of work on approximation algorithms. Alon et al. [5] developed the color coding technique for approximating the number of paths, trees, and tree-width-bounded subgraphs of size $O(\log n)$ in polynomial time. This was improved by Alon et al. [3] to an algorithm with running time $O(k|E|2^k e^k \log(1/\delta)\frac{1}{\varepsilon^2})$, where ε and δ are error and confidence parameters, respectively. Huffner et al. [25] optimized the parameters of the color coding technique, including the choice of the number of colors, and efficient data structures to minimize the running time. Aravind et al. [6] gave results for other special classes of subgraphs, including cliques.

A lot of practical heuristics have also been developed for various versions of these problems, especially for the frequent subgraph mining problem. An example is the "Apriori" method, which uses a level-wise exploration of the template [26, 32], for generating candidates for subgraphs at each level; later these methods are made to run faster by better pruning and exploration techniques, e.g., [23, 32, 55]. Other approaches in relational databases and data mining involve queries for specific labeled subgraphs and have combined relational database techniques with careful depth-first exploration, e.g., [10, 48, 49]. Kuramochi et al. [32] develop techniques for frequent subgraph discovery in ensembles of graphs and apply them to discover new patterns in chemical and biological networks.

As discussed in Sect. 2, other variants of subgraph enumeration problems have been studied extensively. Pržulj et al. [44, 45] develop techniques for computing graphlet distributions in biological networks and use it to characterize the networks.

There has been a lot of work on developing "graph query languages" for specifying various classes of relational queries in complex labeled graphs; examples include GraphLog [12], QGraph [27], GraphDB [21], and a variety of RDF languages [41]. These frameworks are able to use formal language constructs to specify subgraphs. However, the algorithmic techniques for supporting them are generally quite limited and do not scale to large graphs.

Most of these approaches are sequential and generally only scale to fairly small graphs and templates; parallel algorithms are unavoidable for scaling to much larger networks and templates. Several parallel techniques have been developed for variants of subgraph enumeration problems, including [10,42,52,57,58]. We discuss the main challenges in parallelization and some of these techniques in more detail below.

3.2 *Parallel Algorithms: Challenges and Main Techniques*

In general, traditional parallel computing techniques (e.g., those developed for physical problems, such as n-body simulations and molecular dynamics) are not well suited for the motif finding problem in large-scale social networks because of several reasons [36], including the following: (i) a large network may not fit into the memory of a single processor, and thus, it requires partitioning the network; however, social networks are heterogeneous and unstructured, which makes partitioning the network into "independent" partitions difficult [22]; (ii) a subgraph embedding matching the given template may lie over multiple partitions of the network leading to a complex coordination and communication among the processors; (iii) computations on social networks tend to have very low locality because of the irregular structure; and (iv) matching subgraphs involve exploration of the graph structure, which are highly memory-intensive, and there is very little computation to hide the latency to memory accesses.

Two broad approaches for parallelization are distributed-memory architectures with MPI (message-passing interface) and shared-memory architecture OpenMP. In an MPI architecture, multiple processors are interconnected and communicate with each other via message passing. Each processor has its own main memory (distributed memory), and one processor cannot access the main memory of other processors. The computation task is divided among the processors, and the processors coordinate computation by exchanging messages with each other. To facilitate efficient communications among the processors, MPI architecture provides various communication primitives such as broadcast, aggregation (MPI-reduce), etc., in addition to simple *send* and *receive* functions that can be used for communications between any two processors. In Sect. 4.1, we discuss a parallel algorithm using MPI architecture for counting subgraphs and in Sect. 6.2 a parallel algorithm for enumerating maximal cliques. In OpenMP architecture, the processors share the same common main memory. Data are stored in the common memory, and all processors can perform read/write operation on the same data with the help of some locking mechanism. Some parallel algorithms using OpenMP shared memory architecture can be found in [20, 29, 37, 38]. Shared-memory architecture is not suitable for large graphs that do not fit in the main memory of a single computing machine, and the number of processors can also be very limited compared to a distributed memory system. As a result distributed memory parallelization became recent trend to deal with large volume of data. In this chapter, we do not discuss any shared memory algorithm. Interested readers are referred to [20,29,37,38] and references therein for such algorithms.

Another recent distributed memory framework (beside MPI) is Apache's Hadoop [1], which is based on MapReduce [13] principle and provide a generic programming framework for parallel algorithms. Informally, this architecture involves two basic operations: Map and Reduce. A Map function processes data in the form of key/value pairs (k_1, v_1). On completion, the system groups the intermediate pairs by each key k_1 and collects all values associated with it. All values associated with a specific key are then provided as input to a Reduce function. This Map-Reduce sequence can be repeated multiple times. Hadoop-based techniques have been developed for a number of graph problems, including connectivity and diameter [28].

The sequential approaches for subgraph counting problems discussed in Sect. 3.1 are notoriously hard to parallelize as it is very difficult to decompose the task into independent subtasks. It is not clear if candidate generation approaches [23, 32, 55] can be parallelized and scaled to large graphs and computing clusters. Some recently proposed parallel algorithms for finding subgraph embeddings are [10, 57, 58]. For example, the approach of Bröcheler et al. [10] includes an intensive preprocessing step and requires prior knowledge on the probability distribution over input queries. The algorithms presented in [57] and [58] use a color coding approach, which is discussed in detail in Sect. 4.

Two very recent papers [42, 52] develop MapReduce-based algorithms for counting number of triangles in a network. The detail of these two algorithms are given in Sect. 5.

4 Approximate Counting of Trees Using Color Coding

Color coding is a randomized approximation algorithm proposed by Alon et al. [4], to determine $emb(T, G)$, the number of embeddings of template T in G (See Sect. 2 for the definition). First we focus on templates T which are trees or "tree-like" and use the following definition for approximation algorithms.

Definition 1. We say that a randomized algorithm $\mathcal{A}$ produces an (ε, δ) -approximation to $emb(T, G)$, if the estimate Z produced by $\mathcal{A}$ satisfies: $\Pr[|Z - emb(T, G)| > \varepsilon \cdot emb(T, G)] \leq 2\delta$; in other words, $\mathcal{A}$ is required to produce an estimate Z whose difference to the precise value is a very small proportion of it (i.e., $\varepsilon \cdot emb(T, G)$), with high probability.

A potential "divide-and-conquer" approach for counting the number of embeddings of template T in graph G (see, e.g., Fig. 2) might be to partition T into sub-templates T_1 and T_2 (as in Fig. 3), get counts for these sub-templates, and then combine the results. The main issue in this approach is in the last step of combining the number of embeddings of T_1 and T_2. For instance, in Fig. 3, the mapping of nodes a and b to 6 and 2, respectively, is a valid embedding of T_1. Similarly, the mapping of nodes c, d, and e to $3, 6$, and 5, respectively, is a valid embedding of T_2. However, these two embeddings of T_1 and T_2 cannot be combined to get a valid embedding of T because of their overlap on node 6. Such overlaps could be handled by keeping track of additional information but can potentially lead to high overheads. The color-coding algorithm addresses this by a very elegant idea: if we consider "colorful" embeddings (an embedding is colorful if each vertex has distinct color as shown in Fig. 2) in a graph G whose nodes are colored, the combination step

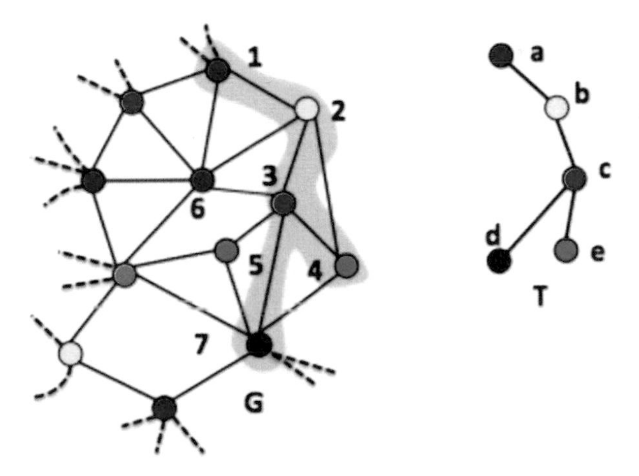

Fig. 2 Example showing the notion of "colorful" embeddings of template T in graph G. The nodes of G are colored with {red, blue, yellow, green, black}. The mapping of nodes a, b, c, d, e to $1, 2, 3, 7, 4$, respectively, is a colorful embedding (highlighted in the figure). The mapping of a, b, c, d, e to $1, 2, 3, 4, 5$, respectively, is not a colorful embedding

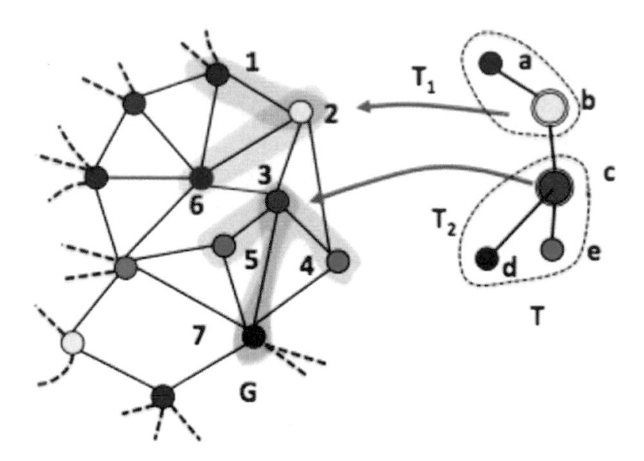

Fig. 3 Illustration of the color coding algorithm: template T is partitioned into T_1 and T_2, with the roots shown with larger circles and a double line. We have $C(2, T_1, S_1 = \{\text{yellow, blue}\}) = 2$ and $C(3, T_2, S_2 = \{\text{red, green, black}\}) = 2$. And the number of colorful embeddings for T with the yellow node mapped to node 2 in G is 4

can be done easily by a dynamic programming procedure, which we describe using the example in Fig. 3. We assume the trees T_1, T_2, and $T_0 = T$ are rooted at nodes b, c, and b, respectively. Let $\rho(T_i)$ denote the root of tree T_i. Let S denote the set of the five colors in Fig. 3. For $S' \subset S$, $|S'| = |T_i|$, let $C(v, T_i, S')$ denote the number of colorful embeddings of tree T_i in G using colors in the set S', with $\rho(T_i)$ mapped to v. For instance, $C(2, T_1, S' = \{\text{blue, yellow}\}) = 2$ because the mappings of nodes b, a to $2, 1$ and $2, 6$, respectively, are valid embeddings. Similarly, $C(3, T_2, S'' = \{\text{red, green, black}\}) = 2$. Finally, observe that $C(2, T_0, S)$, which is the number of colorful embeddings of T_0 rooted at 2, must involve embeddings of T_1 rooted at 2, using a set S_1 of colors, and embeddings of T_2 rooted at a neighbor of 2, using a set S_2 of colors; the embeddings must be enumerated over all possible S_1, S_2 such that $|S_1| = 2$, $|S_2| = 3$, and $S_1 \cap S_2 = \phi$. In this example, the only possible choices of S_1 and S_2 which give nonzero counts are S' and S'', respectively. It follows that $C(2, T_0, S) = 4$. This suggests that counting the number of colorful embeddings could be done by dynamic programming. Further, if the coloring is done randomly, the number of colorful embeddings is proportional to $emb(T, G)$. We discuss the details a bit more formally below; see [4] for additional details. The algorithm involves the following steps:

1. For $i = 1$ to $N = O(\frac{e^k \log 1/\delta}{\varepsilon^2})$ perform the following steps, where $k = |V_T|$ is the number of nodes in the template, and δ and ε are input parameters such that the approximation factor is $1 \pm \varepsilon$ with probability at least $1 - 2\delta$:
   a) (Color coding) Color each node $v \in V(G)$ uniformly at random with a color from $\{1, \ldots, k\}$.
   b) (Counting) Use dynamic programming to count all "colorful" embeddings of T in G, where an embedding H in G isomorphic to T is said to be colorful

(see Fig. 2 and the discussion below), if all nodes in H have distinct colors. Let X_i be the number of embeddings in iteration i.

2. Output an estimate of the actual number of embeddings using the X_is in the following manner: partition the N samples above into $t = O(\log 1/\delta)$ sets, and let Y_j be the average of the jth set. Output the median Z of $Y_1, \ldots, Y_t$.

Alon et al. [4] show that the above (randomized) algorithm gives an estimate of the actual number of embeddings of T in G within a multiplicative factor of $(1 \pm \varepsilon)$, with probability at least $1 - 2\delta$, where ε and δ are arbitrary input parameters. As discussed above, the main insights of the algorithm are as follows: (i) the probability that any given embedding H of T with k nodes is colorful is precisely $\frac{k!}{k^k}$, so that the expected number of colorful embeddings is $emb(T, G) \cdot \frac{k!}{k^k}$ and (ii) for colorful embeddings, there is a simple recurrence relationship, which leads to a dynamic program. As defined above, let $\rho(T)$ denote an arbitrary root of T. Let $T_1, \cdots, T_\ell$ denote the subtrees of $\rho(T)$. Then, following the above discussion of the example in Fig. 3,

$$C(v, T, S) = \sum_{S_1,\ldots,S_\ell, w_1,\ldots,w_\ell \in N(v)} \Pi_{i=1}^{\ell} C(w_i, T_i, S_i), \tag{1}$$

where the summation is over all possible partitions of $S = S_1 \cup \ldots \cup S_\ell$ and over all possible ℓ-tuples of neighbors $w_1, \ldots, w_\ell$ of v. This implies that the number of colorful embeddings can be computed bottom-up. Alon et al. [4] show that this approach leads to a dynamic programming algorithm with a running time of $O(|E| \cdot 2^k \cdot e^k \log 1/\delta/\epsilon^2)$. Huffner et al. [25] show that this algorithm can be speeded up by careful data structures and using potentially more than k colors. Further, [18, 19] extend this approach to count other kinds of subgraphs besides trees, including cycles.

4.1 Color Coding in Parallel

The main issue with the color-coding algorithm is that the memory requirement for the dynamic programming table is $O(nk2^k)$, making it infeasible even for moderate-size graphs to run on a single machine. We now discuss techniques for parallelizing this algorithm, in order to distribute the memory consumption over multiple machines or even use distributed file system to store the intermediate results as in SAHAD. Our general approach is to partition the graph into small parts, compute the quantities $C(v, T, S)$ bottom-up, and run the dynamic programming step in parallel, to compute the recurrence in each stage. The main challenge is to maintain and exchange these quantities across processors, especially at higher levels of the dynamic program. We discuss two different approaches for parallelization.

ParSE: An MPI-based Parallel Algorithm. This approach focuses only on templates T that can be partitioned into sub-templates T_1 and T_2 by removing a cut-edge. In [57], we develop PARSE, an MPI-based parallel algorithm for estimating $emb(T, G)$ for such templates T. The basic idea is that if T_1, T_2 have low diamater (e.g., if they are star graphs or cliques), then $C(v, T_1, S_1)$ and $C(v, T_2, S_2)$ can be determined exactly by a backtracking-based approach quite efficiently. We only solve one recurrence for each edge (u, v) of the form (1). This is done in two steps: (i) We first partition the set V into parts V_p corresponding to each processor p and store the graph G_p consisting of all the edges incident on V_p on processor p. (ii) For each p and each $v \in V_p$, we compute $C(v, T_1, S_1)$ and $C(v, T_2, S_2)$ for possible color sets S_1, S_2 by a local backtracking-based algorithm. (iii) These counts are sent to a designated processor called the master node, which aggregates them (by computing the above recurrence).

The overall running time for this approach has two components: the time to compute the counts for templates T_1 and T_2 locally within each G_p and the time for exchanging the counts and running the aggregation step. The former can be done in time $O(\max_p |V_p|\Delta^{k'})$, where Δ is the maximum node degree in G, and $k' = \max\{|T_1|, |T_2|\}$, and the latter can be done in time $O(mk^{k'})$, giving a time complexity of roughly $O(\frac{n}{Q}\Delta^{k'} + mk^{k'})$ for each iteration (recall that this has to be repeated multiple times, depending on ϵ and δ, as in the sequential algorithm), where Q is the number of processors.

As a result, this approach is able to handle more complex templates than trees, and the template can have more than constant treewidth. However, the limitation is that only one level of the dynamic program is implemented here, and therefore, this approach works if the counts for T_1, T_2 can be determined exactly. We extend this to templates which are trees but with arbitrary diameter by a different approach below.

SAHad: A Hadoop-based Parallel Algorithm. Extending the approach in PARSE to implement all the steps of the dynamic program leads to a very high communication overhead because the recurrence steps are computed only by the master node. Further, the amount of data in each step also increases significantly, making it infeasible for the master node to store all the intermediate colored counts. In [58], we develop, SAHad, an approach based on Hadoop for parallelizing the color-coding step.

We focus on templates T which are trees. To handle trees with higher depth, we first partition T into a sequence of sub-templates $\{T_0, T_1, \ldots, T_r\}$ recursively. As before, we let $\rho(T')$ denote the root of the sub-template T'. Each T_i (which is not a single node) is partitioned into T_i' and T_i'', which are referred to as the children of T_i. Further, we have $\rho(T_i) = \rho(T_i')$, i.e., the root of T_i is the same as one of its children. The basic idea is to solve the above recurrences by transforming it to an appropriate map-reduce operation. We map the count $C(v, T_i, S_i)$ with a key value of u, for each neighbor u of v. As a result, the reduce operation is able to bring together all the quantities of the form $C(v, T_i', S')$ and $C(u, T_i'', S'')$, which allows the recurrence to be solved. The implementation involves a sequence of $r \leq k$ map-reduce phases. Each phase involves running one step of the dynamic program, which involves a total work complexity of $O(|E|2^{2k})$. Therefore, the total work complexity for a single run of this algorithm is $O(k|E|2^{2k})$, and this has to be repeated $e^k \log 1/\delta \frac{1}{\epsilon^2}$ times, as in the sequential algorithm. Thus, the total work complexity of this algorithm is close to the sequential algorithm.

5 Counting Triangles Using MapReduce Framework

A triangle is a special subgraph with three nodes and edges between every pair of nodes, i.e., there are three edges in a triangle. Due to its symmetrical nature and smaller size, special algorithms can be designed that are comparatively faster than any general subgraph count algorithm. Recently, two parallel algorithms [42, 52] have been designed to count triangles in a given network using the MapReduce framework. The algorithm in [42] is a randomized algorithm to find approximate count, whereas [52] presents an algorithm to find the exact number of the triangles.

5.1 An Exact Algorithm Using MapReduce

One brute-force approach to count triangles in a network $G(V, E)$ is to check for all possible triples (u, v, w) with $u, v, w \in V$, whether (u, v, w) forms a triangle, i.e., $(u, v), (v, w), (u, w) \in E$. There are $\binom{n}{3}$ such triples, and thus the algorithm runs in $\Omega(n^3)$ time. A simple and efficient algorithm to count the triangles in a network can be as follows: for each node $v \in V$, find the number of edges among its neighbors, which is the number of triangles containing node v. Let $N(v)$ be the set of the neighbors of v and $T(v)$ be the number of triangles containing node v, as counted above. Then the total number of triangles in the network is $T = \sum_v T(v)/3$. In this method, each triangle is counted three times, once for each node in the triangle. Using an ordering $\prec$ of the nodes, it is possible to improve the efficiency of this algorithm by counting each triangle exactly once as shown below [50]:

```
T ← 0
For each v ∈ V do
  For each u ∈ N(v) and u ≺ v do
    For each w ∈ N(v) and w ≺ u do
      If (u, w) ∈ E then
        T ← T + 1
```

Let $\hat{d}_v$ be the number of neighbors u of v such that $v \prec u$. Assuming, the neighbor lists $N(v)$, for all v, are sorted and a binary search is used to check $(u, w) \in E$, the running time of this algorithm can be shown to be $O(\sum_v \hat{d}_v d_v + \hat{d}_v^2 \log d_{\max})$, where $d_{\max} = \max_v d_v$. Although any ordering of the nodes makes sure that each triangle is counted exactly once, the performance of this algorithm can be improved further by using an ordering based on the degrees of the nodes, with ties broken by node IDs. This degree-based ordering of the nodes can improve running time significantly, especially for a graph with skewed degree distribution. It is easy to see that this running time is minimized when $\hat{d}_v$ values of the nodes are as close to each other as possible, although, for any ordering of the nodes, $\sum_v \hat{d}_v = m$ is invariant. On the other hand, the more $\hat{d}_v$ values of nodes are diverse from each other, the larger the running time becomes. Now notice that in the degree-based ordering, $\hat{d}_v$ is significantly smaller than degree d_v for a node v with larger degree in contrast to that of a node with smaller degree, and thus, skewness and diversity of the $\hat{d}_v$ values are reduced significantly leading to improved running time.

An elegant parallel implementation of the above algorithm can be done using MapReduce framework as shown in [52]. The implementation has two pairs of map and reduce functions. The first pair of map and reduce imposes an ordering on the nodes, to make sure each triangle is counted only once, and generates 2-paths centered on each node. The second pair of map and reduce functions check whether each of such 2-paths forms a triangle with an edge in the network. The map and reduce functions are as follow:

Map 1: Input: $\langle (u,v); \emptyset \rangle$
  If $u \prec v$, then emit $\langle u; v \rangle$

Reduce 1: Input $\langle v; S \rangle$
  For all $u, w \in S$, emit $\langle v; (u,w) \rangle$

Map 2: Input: $\langle v; (u,w) \rangle$ or $\langle (u,v); \emptyset \rangle$
  For input of type $\langle v; (u,w) \rangle$, emit $\langle (u,w); v \rangle$
  For input of type $\langle (u,v); \emptyset \rangle$, emit $\langle (u,v); \$ \rangle$

Reduce 2: Input $\langle u, w; S \rangle$
  If $\$ \in S$ then
    For all $v \in S \cap V$, emit $\langle v; 1 \rangle$

5.2 Approximation Algorithm by Sparsification of the Network

Sampling is a standard technique widely used in designing approximation algorithms. The purpose of using an approximation algorithm is to use less computing resources, such as time and memory space, by sacrificing the quality of the solution. In many cases, such approximation algorithms provide very close approximation to the exact solution while providing significant savings in computing resources. Sparsification of a network is a sampling technique where some randomly chosen edges are retained and the rest are deleted and then computation is performed in the network obtained after sparsification of the original network. Sparsification of a network can save both computation time and memory requirement and provide a way to deal with massive networks that do not fit in the main memory.

Sparsification of networks can be effectively used in counting triangles as shown in [53]. Let $G(V, E)$ and $G'(V, E' \subset E)$ be the networks before and after sparsification, respectively. Network $G'(V, E')$ is obtained from $G(V, E)$ by retaining each edge, independently, with probability p and removing with probability $1 - p$. Now any algorithm can be used to find the exact number of triangles in G'. Let $T(G')$ be the number of triangles in G'. The estimated number of triangles in G is given by $\frac{1}{p^3}T(G')$, which is an unbiased estimation. It is easy to see that the expected value of $\frac{1}{p^3}T(G')$ is T(G), the number of triangles in the original network G: let the triangles in G be arbitrarily numbered as $1, 2, \ldots, T(G)$ and x_i be an indicator random variable that takes value 1 if Triangle i of G survives in G'. A triangle survives if all of its three edges are retained in G'. Then we have $\Pr\{x_i = 1\} = \frac{1}{p^3}$ and, by the linearity of expectation,

$$E\left[\frac{1}{p^3}T(G')\right] = \frac{1}{p^3}\sum_{i=1}^{T(G)} E[x_i] = \frac{1}{p^3}\sum_{i=1}^{T(G)} \Pr\{x_i = 1\} = T(G).$$

The variance of the estimated number of triangles is $\left(\frac{1}{p^3}-1\right)T(G)+2k\left(\frac{1}{p}-1\right)$, where k is the number of pairs of triangles in G with an overlapping edge [53].

This sparsification method can easily be integrated with the MapReduce-based parallel algorithm given in Sect. 5.1 by modifying `Map 1` function as follows:

```
Map 1: Input: ⟨(u, v); ∅⟩
  If u ≺ v, then emit ⟨u; v⟩ with probability p
```

5.3 *Sparsification Using Color Coding*

Another sparsification technique is given in [42] where the accuracy of the estimated number of triangles is improved with the same sparsification rate p as in the algorithm in Sect. 5.2. The main idea is that if two edges of a triangle are retained, the third edge is always retained. However, a restriction is imposed on the choice of p: p must be chosen such that $\frac{1}{p}$ is an integer. For each node v in the network, a color $C(v)$ is chosen uniformly at random from $\frac{1}{p}$ available colors. An edge (u, v) in G is retained in G' if and only if $C(u) = C(v)$. Then the probability that an edge is retained is p, and the probability that a triangle in G survives in G' is $\frac{1}{p^2}$ in contrast to $\frac{1}{p^3}$ for the algorithm in Sect. 5.2. The estimated number of triangles is given by $\frac{1}{p^2}T(G')$. This estimation is also unbiased as we have $E\left[\frac{1}{p^2}T(G')\right] = T(G)$. As the probability of survival of a triangle is larger in this color-coding-based sparsification, variance of the estimation is smaller. However, the additional step of coloring the nodes and its use in sparsification of the edges can require an additional $O(n)$ memory space and increase the complexity of the parallelization of this algorithm.

6 Algorithms to Find Maximal Cliques

A clique is another special type of subgraph where there is an edge between every pair of nodes in the subgraph. In a social network such a clique is a subset of individuals who are more closely and intensely tied to one another than they are to other members of the network. It is quite common that individuals form cliques based on their demographic characteristics such as age, gender, profession, and ethnicity. Thus, a clique in a social network shows a closely knit community or a part of such a community, and the enumerated maximal cliques can be used in detection of social hierarchies. Among other applications of maximal clique enumeration are genome mapping and gene expression analysis.

The problem of finding and counting cliques in a given network can be formulated in the following two forms:

- Find all cliques of a given size. The size of a clique is the number of nodes in the clique.
- Find all maximal cliques. A clique is maximal if it is not contained in another larger clique; that is, a maximal clique cannot be extended by adding another node to the clique.

Notice that a clique is a generalization of a triangle: a clique with three nodes is a triangle. While most of the algorithms for finding triangles can easily be generalized to find all cliques of a fixed size, such generalizations do not work in finding the maximal cliques. Obviously, in a brute-force approach, one can enumerate cliques of various sizes and for each such clique check whether it is contained in another clique. However, it does not provide a scalable solution for a large social network. Like subgraph enumeration, enumerating maximum cliques is an NP-hard problem [33]. Although the problem is NP-hard, running time can be improved significantly over the brute-force approaches by carefully designed algorithms [2, 11, 35, 43] that exploit various properties of maximal cliques. For this clique enumeration problem, backtracking search-based algorithms [2,11,35] seem to be more efficient. A backtracking search for maximal clique provides the opportunity of avoiding unpromising search paths. An efficient backtracking-search-based algorithm, called *Bron–Kerbosch algorithm or BK algorithm*[11], and a parallel implementation of this algorithm are given in [51]. For its simplicity and efficiency, the BK algorithm stands apart among the sequential algorithms for enumerating all maximal cliques in a network. Next we discuss the BK algorithm and its parallelization.

6.1 Bron–Kerbosch Algorithm

Bron–Kerbosch (BK) algorithm is a widely used sequential algorithm for enumerating maximal cliques. Efficiency and usefulness of the BK Algorithm come from its following characteristics:

- This backtracking algorithm uses branch-and-bound techniques to cut off all unpromising search paths. It explores only search paths that guarantee to yield maximal cliques leading to the time complexity proportional to the number maximal cliques in the network. Any algorithm requires a running time in the order of the number of maximal cliques in the graph as it needs to enumerate all of them.
- The memory requirements of the algorithms is polynomial in the size of the graph.
- Its simplicity makes it promising for parallelization.

The basic approach of BK algorithm is as follows: it starts with a single node of the network as a clique and recursively adds one node at a time to form a larger clique. The algorithm maintains three sets of nodes as shown below. If there is an edge (u, v) in the network, we say u is a neighbor of v and vice versa. Let $N(v)$ be the set of all neighbors of v.

- CLQ: the set of nodes in the current clique. This set is extended by adding more nodes, if possible, to form a larger clique.
- CAND: the set of candidate nodes that can be added to CLQ to extend the current clique. By the definition of a clique, any node in CAND must be a common neighbor of every node in CLQ. However, all of the common neighbors of the nodes in CLQ may not be in CAND. Once a node in CAND is explored, it is removed from CAND and added to the NOT set.
- NOT: set of candidate nodes that have already been explored. All extensions of CLQ containing any node in NOT have already been generated. The purpose of maintaining the NOT set will be clear soon.

Initially, CLQ and NOT are empty, and CAND is the set of all nodes in the network. Then the maximal cliques are enumerated by calling the recursive function `EnumerateMaxCliques` shown below. Notice that any common neighbor of the nodes in CLQ must be either in CAND or in NOT, but not in both. When CAND $= \phi$, clique CLQ cannot be extended anymore; however, at this stage, CLQ is not a maximal clique if NOT $\neq \phi$. Moreover, if at any stage there is a node $w \in$ NOT that is a neighbor of all nodes in CAND, further extension of CLQ will never lead to a maximal clique since the NOT set will never be empty with such extensions.

Algorithm EnumerateMaxCliques (CLQ, CAND, NOT)

1. If CAND $= \phi$
   a. If NOT $= \phi$, output CLQ
   b. Return
2. If there is a $w \in$ NOT s.t. $w \in N(x)$ for all $x \in$ CAND
   Return
3. first_$v \leftarrow$ any node in CAND
4. $v \leftarrow$ first_v
5. EnumerateMaxCliques (CLQ $\cup \{v\}$, CAND $\cap N(v)$, NOT $\cap N(v)$)
6. NOT $\leftarrow$ NOT $\cup \{v\}$
7. CAND $\leftarrow$ CAND $\setminus \{v\}$
8. If CAND $\setminus N(\text{first\_}v) \neq \phi$
   a. $v \leftarrow$ any node in CAND $\setminus N(\text{first\_}v)$
   b. Repeat Step 5-8

Notice that extending CLQ by any neighbor of the first node, first_v, selected from CAND in Step 3 of `EnumerateMaxCliques` is not necessary. Any maximal clique containing any such node will be generated by some other extensions. Further, to extend CLQ, the nodes from CAND can be selected in any order. Thus, choosing a node as first_v with the largest $|\text{CAND} \cap N(\text{first\_}v)|$ reduces the number of candidates and the number of recursive branches (Step 5–8).

6.2 A Parallel Implementation of Bron–Kerbosch Algorithm

The independent nature of the subtasks in the BK algorithm allows efficient parallelization of this algorithm. The BK algorithm (see Sect. 6.1) searches for maximal cliques in a depth-first manner. Each call of `EnumerateMaxCliques` creates one node in the search tree, and each node in the search tree is defined by the corresponding CLQ, CAND, and NOT sets. The root of the search tree is (ϕ, V, ϕ), where V is set of nodes in the network, and only a leaf node produces a maximal clique. An example search tree is given in Fig. 4. All descendant leaf nodes, and the corresponding maximal cliques, of node (CLQ, CAND, NOT) depend only on its CLQ, CAND, and NOT sets, and thus, as long as the corresponding CLQ, CAND, and NOT sets are known, each subtree rooted at (CLQ, CAND, NOT) can be explored independently. In the BK algorithm, Step 5–8 are repeated and `EnumerateMaxCliques` is called for each $v \in$ CAND $\setminus N(\text{first\_}v)$. Each such invocation of `EnumerateMaxCliques` creates a branch of the search tree. Now assuming any ordering of the members of the set CAND, for all v, CLQ $\cup \{v\}$, CAND $\cap N(v)$,and NOT $\cap N(v)$ can be computed in advance and these branches of the search tree can be computed independently in parallel.

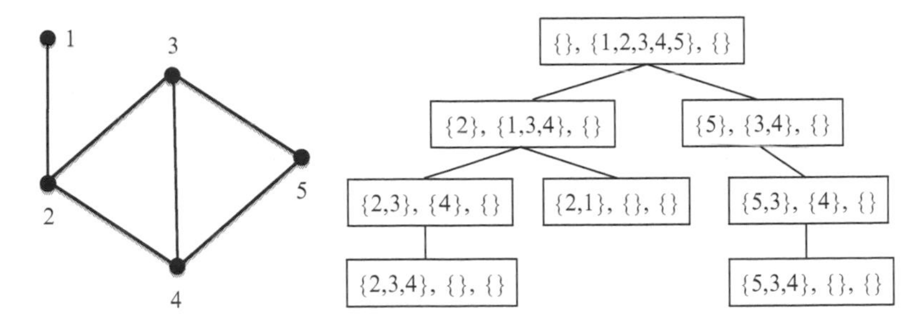

Fig. 4 Execution of BK algorithm: a network and the corresponding search tree. Each node in the search tree shows CLQ, CAND, and NOT sets in this order

Exploiting the above principle, a parallelization of the BK algorithm is given in [51] which achieves almost linear speedup with the number of processors. This parallel algorithm begins with reading the graph data from file by a single processor and copying the entire graph data to the other processors by MPI broadcast. It was experimentally shown in [51] that this MPI broadcast is faster than having each processor read the graph data independently from a single file. Initially, the branches of the root node in the search tree are distributed to the processors. Given that each processor has access to complete graph data, by assuming some ordering of the nodes, each processor can independently decide its initial subtask assignment without any communication. However, different search subtrees may differ greatly in size, and the size of subtree is not known in advance until the subtree is completely traversed. As a result, load balancing needs to be done dynamically. Such dynamic load balancing is done as follows: as soon as a processor becomes idle, it sends a load balance request message to a random target processor. The receiver of the message sends back CLQ, CAND and NOT sets of the root of an unexplored subtree. If the chosen target processor is also idle, the sender tries another random target processor.

The above algorithm achieves very good parallelization. It exhibits a speedup that is almost linear with the number of processors. Experimental results show such speedup is achieved even when thousands of processors are used. The details of the experimental results can be found in [51]. However, its ability to deal with very large networks is limited by the fact that each processor is required to have a copy of the entire network in its local memory.

7 Conclusions and Open Problems

Problems relating to subgraph analysis and queries arise in diverse applications. They are challenging computational problems, and parallel algorithms are the only approach to scale to very large graphs. In this chapter, we discussed some of the key parallel algorithmic issues and techniques for the problem of counting the number of non-induced embeddings of special kinds of subgraphs, including trees and cliques. Despite a lot of active research, this area is still wide open and consists of a number of challenging theoretical and practical problems. We discuss some of these below.

1. An immediate challenge is to scale the basic subgraph enumeration problem to larger graphs and templates and also consider induced subgraphs, which tend to be much harder. Counting can be viewed as a function on the space of possible embeddings, and extension to other kinds of functions, e.g., the average degree of a node over all embeddings, are interesting problems in many applications.
2. Due to advancement of technology, massive real-world datasets are becoming available. As a result, distributed-memory parallel algorithms for subgraph counting problem seem to be necessary to work with massive graphs that do not fit in the main memory of a single computing machine. As discussed in this chapter, such distributed memory parallel algorithms exist for only few special types of subgraphs such as trees, cliques, subgraphs with small diameter, or a subgraph that can be divided into two smaller subgraphs by removing a single cut-edge. Developing an efficient distributed-memory algorithm for any arbitrary template subgraph is still an open problem. Research effort need to be directed toward obtaining such an algorithm.
3. The parallelization of the BK algorithm, discussed in Sect. 6.2, for enumerating maximal cliques requires that each computing machine has a copy of the entire graph. An interesting open problem is to develop a parallel algorithm that does not require each computing machine to have the entire graph, i.e., each computing machine works with only a small part of the graph to be able to deal with massive graphs.
4. Many applications involve dynamic and uncertain graphs, in which edges/nodes might be known to be in the graph with some probability. Extending subgraph enumeration to such settings is a very promising direction.
5. Finally, many applications such as the semantic Web involve broader relational queries. As discussed earlier, a number of graph query languages and RDF approaches have been proposed for such problems. Developing efficient algorithmic tools to support such a formalism is one of the main challenging problems in this area.

Acknowledgement This work has been partially supported by NSF Grant CNS-0626964, SES-0729441, CNS-0831633, and OCI-0904844, and DTRA Grant HDTRA1-0901-0017 and HDTRA1-07-C-0113.

References

[1] A. hadoop, Code and documentation are available at http://developer.yahoo.com/hadoop/
[2] E. Akkoyunlu, The enumeration of maximal cliques of large graphs. SIAM J. Comput. **2**(1), 1–6 (1973)

[3] N. Alon, P. Dao, I. Hajirasouliha, F. Hormozdiari, S. Sahinalp, Biomolecular network motif counting and discovery by color coding. Bioinformatics **24**(13), i241 (2008)
[4] N. Alon, P. Dao, I. Hajirasouliha, F. Hormozdiari, S.C. Sahinalp, Biomolecular network motif counting and discovery by color coding. Bioinformatics **24**(13), 241–249 (2008)
[5] N. Alon, R. Yuster, U. Zwick, Color-coding. J. ACM **42**(4), 856 (1995)
[6] V. Aravind, V. Raman, Approximate counting of small subgraphs of bounded treewidth and related problems. Electronic Colloquium on Computational Complexity (ECCC)(031) (2002)
[7] E. Bloedorn, N.J. Rothleder, D. DeBarr, L. Rosen, Relational graph analysis with real-world constraints: An application in irs tax fraud detection, in *AAAI*, 2005
[8] P. Bonzini, L. Pozzi, Polynomial-time subgraph enumeration for automated instruction set extension, in *Proceedings of the International Conference on Design, Automation and Test in Europe*, 2007
[9] C. Borgelt, M.R. Berhold, Mining molecular fragments: Finding relevant substructures of molecules, in *ICDM*, 2002
[10] M. Bröcheler, A. Pugliese, V. Subrahmanian, Cosi: Cloud oriented subgraph identification in massive social networks, in *Proceedings of the International Conference on Advances in Social Networks Analysis and Mining* (ASONAM) (2010)
[11] C. Bron, J. Kerbosch, Finding all cliques of an undirected graph. Comm. ACM **16**(9), 575–577 (1973)
[12] M.P. Consens, A.O. Mendelzon, Expressing structural hypertext queries in graphlog, in *Proceedings of the 2nd International Conference on Hypertext*, 1989
[13] J. Dean, S. Ghemawat, Mapreduce: simplified data processing on large clusters, in *Proceedings of the Sixth Symposium on Operating System Design and Implementation (OSDI)*, 2004
[14] M. Deshpande, M. Kuramochi, N. Wale, G. Karypis, Frequent substructure-based approaches for classifying chemical compounds. IEEE Trans. Knowl. Data Eng. **17**(8), 1036–1050 (2005)
[15] F. Eichinger, K. Bohm, M. Huber, Mining edge-weighted call graphs to localise software bugs, in *Proc. European Conf. on Machine Learning and Principles and Practice of Knowledge Discovery in Databases (ECML PKDD)*, 2008
[16] F. Eisenbrand, F. Grandoni, On the complexity of fixed parameter clique and dominating set. Theoret. Comput. Sci. **326**(1–3), 57–67 (2004)
[17] L. Getoor, C.P. Diehl, Link mining: a survey. SIGKDD Explor. Newslett. **7**, 3–12 (2005)
[18] M. Gonen, D. Ron, Y. Shavitt, Counting stars and other small subgraphs in sublinear time, in *ACM-SIAM Symposium on Discrete Algorithms (SODA)*, 2010
[19] M. Gonen, Y. Shavitt, Approximating the number of network motifs, in *The 6th Workshop on Algorithms and Models for the Web Graph (WAW)*, 2009
[20] A. Grama, V. Kumar, State of the art in parallel search techniques for discrete optimization problems. IEEE Trans. Knowl. Data Eng. **11**(1), 28–35 (1999)
[21] R.H. Guting: Graphdb: Modeling and querying graphs in databases, in *Proceedings of the 20th international Conference on Very Large Data Bases*, 1994
[22] B. Hendrickson, J. Berry, Graph analysis with high-performance computing. Comput. Sci. Eng. **10**(2), 14–19 (2008)
[23] J. Huan, W. Wang, J. Prins, J. Yang, Spin: Mining maximal frequent subgraphs from graph databases, in *ACM KDD*, 2004
[24] J. Huang, D. Abadi, K. Ren, Scalable sparql querying of large rdf graphs. Proc. VLDB Endowment **4**(11), 1123–1134 (2011)
[25] F. Hüffner, S. Wernicke, T. Zichner, Algorithm engineering for color-coding with applications to signaling pathway detection. Algorithmica **52**(2), 114–132 (2008)
[26] A. Inokuchi, T. Washio, H. Motoda, An apriori-based algorithm for mining frequent substructures from graph data, in *ECML-PKDD*, 2000
[27] B.I. Jensen, H. Blau, N. Immerman, D. Jensen, A visual language for querying and updating graphs. Technical Report, University of Massachusetts Amherst (2002)

[28] U. Kang, C. Tsourakakis, A. Appel, C. Faloutsos, J. Leskovec, Hadi: Fast diameter estimation and mining in massive graphs with hadoop. Technical Report, CMU-ML-08-117, Carnegie Mellon University (2008)
[29] P.N. Klein, S. Subramanian, A randomized parallel algorithm for single-source shortest paths. J. Algorithm **25**(2), 205–220 (1997)
[30] M. Kowaluk, A. Lingas, E. Lundell, Counting and detecting small subgraphs via equations and matrix multiplication, in *ACM SODA*, 2011
[31] M. Kuramochi, G. Karypis, Frequent subgraph discovery, in ICDM, 2001
[32] M. Kuramochi, G. Karypis, Finding frequent patterns in a large sparse graph. Data Min. Knowl. Discov. **11**(3), 243–271 (2005)
[33] E. Lawler, J. Lenstra, A. Kan, Generating all maximal independent sets: Np-hardness and polynomial-time algorithms. SIAM J. Comput. **9**(3), 558–565 (1980)
[34] J. Leskovec, A. Singh, J. Kleinberg, Patterns of influence in a recommendation network. in PAKDD, 2006
[35] E. Loukakis, A new backtracking algorithm for generating the family of maximal independent sets of a graph. Comput. Math. Appl. **9**(4), 583–589 (1983)
[36] A. Lumsdaine, D. Gregor, B. Hendrickson, J. Berry, Challenges in parallel graph processing. Parallel Process. Lett. **17**(1), 5–20 (2007)
[37] K. Madduri, D. Bader, J. Berry, J. Crobak, An experimental study of a parallel shortest path algorithm for solving large-scale graph instances, in *Workshop on Algorithm Engineering and Experiments (ALENEX)*, 2007
[38] K. Madduri, D. Ediger, K. Jiang, D. Bader, D. Chavarra-Miranda, A faster parallel algorithm and efficient multithreaded implementations for evaluating betweenness centrality on massive datasets, in *Proceedings of the 3rd Workshop on Multithreaded Architectures and Applications (MTAAP)*, 2009
[39] E.K. Maxwell, G. Back, N. Ramakrishnan, Diagnosing memory leaks using graph mining on heap dumps, in *Proceedings of the 16th ACM SIGKDD International Conference on Knowledge Discovery and Data Mining* (ACM, New York, NY, 2010), pp. 115–124. DOI 10.1145/1835804.1835822. URL http://doi.acm.org/10.1145/1835804.1835822
[40] R. Milo, S. Shen-Orr, S. Itzkovitz, N. Kashtan, D. Chklovskii, U. Alon, Network motifs: simple building blocks of complex networks. Science **298**(5594), 824 (2002)
[41] Resource description framework (RDF) (2004) Documentations are available at http://www.w3.org/RDF/
[42] R. Pagh, C. Tsourakakis, Colorful triangle counting and a mapreduce implementation. Inform. Process. Lett. **112**(7), 277–281 (2011)
[43] P. Pardalos, J. Xue, The maximum clique problem. J. Global Optim. **4**(3), 301–328 (1994)
[44] N. Pržulj, Biological network comparison using graphlet degree distribution. Bioinformatics **23**(2), e177 (2007)
[45] N. Pržulj, D. Corneil, I. Jurisica, Efficient estimation of graphlet frequency distributions in protein-protein interaction networks. Bioinformatics **22**(8), 974 (2006)
[46] S. Raghavan, H. Garcia-Molina, Representing web graphs, in *ICDE*, 2003
[47] K. Rohloff, R. Schantz, Clause-iteration with mapreduce to scalably query data graphs in the shard graph-store, in *Proceedings of the Fourth International Workshop on Data-Intensive Distributed Computing*, 2011
[48] R. Ronen, O. Shmueli, Evaluating very large datalog queries on social networks, in *ACM EDBT*, 2009
[49] S. Sakr, Graphrel: A decomposition-based and selectivity-aware relational framework for processing sub-graph queries, in *DASFAA*, 2009
[50] T. Schank, Algorithmic aspects of triangle-based network analysis. Ph.D. thesis, Universitat Karlsruhe (TH) (2007)
[51] M. Schmidt, N. Samatova, K. Thomas, B. Park, A scalable, parallel algorithm for maximal clique enumeration. J. Parallel Distr. Comput. **69**(4), 417–428 (2009)
[52] S. Suri, S. Vassilvitskii, Counting triangles and the curse of the last reducer, in *Proceedings of the 20th International Conference on World Wide Web* (WWW), 2011

[53] C. Tsourakakis, U. Kang, G. Miller, C. Faloutsos, Doulion: Counting triangles in massive graphs with a coin, in *Proceedings of the 15th ACM SIGKDD International Conference on Knowledge Discovery and Data Mining* (KDD), 2009
[54] V. Vassilevska, R. Williams, Finding, minimizing, and counting weighted subgraphs, in *ACM STOC*, 2009
[55] X. Yan, X.J. Zhou, J. Han, Mining closed relational graphs with connectivity constraints, in *KDD*, 2005
[56] Z. Zeng, J. Wang, L. Zhou, G. Karypis, Out-of-core coherent closed quasi-clique mining from large dense graph databases. ACM Trans. Database Syst. **32**(2), 13 (2007)
[57] Z. Zhao, M. Khan, V.S.A. Kumar, M. Marathe, Subgraph enumeration in large social contact networks using parallel color coding and streaming, in *39th International Conference on Parallel Processing (ICPP)*, pp. 594–603, 2010
[58] Z. Zhao, G. Wang, A.R. Butt, M. Khan, V.S.A. Kumar, M.V. Marathe, Sahad: Subgraph analysis in massive networks using hadoop, in *Proceedings of the 26th IEEE International Parallel and Distributed Processing Symposium (IPDPS)*, 2012

Part III
Diffusion, Spreading, Mobility, and Transport

A Dynamical Network View of Lyon's Vélo'v Shared Bicycle System

Pierre Borgnat, Céline Robardet, Patrice Abry, Patrick Flandrin, Jean-Baptiste Rouquier, and Nicolas Tremblay

1 Introduction

A current challenge in the study of complex networks is to devise and validate methods that are not limited to static or growing complex networks but that are adaptable to dynamical aspects of complex networks. Among all the complex networks offered by human activities, the transportation networks raise some unique challenges by themselves. The classical methods in transportation research rely on household surveys of movements or on the direct observation of movements by sampling some places in the city. However, thanks to the developments of information technologies, more and more systems of transportation offer digital footprints of population movements and enable their study as complex systems: public transportations in subway (thanks to individual digital subscription cards) [1, 2], railways [3, 4], or air transportation thanks to database of flights [5], with applications ranging from urban planning to epidemiology [6]. The issues we now face are to understand how to cope with these new, large-scale datasets and which methods are useful so as to obtain some insights on the dynamics of people's moves.

We will review in this chapter the progresses made in this direction in the particular case of shared bicycle systems. As modern cities are more and more overcrowded by cars in their centers, alternative means of transportation have been

P. Borgnat (✉) • P. Abry • P. Flandrin • N. Tremblay
Laboratoire de Physique, ENS de Lyon, CNRS, 46 allée d'Italie 69007 Lyon, France
e-mail: Pierre.Borgnat@ens-lyon.fr; Patrice.Abry@ens-lyon.fr; Patrick.Flandrin@ens-lyon.fr; Nicolas.Tremblay@ens-lyon.fr

C. Robardet
LIRIS, INSA-Lyon, CNRS, Bâtiment Blaise Pascal, 69621 Villeurbanne, France
e-mail: Celine.Robardet@insa-lyon.fr

J.-B. Rouquier
Eonos, 53 rue de la Boétie, 75008 Paris, France
e-mail: jrouquie@gmail.com

A. Mukherjee et al. (eds.), *Dynamics On and Of Complex Networks, Volume 2*, Modeling and Simulation in Science, Engineering and Technology, DOI 10.1007/978-1-4614-6729-8_13, © Springer Science+Business Media New York 2013

developed. Among them, shared bicycle programs have gained a renewal in interest in the past 10 years, thanks to the possibility of using fully automated rental systems that are available 24 hours a day, 7 days a week. This innovation led to the growth of different systems in many of the major cities in Europe, e.g., Vélib' in Paris [7, 8] and Vélo'v in Lyon [9, 10], Bicing in Barcelona [11, 12], OYBike in London [13], and Bicikelj in Ljubljana [14] to cite only a few that have attracted quantitive analyses of their data. Indeed, the full automation of the systems creates digital footprints which provide a complete view of all the trips made with these bicycles. Such datasets would have been impossible to obtain before, although it gives, by definition, only a partial view of all the movements done in a city. Despite that, these data are interesting to test ideas related to the study of dynamical networks.

The contribution of this chapter will be first to review previous studies on shared bicycle systems from the point of view of information sciences. We discuss some of the general features that are displayed by these systems. Then, we focus on Vélo'v—the shared bicycle system that is deployed in the city center of Lyon, France's second largest urban community and that was the largest scale system of this type when launched in May 2005. Our main purpose is to discuss an adaptation of community detection methods to dynamical networks and aggregation in dynamical networks. Here, the Vélo'v network is inherently dynamical, in that the network appears only because there are bicycle trips connecting the stations.

The chapter is organized as follows. Section 2 recalls general facts about shared bicycle systems and about their global dynamics (cycle over the week and nonstationarity over several months). Section 3 shows how such a system can be seen as an instance of a dynamical complex network. Section 4 discusses how a spatial aggregation of the Vélo'v complex network can be obtained by adapting methods for community detection. Then, Sect. 5 proposes a typology of the dynamics in the Vélo'v network by estimating a similarity graph and detecting communities. We conclude in Sect. 6.

2 General Dynamics of Vélo'v System

The Vélo'v system is basically a bicycle rental system consisting of 343 automated stations each composed of several stands from where a Vélo'v bicycle can be taken or put back at the end of the trip. The stations are spread out in the city with the objective that, in the center, the distance to a station is no more than 300m. There are around 4000 Vélo'v bicycles available at these stands, and, as the system is automated, a bicycle can be returned at any free stand, usually in a different station than the one it was taken from. The functioning is all automated and works 24 hours a day, all year long. People use long-term or short-term (obtained with a credit card) subscription card to rent a bicycle. This made possible the collection of all the data pertaining to the trips made with these bicycles, without sampling as was previously the rule in transportation studies. This also made possible the display of the state of each station through the Web [15]. Thanks to JCDecaux—Cyclocity and the Grand Lyon, we had access to the anonymized version of the trips

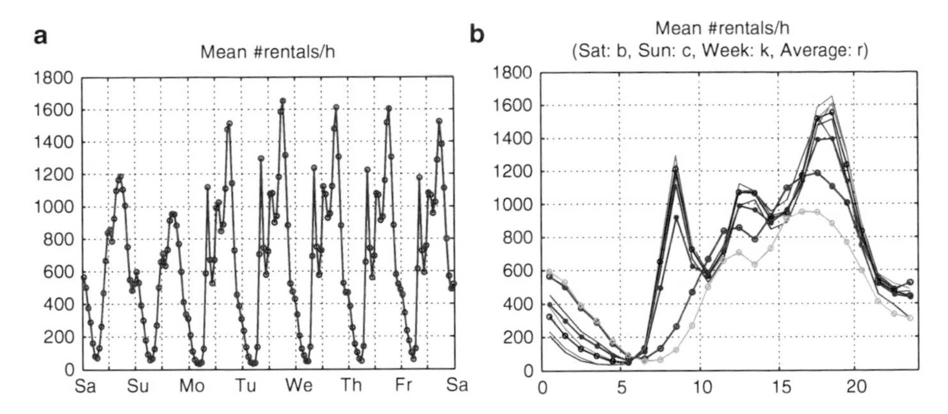

Fig. 1 Weekly pattern of the number of rentals made per hour, summed over all the Vélo'v stations: (**a**) pattern for the whole week and (**b**) superposition of individual day patterns in the week. The average over the 5 ordinary weekdays from Monday to Friday is the thick black curve with circles. The five curves in thin lines that look alike are each for one weekday: Monday in blue, Tuesday in dark green, Wednesday in red, Thursday in cyan, and Friday in purple. Saturday (*thick blue curve with circles*) and Sunday (*thick cyan curve with circles*) reveal a different structure, especially for the peak hours. The average over all days is the thick red curves and is dominated by the weekdays. That is a reason to study separately weekdays and weekends

made with Vélo'v. The dataset consists of a log of all the rentals, with the station and the time of departure and the destination station and the time of arrival. For privacy concerns, there is no information about users. Nevertheless, about the trips, the dataset is complete.

This dataset of the trips was instrumental in studying the general dynamics of the system. A first point was to understand how the size of the Vélo'v system and its popularity increased along the first years of the program. As shown in [10], there has been a clear nonstationary growth of its popularity, in terms of long-term subscribers and in terms of number of trips made (currently the average is more than 20,000 trips per day). Moreover the system is still planned to grow [16]. Also, the number of rentals reveal a general multiplicative trend at the scale of several months that can be estimated using recent data-driven statistical methods [17].

A basic feature of transportation system studies is to discover its time pattern of use: When is it really used? What are the peak hours? Is it a means of transportation for the ordinary weekdays or the weekends? Using periodic averaging over the week combined to detrending of the nonstationary behavior, we were able to estimate the mean pattern of total rentals along the week, as shown in Fig. 1 [10, 18, 19]. It reveals that Vélo'v is used first for ordinary transport on working days. Its activity peaks are during the morning, the lunch time, and the evening, and this is characteristic of a system used to go to work and then to come back, with the lunch break in the middle. In a French city such as Lyon, usual schedule for work is to begin around 9am and finish between 5pm and 7pm. The lunch break is spread out between 11:30am and 2pm. Students at universities take classes between (at most) 8am and 6pm with a lunch break at the same time as workers. Finally, shops are usually open from 9am or 10am to around 7pm. All the peaks of activity detected are thus compatible with this lifestyle. Also all the weekdays are similar in that respect, as shown on Fig. 1b. However, a second type of use exists during the weekends, as seen on this second plot. The peaks are less sharp and more spread out around noon and during the afternoon. This is compatible with leisure activities. Finally, one can see that just after midnight each day, a small bump that can be related to the closure of the public transports right after midnight (this even causes local maxima on Friday and Saturday nights). This type of pattern is reminiscent to uses of ordinary public transports. An exception is the nocturnal activity when public transports are closed. Also, it is compatible with similar patterns for shared bicycle systems in other cities: Paris [7, 13], Barcelona [11, 12], and Ljubljana [14].

The prediction of the number of rentals at a given day and time was addressed in [10, 18, 19] using statistical time-series analysis. The global number of rentals, summed over the city, can be predicted on an hourly basis if one takes into account several important factors: the weather (temperature and rain), the holiday periods, and the existence of a correlation over 1 h. The first two features (temperature and rain) account for most of the nonstationary evolution over the year of the average rentals. When zooming in at finer timescales, the 1-h correlation reflects that one decides to use a bicycle depending on the conditions seen during the previous minutes and a rain condition affects the decision on this hourly scale—as it could be expected.

Finally, several empirical studies of data of community shared bicycles revealed various features such as the most frequent paths taken by these bicycles or the advantages of using a bicycle as compared to a car [14, 20], the distribution of durations and lengths of the trips [10], or the usual speed of the bikers depending on the time of the day [10, 20]. All these features could help to design an agent-based model of shared bicycle systems. Note that their properties should be heterogeneous because the distributions found are usually with long tails. For instance, though the median duration of a trip with Vélo'v is 11 min and the average is little bit less than 30 min, there are rentals lasting more than 2 h, and the distribution has a tail that is roughly a power law [10].

All these studies give a good empirical description of the global features of these systems along time or along space. However, the challenge is to design a joint analysis in space and time.

3 Vélo'v as a Complex Network

From Individual Trips to a Network of Stations

As already proposed in [10, 21], it is worthwhile to study the Vélo'v system from a complex network perspective. Complex networks are usually employed when studying real-world dataset including some relational properties. In the context of shared bicycle systems, a network arises when looking at stations as nodes of a complex network. Let us define $\mathcal{N}$ the set of stations. Each node $n \in \mathcal{N}$ is at a specific geographical place in the city. Going from one station to any other is theoretically possible and around half of all theoretically possible trips have been done at least once. However some preferred trips appear in the data, and the they are not necessarily local in space. That is the reason why a representation of the Vélo'v system as a weighted network makes sense.

Our proposition is that the relation between the nodes of the Vélo'v network—the stations—is created by the trips made from one station to another. The greater the number of trips made, the more linked the stations are. Let us define $\mathcal{D} = \{(n, m, \tau)\}$ as the set of individual trips going from station $n \in \mathcal{N}$ to station $m \in \mathcal{N}$ at time τ. The Vélo'v network is defined as $\mathcal{G} = (\mathcal{N}, \mathcal{E}, T)$ where the set of possible edges is $\mathcal{E} \subseteq \mathcal{N} \times \mathcal{N}$ and T is a function defining a weighted adjacency matrix varying in time. Let us define $\mathcal{T}$ as a set containing the times t of interest and $\mathcal{S}$ a set of timescales Δ for aggregation. The function $T : \mathcal{E} \times \mathcal{T} \times \mathcal{S} \to \mathbb{N}$ is obtained by the following equation:

$$T[n, m](t, \Delta) = \#\{(n, m, \tau) \in \mathcal{D} \mid t \leq \tau < t + \Delta\} \tag{1}$$

where # is the cardinal of the set. The result $T[n, m](t, \Delta)$ can be seen as an adjacency matrix of the weighted, directed graph that represents a snapshot of the Vélo'v network. On each edge, the weight is then the number of bicycles going from the station n to station m between times t and $t + \Delta$. More generally, the network $\mathcal{G}$ is a dynamical network and issues arise when we need to deal both with its spatial nature (the nodes and the edges) and with its time evolution as obtained when varying t or the timescale Δ.

In the rest of the article, we follow the same ideas as in [10,21], but we propose a more unified and streamlined approach to study both space and time aggregation for the Vélo'v complex network. Other general ideas on how to cope with time evolving networks can be found in other chapters of this book and in the review [22].

Timescales and Aggregation in Time of Vélo'v Networks

First we take into account the cyclic nature of the network: the same pattern repeats itself each week and we estimate T using periodical averaging. For that, let us decide on a timescale Δ and then define a period $P = p\Delta,\ p \in \mathbb{N}$. Let us set $\mathcal{T}_P = \{k\Delta \bmod P; k \in \{0, \ldots, p-1\}\}$ so that when applying Eq. (1), the first interval in which the trips are counted is $[0 \quad \Delta] \bmod P$. The periodical estimation of the Vélo'v network is obtained by dividing T applied on this choice of $\mathcal{E} \times \mathcal{T}_P \times \mathcal{S}$ by the number of periods P in the data. As the main period in the data is the week [19], we let P be equal to 1 week in the following.

Usually, a specific value for time aggregation Δ is used. However, varying the timescales Δ for the analysis would be of potential interest, e.g., as it was done for computer network analysis [23]. For complex networks, the importance of considering different time window lengths for aggregating networks or the simpler idea that the analyses depend on the observation time have already been put forward in various contexts: to study cattle mobility [24] and communication network [25] or for information spreading in such network [26] to cite a few relevant references. For Vélo'v, given that the average duration of a rental is less than half an hour, aggregating in time over duration Δ larger than that smooths out most of the fluctuations due to individual trips. We classically use $\Delta = 1$h (as in Fig. 1) or 2h. For simplicity, $P = 1$ week and $\Delta =$2h for all the results displayed hereafter.

First, it is possible to reduce the dimension in time of this evolving network by using a principal component analysis in time, as in [10, 21]. It turns out that the principal components display peaks in their time evolution that correspond exactly to the different peaks already commented for the global number of rentals, as shown in Fig. 1. In the following, we will keep in the the dynamical adjacency matrix $T[n, m](t, \Delta)$, only the 19 peaks of activity in time, as given by the global behavior as well as the principle components: every ordinary days around 8am, 12am, and 5pm and each of the two weekend days around 12am and 4pm. We note $\mathcal{T}_P^*$ this set of peak moments and $\#\mathcal{T}_P^* = 19$. The representation of the series of snaphots of the Vélo'v network at peak activity is then $T[n, m](t, \Delta)$ with $t \in \mathcal{T}_P^*$. Another reason for keeping these 19 peaks is that classical approaches in transportation research are often focusing on the activity peaks and doing the same way will help future comparison of results.

Second, at large timescales, there are several manners to aggregate the network. Classical aggregation is to sum over time the different snapshots and this focuses on the strong exchanges between stations at the scale of the aggregation time (typically the week here). In a sense, it can be viewed as going from one timescale to a larger one (hence giving a crude version of multiresolution analysis). Using the 19 peak activity times $\mathcal{T}_P^*$, an aggregated view over the whole week is obtained as $\sum_{t \in \mathcal{T}_P^*} T[n, m](t, \Delta) \triangleq < T[n, m] >_{\mathcal{T}_P^*}$ where $< \cdots >_{\mathcal{T}_P^*}$ stands for this particular aggregation in time. A question in the following is how the different snapshots are similar to or different from the aggregated network over the week.

4 Aggregation in Space for the Vélo'v Network

The series of snapshots of graphs convey detailed information, yet this is too much information for modeling. However, aggregating over all the nodes as done in Sect. 2 does not give enough details. We would like to aggregate on intermediate scales for the nodes in the network. There are two classical approaches to aggregation in a network: find clusters of nodes that are strongly linked together, also called communities [27], or use multiscale harmonic decomposition over a graph (e.g., [28]). Here, we explore the spatial aggregation that is obtained by

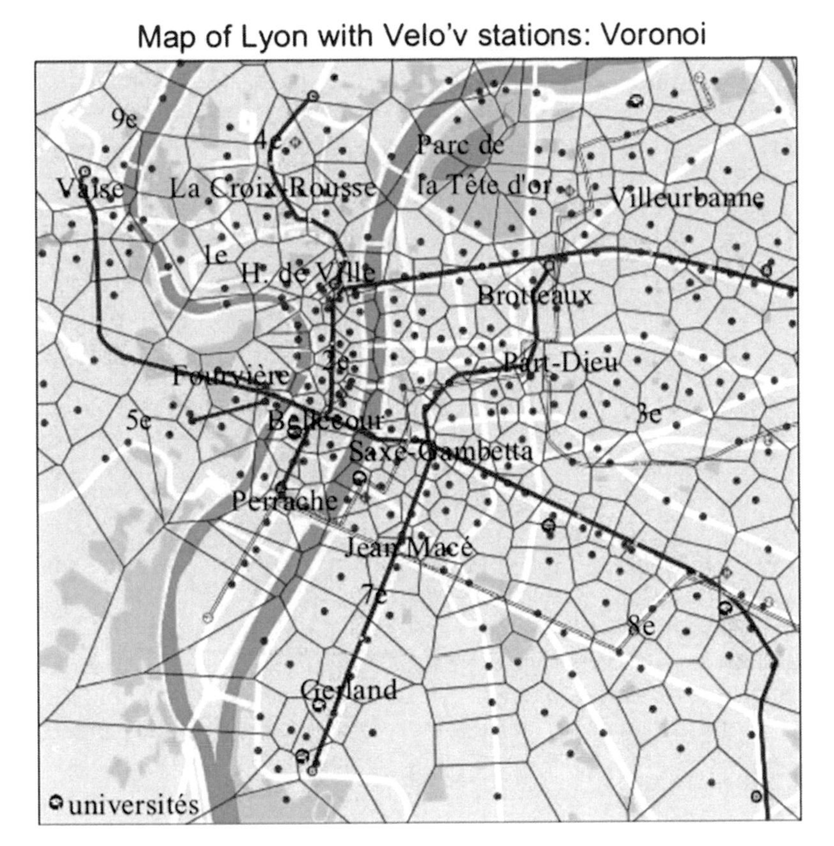

Fig. 2 Map of the cities of Lyon and Villeurbanne with the Vélo'v stations and their Voronoi diagram [29]. Each dot is a Vélo'v station surrounded by its own Voronoi cell. One sees that the city is well covered, with a higher density of stations in the center. Major roads are in white; main public transport lines are in red (subways) and gray (tramways). Parks are in green and the two rivers are in blue (the Saône on the west side, the Rhône on the east side). Names of the different parts of the city are given, as well as names of main hubs of transportations: Part-Dieu and Perrache for the main train stations; Vaise and Jean Macé for secondary train stations; Bellecour, Hôtel de Ville, Brotteaux (including Charpennes), and Saxe-Gambetta for other important hubs of the public transport system. Gerland, Croix-Rousse, and Villeurbanne are other parts of the city that will be discussed afterwards. Finally, the locations of the downtown university campuses are shown

looking at communities of stations. So as to compare with the urban organization of the city, a map of the city is in Fig. 2 that shows the main lines of transportations and provides the places and names of the most important hubs for public transportation. By referring to this map, the reader will follow with greater ease the comments about the spatial aggregation proposed by community aggregation in this section.

Aggregation of Network by Communities

At a given timescale and instant, we propose to aggregate the network in space over its communities of nodes. A community is often defined as a subset of nodes that are strongly linked together inside the network (see the review in [27]). We adopt the modularity as a metrics to find communities. Modularity was first proposed in [30] and extended in [31] to the case of directed networks. Assume that t and Δ are set to specific values and use the adjacency matrix $T[n,m](t,\Delta)$ obtained that way. Modularity is defined as

$$Q = \frac{1}{2W} \sum_{\{n,m\} \in \mathcal{N} \times \mathcal{N}} \left[T[n,m] - \frac{\sum_{j \neq n} T[j,n] \cdot \sum_{k \neq m} T[m,k]}{2W} \right] \delta_{c_n, c_m} \quad (2)$$

where $W = \sum_{n,m} T[n,m]$ is the total weight of the network and c_n is the partition index of the group in which n belongs. Modularity is a number between -1 and $+1$. If there is a community structure in the graph and the index c_n reflects this structure (by taking a different value for each community), Q should be large, typically larger than 0.4. Conversely, if one finds a partition index c_n for which Q is large enough, it tells that there are communities of nodes. As a consequence, finding communities is possible by maximizing Q over the set of $\{c_n, n \in \mathcal{N}\}$ of possible partitions of nodes. However, this task is hard: the complete maximization is NP-complete [32] and many approximations such as the one in [33], have a tendency to propose too big communities. In this work, we use the fast, hierarchical, and greedy algorithm proposed in [34] (called the Louvain algorithm) as a simple way to find relevant communities. It is reviewed in [27] that modularity is a good metrics to find communities and that this algorithm works correctly as compared to other methods. Nevertheless, the results shown hereafter are not specific of this choice of algorithm to find communities in a network, and for instance the *infomap* algorithm of [35] would work as well.

Communities in Vélo'v Networks

First, the method is applied to the time-aggregated network $< T[n,m] >_{T_P^*}$. Figure 3 shows the community structure obtained by approximate maximization of the modularity Q by the Louvain algorithm. Four communities appear in this network and they are displayed on the figure. The main feature is that the obtained communities, when shown on a map using the GPS coordinates of each Vélo'v station, are easily grouped on a geographical basis. This can be surprising as the partitioning in communities is blind to any geographical consideration. Anyway, one recognizes in the proposed communities a partition of Lyon City that reflects its general organization. The center of the city is spread out between the Presqu'île (between the two rivers, the Saône on the west and the Rhône on the east) and Part-Dieu (transport hub comprising the main railway station and a subway station) (blue community); the northeast part contains the 6th district and Villeurbanne which are well connected together with a major science university campus in the north

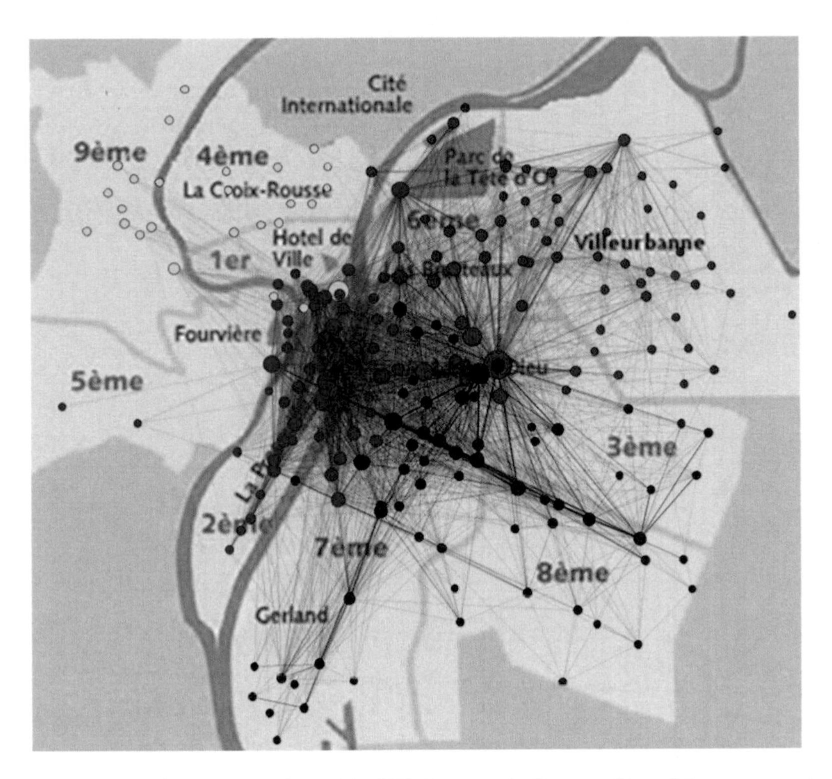

Fig. 3 Aggregation in space and time of the Vélo'v network: Communities of the average network $< T[n,m] >_{T_P^*}$ are shown on top of a simplified map of Lyon. Each community has its own color. The size of one node is proportional to the number of trips made to and from this station; the width of each edge is proportional to the number of trips made between two stations. For the sake of clarity, the undirected version of the graph is shown and stations with small connectivity (degree at most 2) and edges with low activity (at most one trip per week) are not shown

(red community); the south and southeast parts are organized along two major roads (one from Gerland to Saxe-Gambetta and then Part-Dieu and a second along the limit between the 3rd and the 8th district to Saxe-Gambetta) (black community on the map). Finally, the north (Croix-Rousse) and northwest (Vaise, 9th district) are separate from other parts because Croix-Rousse is on the top of a high hill and Vaise is accessible only along the Saône river between Croix-Rousse hill and Fourvière hill (5th district); this creates a fourth separate community. This good matching of communities found by modularity and of the geographical partition of the city in large zones was first discussed in [10]. It even holds if one uses a smaller scale for communities, by keeping the hierarchical structure that is obtained thanks to the Louvain algorithm.

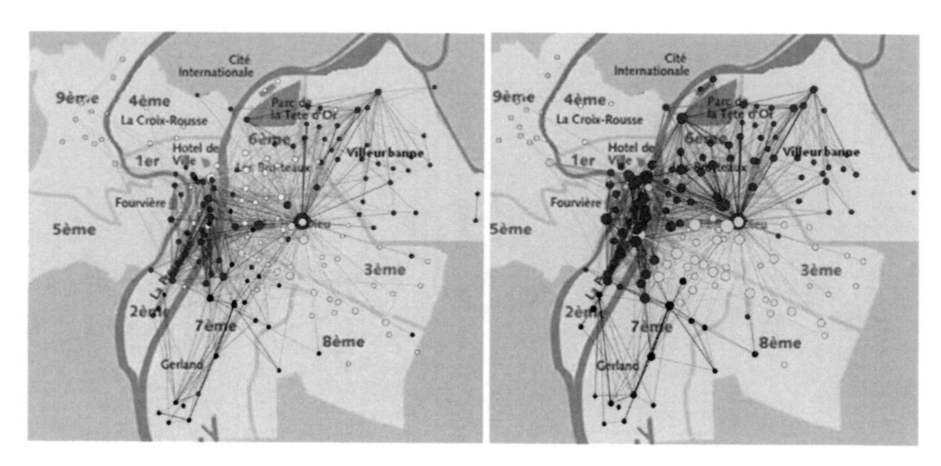

Fig. 4 Aggregation in space of snapshots of the Vélo'v network. *Left*: Monday am (7am–9am). *Right*: Monday pm (4pm–6pm). Each community has an arbitrary color. The size of each node is proportional to its incoming flow added to its outgoing flow. Note that the number of trips is usually larger during the afternoon (as was already seen on Fig. 1, the peak for two hours in the afternoon is 1.5 times higher than the one in the morning)

Communities in Vélo'v Networks During the Weekdays

Communities can also be looked at for individual snapshots $T[n,m](t,\Delta)$, with $t \in \mathcal{T}_P^*$, of the Vélo'v network. Figure 4 displays the community structure obtained by the Louvain algorithm for two snapshots taken during ordinary job days (here, on Mondays). The community structure at a given time in the week still matches a geographical partitioning of the city. Because the timescale is finer, with less aggregation in time, more details are apparent and some specific Vélo'v stations are not in the same community as their surrounding stations. For instance, the northern community (Villeurbanne) which includes a major university campus, includes also a station on the banks of the Rhône where there is also a university. Also, there are more communities (6 here) than in the average network. One new community groups together the Croix-Rousse hill with the Hôtel de Ville and the 6th district which contain the closest downhill subway stations. The comparison of the maps in the morning (on the left) and at the end of the afternoon (on the right) is interesting: it shows that most of the communities are left unchanged. This indicates that Vélo'v, like other ordinary transportation means, is used for commuting (from home to work in the morning and back in the late afternoon). However, the community grouping the Croix-Rousse hill and the 6th district is not present anymore: it is not hard to figure out that people living on the Croix-Rousse will not use Vélo'v bicycles (which are heavy bicycles) to go up the hill back to their home. Apart from that, more than 90% of the remaining nodes did not change community. Note that the same results are obtained for other ordinary weekdays.

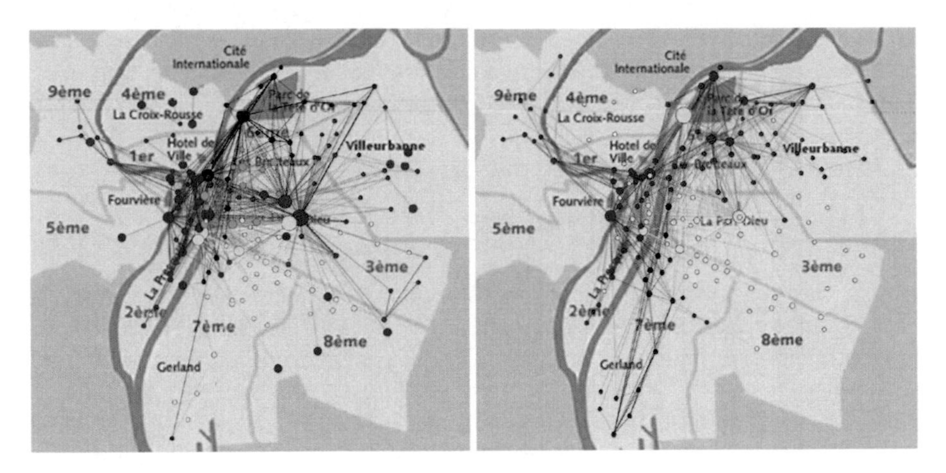

Fig. 5 Aggregation in space of snapshots of the Vélo'v network. *Left*: Saturday pm (3pm–5pm). *Right*: Sunday pm (3pm–5pm). Each community has an arbitrary color

Communities in Vélo'v Networks During the Weekends

When turning to the analysis of the weekend uses of Vélo'v, the features change a little bit as shown in Fig. 5. A first point is that the most active stations during the weekends are not always the same than during the ordinary days. Major transportation hubs are unchanged (Part-Dieu, downhill of Fourvière, Hôtel de Ville,...), yet new active stations appear near places for shopping (in the Presqu'île) and all around the large and green city park of la Tête d'Or in the north.

On Sunday (on the right), the communities could be reminiscent of the time-averaged ones excepted on two points. First, Vaise (9th district) is grouped with the Presqu'île community, possibly because the paths along the river are a pleasant leisure trip. Second, the community ranging from Part-Dieu to the 3rd and 8th districts contains a station which is most active during the weekends: the station at the entrance of the city park of la Tête d'Or. Also some stations along the Rhône river are grouped in the same community. Here again, this is not a great surprise as the park of la Tête d'Or is a main destination for Sunday's outdoor activities. This community connects this park to other places that are either hubs of transportation (like Part-Dieu and Saxe-Gambetta) or other subway stations.

For Saturdays (on the left), the situation is more complex and does not reflect easily a simple geographical partition of the city. A community organized around the park of la Tête d'Or and grouping many stations around the park and on the river banks (having an easy access to the park thanks to bicycle paths on the river banks) is clearly visible. A surprising feature is that the periphery (Vaise, Croix-Rousse but also the most eastern parts of the city) is grouped in the community of the city center (in blue). This is a clue showing that some people uses Vélo'v for longer distance trips on Saturdays than on ordinary days.

This last aspect is one example of the fine-scale analyses that are made possible by aggregating the network in a meaningful manner. It helps finding some unexpected structure that could be probed with more details in the complete dataset.

5 Typology of Dynamics of the Vélo'v Network

Another method for aggregation of nodes is possible: we now want to group two nodes if they have the same usage pattern, whereas in the previous section we grouped nodes exchanging many bicycles. Such an aggregated view is different from summing up the individual snapshots. With the objective of proposing a streamlined methodology for aggregating networks in space and/or time, we will show how the notion of communities can be tailored to group nodes with similar behaviors. For that, the idea is to first build a new similarity network from the dynamical network, before finding communities of similar nodes.

Similarity Graph for the Dynamics

The principle is to quantify the resemblance over time of the different flows between stations. For that, one considers each snapshot to be one observation of the network and then builds a similarity matrix between stations based on these observations. Given a station $n \in \mathcal{N}$, two feature vectors characterize its activity: the incoming flows $F^{in}[n](t) = \sum_{i \in \mathcal{N}} T[i,n](t,\Delta)$ and the outgoing flows $F^{out}[n](t) = \sum_{j \in \mathcal{N}} T[n,j](t,\Delta)$ where $t \in \mathcal{T}_P$ and Δ is constant. For a given pair of stations $[n,m]$, the feature vectors can be computed to quantify if these activity patterns look alike or not. A general approach relies on the choice of a distance d between features (see for instance [36] for many possible distances or [27] for application on graphs), leading to distances between activities of stations n and m:

$$D^{in}[n,m] = d(F^{in}[n], F^{in}[m]) \quad \text{and} \quad D^{out}[n,m] = d(F^{out}[n], F^{out}[m]). \qquad (3)$$

For dealing with observations at different times $t \in \mathcal{T}_P$, it is natural to use a correlation distance over the various observations. The empirical estimator of correlations reads as

$$D^{in}[n,m] = \frac{1}{\#\mathcal{T}_P} \sum_{t \in \mathcal{T}_P} \tilde{F}^{in}[n](t)\tilde{F}^{in}[m](t) \qquad (4)$$

where $\tilde{F}^{in}[n]$ is the centered and normalized version for each n of $F^{in}[n]$ (and respectively for $\tilde{F}^{out}[n]$) and $\mathcal{T}_P$ is a set of times of interest.

When looking at individual snapshots of the Vélo'v network, we have commented that the behaviors during the weekdays are roughly unchanged from one day to another. It makes sense to take as the set of relevant times $\mathcal{T}_P$ the 15 peaks of activity of the weekdays that were already used previously: 8am, 12am, and 5pm, with $\Delta = 2$h. We obtain two correlation matrices of size $\#\mathcal{N} \times \#\mathcal{N}$ that we note ρ^{in}_{week} and ρ^{out}_{week}. For the weekends, the behavior of the stations is different and we compute separately a correlation for the features during the weekends. Using for $\mathcal{T}_P$ the times 12am and 4pm of Saturday and Sunday and a timescale $\Delta = 2$h, two other correlation matrices are obtained: $\rho^{in}_{\text{w.-end}}$ and $\rho^{out}_{\text{w.-end}}$. Note that one could use not only the peak activities but the whole temporal features over these days (like the one reported in Fig. 1 for the global system). However, it is less important if during the low activity parts of the days, two stations have the same behavior (which could be no activity at all and that would have no real relevance).

Remember that the goal is to compare the behaviors of stations, hence the choice of looking at in–in or out–out correlations. An alternative would be to study in–out correlations between pairs of stations; this metrics would describe whether two stations are well connected in the meaning that bicycles leaving one station have a good chance to arrive at another station. However, this metrics appear to be less interesting: first, the mere study of the flows connecting two stations, as studied in Sect. 4, gives already a picture of how well two stations are connected in this acceptance, and this new study would be somewhat redundant; second, the flows leaving a given station are usually really spread between many other stations: the statistical confidence on estimated in-out correlations is low and we leave their study to further work.

For the Vélo'v network, the situation is that many nodes ($\#\mathcal{N} = 334$) but only a few observations on one dataset because $\#\mathcal{T}_P$ is 15 for weekdays and 4 for weekends. Recent theoretical studies about *correlation screening* [37] have shown that, even under the null hypothesis of no correlations between the node features, one should expect large estimated values if using the empirical correlation for large $\#\mathcal{N}$ and small $\#\mathcal{T}_P$. The number of false discovery of nonzero correlations can then be really large. In [37], expressions are given to estimate the threshold under which false discovery becomes dominant. As a consequence, if one wants to build a network of similarity between nodes based on correlation for a small number of observations (as it is often the case), it is expected that a thresholding operation is needed on the correlation to reduce the number of false discoveries. Using [37] and the specific values for the Vélo'v network, a threshold in correlation of 0.8 is reasonable to obtain some statistical confidence of discovering real correlations. Figure 6 shows the histogram of the values of the correlation matrix ρ^{out}_{week} outside the diagonal. As expected, the distribution is broad in $[-1, 1]$. Still, a maximum in the probability of finding large correlations occurs around 0.85 that is not predicted by a null hypothesis of uncorrelated node features. This is a sign of existing similarities in the nodes' activities in the network.

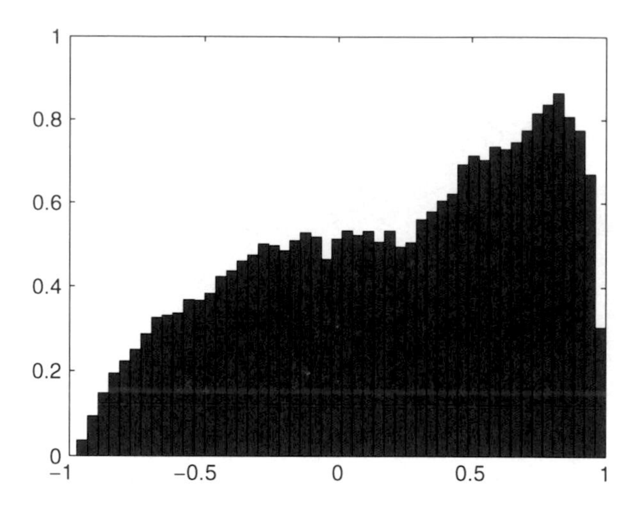

Fig. 6 Distribution of the empirical correlations ρ^{out}_{week} values

We build a similarity graph for the nodes by thresholding the 4 correlation matrices and summing up the thresholded correlations. The weight $S[n,m]$ on each edge of the similarity graph is then

$$S[n,m] = \sum_{\substack{dir.=\{in,out\} \\ time=\{\text{week},\text{w.-end}\}}} \max(\rho^{dir.}_{time}, \eta) - \eta, \tag{5}$$

where η is the threshold. Based on the analysis before, we set $\eta = 0.8$ (though the results are not sensitive to small changes of η). An important remark is that the threshold is applied directly on ρ, not on its absolute value: for the Vélo'v network, negative correlations are discarded because stations with opposite activities would not be in a group of similar behavior. On the contrary, it could be used as to detect stations whose behaviors are far remote, but this is not our objective here.

As a consequence, the similarity weights $S[n,m]$ are between 0 and 4. If two nodes are never similar, neither during the week nor during the weekends, both for incoming and leaving flows, the weight is 0 and the nodes (i.e., Vélo'v stations) are not connected in the similarity graph. If the nodes have similar behavior along time for some of the features, the weight will increase by being higher than η, 2η, 3η, or 4η if they are similar for one, two, three, or all of the feature correlation matrices. Using thresholding before summing the correlations allows us to escape the poor estimation in correlation screening.

Communities of Dynamical Activities

Given the similarity graph, quantifying if the activities of two stations look alike along time or not, it is possible to build a typology of the stations by grouping them according to these correlations. This is simply framed as a problem of detecting communities in the similarity graph of weights S. The same method of community

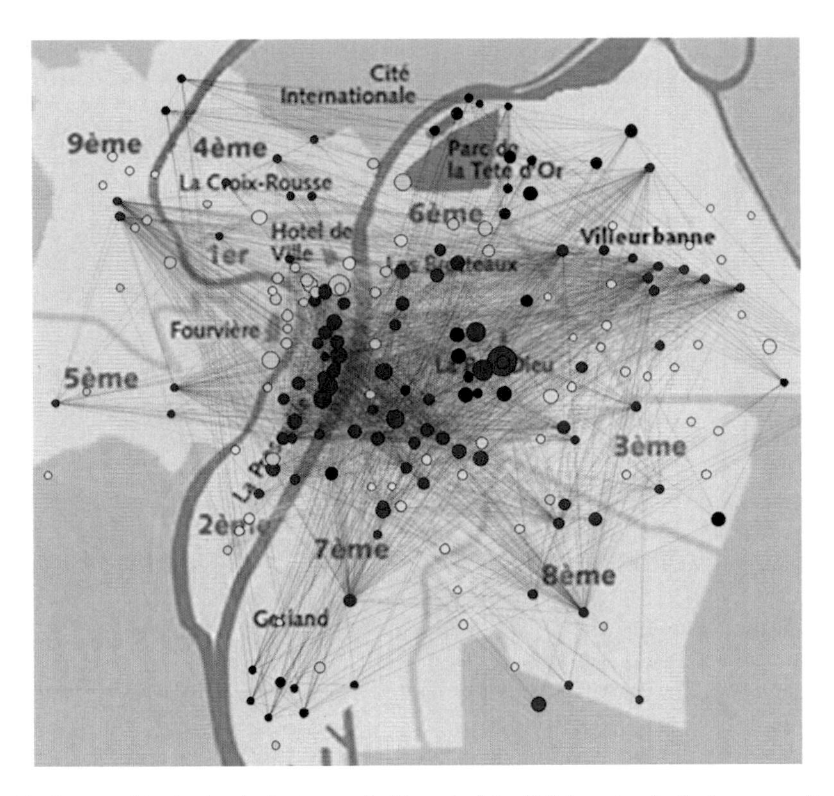

Fig. 7 Communities in the similarity graph $S[n,m]$ of the Vélo'v network. Each community (in an arbitrary color) reflects a specific pattern of usage of the station during the week. Edges are nonzero weights of similarity $S[n,m]$ and nodes have a size proportional to the number of trips made to and from this station

detection, using modularity and the Louvain algorithm, is used on the weighted similarity graph. It provides communities of stations that share, by design of S, their pattern of activity in time. Each community is a type of dynamical activity in the Vélo'v network. The set of communities can be seen as a typology of the different dynamics at work in the network. Figure 7 shows the obtained typology on the Lyon map.

As compared to the previous communities obtained for space aggregation, the similarity communities cannot be matched on a simple geographical partitioning of the city. However, it can be interpreted as a kind of segmentation of the city in various zones of activity. For instance, the community in black groups most of the nodes from the university campus (Villeurbanne and near the park in the north, Gerland in the south, the medicine university in the east, and the university on the banks of the Rhône) and parts of the city with many companies – places to which people commute. The community of Part-Dieu (in red) includes many places of major subway or tramway hubs. Another (in dark blue) is spread out in the city center and has extensions along the stations of the subway lines crossing the center at Bellecour. Finally, the two remaining communities (in light colors) group parts of the cities that are in the east (mostly residential area) or near the Saône river banks in the center (where there are many shops, especially active during the weekends).

All these communities, found through the similarity patterns in the dynamical Vélo'v network and thanks to information science methods, offer a typology of the different neighborhoods in Lyon that will be compared in the future to social science studies.

6 Conclusion

Studying the Vélo'v shared bicycle system as a dynamical network, we have discussed in this chapter how the methods for complex networks can be used to understand part of the dynamics of movements in a city. A key point of the study is the existence of digital footprints left by the use of such automated systems. The main point of this article, after a review of general results on community shared bicycle systems, is to show that community detection thanks to modularity maximization offers a way to create space and/or time aggregation of a network representation of the system. This method for aggregation in space would make possible the modeling of the Vélo'v systems in two levels: a first level inside each community (each having many trips inside the group) and a second level between the communities. Separating the modeling in two levels like that is a good step to reduce the dimensionality of possible modeling of the number of trips made with bicycles in space and time. Also, the aggregated networks that were obtained are interesting as they enable future comparisons of results derived by these complex network methods to results obtained more traditionally by economical and social studies of the city and its transportation systems.

Acknowledgements JCDecaux—Cyclocity and the Grand Lyon Urban Community are gratefully acknowledged for having made Vélo'v data available to us. The authors thank Eric Fleury, Antoine Scherrer, Pablo Jensen, Luc Merchez, Charles Raux, Alain Bonnafous, and Marie Vogel for interesting discussions on the study of Vélo'v. This work has been partially supported by IXXI (Institut des Systèmes Complexes—Complex Systems Institute) Rhône-Alpes in Lyon.

References

[1] V. Latora, M. Marchiori, "Is the Boston subway a small-world network?" Phys. A **314**, 109–113 (2002)
[2] C. Roth, S.M. Kang, M. Batty, M. Barthelemy, "Structure of urban movements: polycentric activity and entangled hierarchy flows". PLoS One **6**(1), e15923 (2011)
[3] P. Sen, S. Dasgupta, A. Chatterjee, P.A. Sreeram, G. Mukherjee, S.S. Manna, "Small-world properties of the Indian railway network". Phys. Rev. E **67**, 036106 (2003)

[4] S. Ghosh, A. Banerjee, N. Sharma, S. Agarwal, N. Ganguly, "Statistical analysis of the Indian Railway network: a complex network approach". Acta Phys. Pol. B Proc. Supp. **4**(2), 123–137 (2011)
[5] A. Barrat, M. Barthélemy, A. Vespignani, "The effects of spatial constraints on the evolution of weighted complex networks". J. Stat. Mech. **5**, P05003 (2005)
[6] A. Barrat, M. Barthélemy, A. Vespignani, *Dynamical Processes on Complex Networks* (Cambridge University Press, Cambridge, 2008)
[7] F. Girardin, "Revealing Paris Through Velib' Data" http://liftlab.com/think/fabien/2008/02/27/revealing-paris-through-velib-data/ (2008)
[8] R. Nair, E. Miller-Hooks, R. Hampshire, A. Busic, "Large-scale bicycle sharing systems: analysis of Vélib". Int. J. Sustain. Transport **7**(1), 85–106 (2013)
[9] http://www.velov.grandlyon.com/
[10] P. Borgnat, C. Robardet, J.-B. Rouquier, P. Abry, E. Fleury, P. Flandrin, "Shared bicycles in a city: a signal processing and data analysis perspective". Adv. Complex Syst. **14**(3), 415–438 (2011)
[11] J. Froehlich, J. Neumann, N. Oliver, "Measuring the pulse of the city through shared bicycle programs", in *International Workshop on Urban, Community, and Social Applications of Networked Sensing Systems - UrbanSense08*, Raleigh, North Carolina, USA, November 4, 2008
[12] J. Froehlich, J. Neumann, N. Oliver, "Sensing and predicting the pulse of the city through shared bicycling", in *Twenty-First International Joint Conference on Artificial Intelligence (IJCAI-09)*, pp. 1420–1426, Pasadena, California, USA, July 11–17, 2009
[13] N. Lathia, S. Ahmed, L. Capra, "Measuring the impact of opening the London shared bicycle scheme to casual users". Transport. Res. C Emerg. Tech. **22**, 88–102 (2012)
[14] D. Lorand, M. Dunja, Bicikelj: environmental data mining on the bicyle, in *Proceedings of the ITI 2012 34th International Conference on Information Technology Interfaces (ITI)*, Dubrovnik, 2012, pp. 331–336
[15] http://www.velov.grandlyon.com/Plan-interactif.61.0.html
[16] http://www.grandlyon.com/PDU.55.0.html
[17] A. Moghtaderi, P. Flandrin, P. Borgnat, "Trend filtering via empirical mode decompositions". Comput. Stat. Data Anal. **58**, 114–126 (2013)
[18] P. Borgnat, A. Abry, P. Flandrin, "Modélisation statistique cyclique des locations de Vélo'v à Lyon", in *Symposium GRETSI-09*, Dijon, FR, Sept., 2009
[19] P. Borgnat, A. Abry, P. Flandrin, J.-B. Rouquier, "Studying Lyon's Velo'V: a statistical cyclic model", in *Proceedings of ECCS'09*, Warwick, Sept., 2009
[20] P. Jensen, J.-B. Rouquier, N. Ovtracht, C. Robardet, Characterizing the speed and paths of shared bicycles in Lyon. Transport. Res. D Transport Environ. **15**(8), 522–524 (2010)
[21] P. Borgnat, E. Fleury, C. Robardet, A. Scherrer, "Spatial analysis of dynamic movements of Vélo'v, Lyon's shared bicycle program", in *Proceedings of ECCS'09*, Warwick, Sept., 2009
[22] P. Holme, J. Saramäki, "Temporal networks". Phys. Rep. **519**(3), 97–125 (2012)
[23] P. Abry, R. Baraniuk, P. Flandrin, R. Rieidi, D. Veitch, "Wavelet and multiscale analysis of network traffic". IEEE Signal Proc. Mag. **3**(19), 28–46 (2002)
[24] P. Bajardi, A. Barrat, F. Natala, L. Savini, V. Colizza, "Dynamical patterns of cattle trade movements". PLoS ONE **6**(5), e19869 (2011)
[25] G. Krings, M. Karsai, S. Bernhardsson, V. Blondel, J. Saramäki, "Effects of time window size and placement on the structure of an aggregated communication network". EPJ Data Sci. **1**(4), 1–16 (2012)
[26] G. Miritello, E. Moro, R. Lara, "Dynamical strength of social ties in information spreading". Phys. Rev. E **83**(4), 045102 (2011)
[27] S. Fortunato, "Community detection in graphs". Phys. Rep. **486**, 75–174 (2010)
[28] D. Hammond, P. Vandergheynst, R. Gribonval, "Wavelets on graphs via spectral graph theory". Appl. Comput. Harmonic Anal. **30**(2), 129–150 (2011)
[29] F.R. Preparata, M.I. Shamos, "*Computational Geometry: An Introduction*" (Springer, New York, 1985)

[30] M. Newman, M. Girvan, "Finding and evaluating community structure in networks". Phys. Rev. E **69**, 0206113 (2004)
[31] E. Leicht, M. Newman, "Community structure in directed networks". Phys. Rev. Lett. **100**, 118703 (2008)
[32] U. Brandes, D. Delling, M. Gaertler, R. Görke, M. Hoefer, Z. Nikoloski, D. Wagner, "On modularity clustering". IEEE TKDE **20**(2), 172–188 (2008)
[33] A. Clauset, M. Newman, C. Moore, "Finding community structure in very large networks". Phys. Rev. E **70**, 066111 (2007)
[34] V. Blondel, J.-L. Guillaume, R. Lambiotte, E. Lefebvre, "Fast unfolding of communities in large networks". J. Stat. Mech. P10008 (2008)
[35] M. Rosvall, C.T. Bergstrom, "Maps of information flow reveal community structure in complex networks". PNAS **105**(4), 1118–1123 (2008)
[36] M. Basseville, "Distance measures for signal processing and pattern recognition". Signal Process. **18**(4), 349–369 (1989)
[37] A.O. Hero, B. Rajaratnam, "Large scale correlation screening". J. Am. Stat. Assoc. **106**(496), 1540–1552 (2011)

Generalized Voter-Like Models on Heterogeneous Networks

Paolo Moretti, Andrea Baronchelli, Michele Starnini, and Romualdo Pastor-Satorras

1 Introduction

The study of dynamical ordering phenomena and consensus formation in initially disordered populations is a central problem in the statistical physics approach to social and natural sciences, whether the broad idea of consensus is referred to opinions, voting intentions, language conventions, social habits, or inherited genetic information. Indeed, people tend to align their opinions [1], segregated populations gradually lose their genetic diversity [2], and different social groups spontaneously develop their own seemingly arbitrary traits of dress or jargon [3]. Understanding how global order can emerge in these situations, self-organized by purely local interactions, represents an important theoretical and practical problem. Our ability to grasp these issues has been mainly driven by the analysis of simple statistical models, which capture the essential ingredients of copying/invasion local dynamics. The simplest of those copying/invasion processes are the voter model [4] and the Moran process [5], which focus principally on social [1] and evolutionary [6] dynamics, respectively.

P. Moretti (✉)
Departamento de Electromagnetismo y Física de la Materia and Instituto "Carlos I" de Física Teórica y Computacional, Universidad de Granada, Facultad de Ciencias, Fuentenueva s/n, E 18071 Granada, Spain
e-mail: pmoretti@onsager.ugr.es

A. Baronchelli
Department of Physics, College of Computer and Information Sciences, Bouvé College of Health Sciences, Northeastern University, Boston MA02120, USA
e-mail: a.baronchelli@neu.edu

M. Starnini • R. Pastor-Satorras
Departament de Física i Enginyeria Nuclear, Universitat Politècnica de Catalunya, Campus Nord B4, 08034 Barcelona, Spain
e-mail: michele.starnini@upc.edu; romualdo.pastor@upc.edu

A. Mukherjee et al. (eds.), *Dynamics On and Of Complex Networks, Volume 2*, Modeling and Simulation in Science, Engineering and Technology, DOI 10.1007/978-1-4614-6729-8_14,

These two models describe a population as a set of agents, each one carrying a state (opinion, trait, genome) represented by a binary variable $\sigma = \pm 1$. At each time step, an ordered pair of adjacent agents (i, j) is selected at random. In the voter model, as a paradigm of copying processes, the system is updated as $\sigma_i := \sigma_j$, the first agent copying the state of its neighbor. The voter model can thus be conceived a very simplistic model of opinion formation in society in which individuals select their beliefs by the (admittedly not very realistic) procedure of just imitating one of their neighbors. On the other hand, in the Moran process it is the neighboring agent the one who copies the state of the first agent, $\sigma_j := \sigma_i$, or, from another perspective, the state of the first one "invades" the neighbor in contact. The Moran process represents thus a simple approximation to the evolutionary dynamics of an asexual population, constrained to have a fixed size [2].

In finite, initially disordered systems, and in the absence of bulk noise (i.e., agents spontaneously changing their state [1]), stochastic copying/invasion dynamics lead to a uniform state with all individuals sharing the same value σ, the so-called consensus. Being the two final states symmetrical in principle, the dynamical evolution towards either state will only depend on the initial configuration. In order to characterize how the consensus is reached, two quantities are usually considered. The exit probability $E(x)$ is defined as the probability that the final state corresponds to all agents in the state $+1$ when starting from a homogeneous initial condition with a fraction x of agents in state $+1$. Accordingly, the consensus time $T_N(x)$ is defined as the average time required to reach consensus, independently of its value, in a system of N agents.

Voter-like processes were originally considered in regular topologies. In this case, the voter model remarkably is one of the few stochastic nonequilibrium models that can be exactly solved in any number of dimensions [7, 8]. As it turns out, its symmetries imply the equality of probabilities of the spin flips $P(+- \to ++) = P(+- \to --)$, and therefore the average conservation of magnetization $m = \sum_i \sigma_i / N$, which in turn implies that the exit probability takes the linear form $E(x) = x$. On the other hand, the consensus time can be seen to scale with the number N of agents as $T_N \sim N^2$ in $d = 1$, $T_N \sim N \ln(N)$ in $d = 2$, and $T_N \sim N$ for $d > 2$ [9].

The results of the voter and other copying/invasion processes in the context of social and evolutionary dynamics acquire a larger relevance when they are considered in systems endowed with realistic, nontrivial topologies. In fact, the strong heterogeneity of social and ecological substrates is better encoded in terms of a complex network [10] rather than by a homogeneous d-dimensional lattice. Therefore, a large theoretical effort has been recently devoted to uncover the effects of a complex topology on the behavior of the voter model as well as of general dynamical processes [11,12]. At the most basic level, while voter and Moran models are equivalent at the mean-field level for regular topologies, if the connection pattern is given by a complex network they behave differently, since the order in which the interacting agents (i, j) are selected becomes relevant [13, 14]. Additionally, the heterogeneity in the connection pattern, as measured by the degree distribution $P(k)$, defined as the probability that a randomly chosen agent is connected to k other agents (has degree k), plays a relevant role in the scaling of the consensus time, especially for scale-free networks [15] with a degree distribution scaling as a power law, $P(k) \sim k^{-\gamma}$ [13, 14, 16]. A theoretical understanding of the behavior of the voter model on heterogeneous networks was finally put forward by Redner and coworkers [17–19], who showed that for the voter model the consensus time scales with the system size as $T_N \sim N\langle k\rangle^2/\langle k^2\rangle$, i.e., inversely proportional to the second moment of the degree distribution $\langle k^2\rangle = \sum_k k^2 P(k)$. In scale-free networks with degree exponent $\gamma < 3$, the second moment $\langle k^2\rangle$ diverges and becomes size dependent, implying thus a sublinear growth of the consensus time, as previously observed in numerical simulations [13, 16], and contrarily to the homogeneous mean-field expectation.

When discussing the role of heterogeneity in social or evolutionary contexts, a relevant question is whether the complexity of the substrate alone is able to encode the heterogeneity of a realistic dynamical process of social or environmental relevance. The common objection to extremely simple models is that in reality individuals behave and relate to their peers in different ways, i.e., they are *heterogeneous* both in the way in which they interact with others and in the way in which they react to these interactions. For example, in a social context, it could be the case that some agents are more reluctant to change their opinion (*zealots*), while some agents can assign different importance to the opinion of their different neighbors (i.e., close friends can be more trusted than casual acquaintances). Lately, different variations of voter-like models have been put forward, in an effort to take into account the intrinsic variability of agents and their individual propensity to interact with peers [18, 20–24].

In this chapter, we describe a generalization of the voter model on complex networks that encompasses different sources of degree-related heterogeneity and that is amenable to direct analytical solution by applying the standard methods of heterogeneous mean-field theory [11, 12]. Our formalism allows for a compact description of previously proposed heterogeneous voter-like models and represents a basic framework within which we can rationalize the effects of heterogeneity in voter-like models, as well as implement novel sources of heterogeneity, not previously considered in the literature.

2 A Generalized Heterogeneous Voter-Like Model on Networks

We consider a generalized heterogeneous voter-like model, given as a stochastic process on networks, and defined by the following rules [25]:

- Each vertex i has associated a given *fitness* f_i.
- A source vertex i is selected at random, with a probability $f_i/\sum_j f_j$, i.e., proportional to its fitness f_i.
- A nearest neighbor j of i is then selected at random.
- With probability $Q(i, j)$, i copies the state of vertex j with. Otherwise, nothing happens.

The fitness function f_i affects the probability that the given node i is chosen to initiate the opinion-update process at a given time t [18]. In the case of copying dynamics, it measures the propensity of a given node to change its state. Once the individual i is chosen and a neighbor j is selected, the probability $Q(i, j)$ measures how probable the actual update is, introducing a weight in the adjacency relation between the two individuals [24]. We can easily check that the standard voter model is recovered by setting $f_i = Q(i, j) = 1$, while the Moran process corresponds to $f_i = 1$ and $Q(i, j) = k_i/k_j$. Other variations of copying/invasion dynamics can be recovered by the appropriate selection of the f_i and $Q(i, j)$ functions.

Recently, a generalized formalism for the class of copying/invasion voter-like models on networks has been proposed [26, 27] in which the process is identified by the copying rate C_{ij}, encoding the full structure of the contact network and the stochastic update rules, and that in our case takes the form

$$C_{ij} = \frac{f_i}{\sum_p f_p} \frac{a_{ij}}{k_i} Q(i, j), \tag{1}$$

where a_{ij} is the adjacency matrix of the network, taking value 1 if vertices i and j are connected by an edge, and zero otherwise. Within this formalism, exact results can be obtained, but at the expense of computing the spectral properties of the copying rate matrix C_{ij}, a highly nontrivial task unless the matrix C_{ij} has a relatively simple form.

Here we follow a different path, applying to our model the technique of heterogeneous mean-field theory, which leads to simple estimates for central properties such as the exit probability and the consensus time in a rather economical way. While this technique is known to be not exact in several instances, it is still nevertheless able to account with a reasonable accuracy for the results of direct numerical simulations of the model.

3 Heterogeneous Mean-Field Analysis

The analytical treatment of voter-like models on complex networks is made possible by the heterogeneous mean-field (HMF) approach, which has traditionally provided a powerful analysis tool for dynamical processes on heterogeneous substrates [11, 12]. Two main assumptions are made: (*i*) Vertices are grouped into degree classes, that is, all vertices in the same class share the same degree and the same dynamical properties; (*ii*) The real (*quenched*) network is coarse grained into an *annealed* one [11], which disregards the specific connection pattern and postulates that the class of degree k is connected to the class of degree k' with conditional probability $P(k'|k)$ [28]. In general, the HMF approach allows for simple analytic solutions. In the case of voter-like models it has proved remarkably powerful in estimating the quantities of interest, showing reasonable agreement with numerical simulations in real quenched networks [17–19, 21, 24].

Following the standard HMF procedure, we work with the degree-class average of the fitness function and microscopic copying rate. Averages are taken over the set of vertices with a given fixed degree, i.e.,

$$f_i \to \frac{1}{NP(k)} \sum_{i \in k} f_i \equiv f_k, \tag{2}$$

$$Q(i,j) \to \frac{1}{NP(k)} \frac{1}{NP(k')} \sum_{i \in k} \sum_{i \in k'} Q(i,j) \equiv Q(k,k'), \tag{3}$$

where $i \in k$ denotes a sum over the degree class k and $P(k)$ is the network's degree distribution. Thus, f_k represents the fitness of individuals of degree k, assumed to be the same, depending only on degree, for all of them, while $Q(k,k')$ represents the probability that a vertex of degree k copies the state of a vertex of degree k'. Analogously, the contact pattern is transformed according to the method described in [29], obtaining

$$\frac{a_{ij}}{k_i} \to \frac{[NP(k)]^{-1} \sum_{i \in k} \sum_{j \in k'} a_{ij}}{[NP(k)]^{-1} \sum_{i \in k} \sum_{r} a_{ir}} \equiv P(k'|k). \tag{4}$$

Disregarding the *microscopic* details of the actual contact pattern, our generalized voter model is thus defined in terms of the *mesoscopic* copying rate

$$C(k,k') \equiv \frac{f(k)}{\langle f(k) \rangle} P(k'|k) Q(k,k'). \tag{5}$$

Here and in the following, we adopt the convention $\langle \cdot \rangle = \sum_k P(k)(\cdot)$. In order to provide a quantitative measure of the ordering process, we shall consider the time evolution of the fraction of vertices of degree k in the state $+1$, x_k. Transition rates for x_k will be given by the probability $\Pi(k;\sigma)$ that a spin in state σ at a vertex of degree k flips its value to $-\sigma$ [17, 19, 24] in a microscopic time step. It is easy to show that, from the definition of the generalized voter model, such probabilities can be written as

$$\Pi(k;+1) = x_k P(k) \sum_{k'} (1 - x_{k'}) C(k,k') \tag{6}$$

$$\Pi(k;-1) = (1 - x_k) P(k) \sum_{k'} x_{k'} C(k,k'), \tag{7}$$

thus leading to the rate equation [24]

$$\dot{x}_k = \frac{\Pi(k;-1) - \Pi(k;+1)}{P(k)} \equiv \sum_{k'} C(k,k')(x_{k'} - x_k). \tag{8}$$

Given the very broad definition of the interaction rate $Q(k,k')$, a solution to the problem at hand cannot be easily provided in closed form, unless further assumptions are made. We will thus make the following additional assumptions:

(i) Dynamics proceed on uncorrelated networks, i.e., [30]

$$P(k'|k) = \frac{k'P(k')}{\langle k \rangle}. \tag{9}$$

(ii) The interaction rate can be factorized as

$$Q(k,k') = a(k)b(k')s(k,k'), \tag{10}$$

where $s(k,k')$ is any symmetric function of k and k'.

This simplified form encompasses a broad range of voter-like dynamical processes, including most previously proposed models and a variety of novel applications, with the remarkable advantage of being promptly solvable, as it will become clear in the rest of this chapter.

In order to provide a general solution in the most compact notation possible, we rewrite the rate equation as

$$\dot{x}_k = \sum_{k'} P(k')\Gamma(k,k')(x_{k'} - x_k), \tag{11}$$

where we have defined $\Gamma(k,k') = u(k)v(k')s(k,k')$ and

$$u(k) = \frac{a(k)f(k)}{\langle f(k) \rangle}, \qquad v(k') = \frac{b(k')k'}{\langle k \rangle}. \tag{12}$$

We analyze the behavior of the linear process at hand in the canonical way, by first determining the inherent conservation laws [17–19]. We define a generic integral of motion $\omega[x_k(t)]$ for Eq. (11) such that $d\omega/dt = 0$. By definition of total time derivative

$$\frac{d\omega}{dt} = \nabla_{\mathbf{x}}\omega \cdot \dot{\mathbf{x}} = \sum_k \frac{\partial \omega}{\partial x_k}\dot{x}_k = 0. \tag{13}$$

In analogy with previous results [17–19], we look for conserved quantities that are linear in x_k imposing $\partial\omega/\partial x_k = z_k$ independent of x_k, so that conserved quantities will be given by

$$\omega = \mathbf{z} \cdot \mathbf{x} = \sum_k z_k x_k, \tag{14}$$

where z_k is any solution of $\sum_k z_k \dot{x}_k = 0$ and $\dot{x}_k$ is given by Eq. (11). It is easy to prove that $z_k \propto P(k)v(k)/u(k)$ always satisfies the above condition, so that a conserved quantity is found up to multiplicative factors and additive constants.

Upon imposing $\sum_k z_k = 1$ as one of the possible normalization conditions, the conserved quantity becomes

$$\omega = \mathbf{z} \cdot \mathbf{x} = \frac{\langle v(k)/u(k)\, x_k \rangle}{\langle v(k)/u(k) \rangle}. \tag{15}$$

In analogy with the simplest definition of the voter model [17], the conserved quantity bears all the information required to calculate the exit probability E, which we previously introduced as the probability that the final state corresponds to all spins in the state $+1$. In the final state with all $+1$ spins, we have $\omega = 1$, while $\omega = 0$ is the other possible final state (all -1 spins). Conservation of ω implies then $\omega = E \cdot 1 + [1 - E] \cdot 0$; hence,

$$E = \omega = \frac{\langle v(k)/u(k)\, x_k \rangle}{\langle v(k)/u(k) \rangle}. \tag{16}$$

Starting from a homogeneous initial condition, with a given density x of randomly chosen vertices in the state $+1$, we obtain, since $\omega = x$,

$$E_h(x) = x, \tag{17}$$

completely independent of the defining functions a, b, and s, and taking the same form as for the standard voter model [1]. On the other hand, if the initial condition corresponds to a single seed, that is, an individual $+1$ spin in a vertex of degree k,

$$\omega = E_1(k) = \frac{v(k)/u(k)}{N \langle v(k)/u(k) \rangle}, \tag{18}$$

which does not depend on the functional form of the symmetric interaction term $s(k, k')$.

Equation (11) predicts that the set of variables of x_k rapidly converge to a steady state. It is easy to see that any choice of x_k that is constant in k is a solution to the steady state condition $\dot{x}_k = 0$. This solution is unique and does not depend on initial conditions if the square matrix $P(k')\Gamma(k, k')$ is irreducible and primitive (it certainly is when working with positive rates, which we will do in the following) [31]. If we call the steady state x^∞, then it is easy to prove that

$$\omega = \sum_{k'} z_{k'} x_{k'} = x^\infty, \tag{19}$$

that is, the steady state value for x_k equals the conserved quantity. Such result is well known in simpler formulations of the voter model and becomes crucial in the computation of the consensus time, even in our generalized case.

As we noted above, the convergence to the steady state distribution occurs on very short time scales. As soon as the steady state is reached, stochastic fluctuations become relevant and the system begins to fluctuate diffusively around this value, until consensus is reached in one of the two symmetric states. Such fluctuations are integral to finite systems and occur at long time scales, making this time-scale separation possible in large enough systems. In the light of such considerations, the average consensus time $T_N(\mathbf{x})$ for a system in a generic steady state $\mathbf{x}$ can be derived extending the well-known recursive method to our general case [19]. At a given time t, $T_N(\mathbf{x})$ must equal the average consensus time at time $t + \Delta t$ plus the elapsed time $\Delta t = 1/N$ that is, in our notation,

$$T_N(\mathbf{x}) = \bar{\Pi}\, T_N(\mathbf{x}) + \sum_{k,s} \Pi(k;s) T_N(\mathbf{x} + \mathbf{\Delta x}^{(k)}) + \Delta t, \tag{20}$$

where $\bar{\Pi} = 1 - \sum_{k,s} \Pi(k;s)$ is the probability that no state change occurs, while the sum is the weighted average over possible state updates $\mathbf{x} \to \mathbf{x} + \mathbf{\Delta x}^{(k)}$. The variation $\mathbf{\Delta x}^{(k)}$ is a vector whose all components are zero except for the k-th, which equals the update-unit $\Delta_k = [NP(k)]^{-1}$. Expanding to second order in Δ_k, taking $x_k = \omega$ as the initial state, and changing variables such that $\partial/\partial x_k = z_k \partial/\partial \omega$, we obtain the backward Kolmogorov equation

$$-1 = \frac{\mathbf{z}^T \Gamma \mathbf{z}}{N} \omega(1-\omega) \frac{\partial^2 T_N}{\partial \omega^2} \tag{21}$$

leading to

$$T_N = -N_{\mathrm{eff}}[\omega \ln \omega + (1-\omega)\ln(1-\omega)] \tag{22}$$

where we have defined the effective system size $N_{\mathrm{eff}} = N/\sum_{k,k'} z_k \Gamma(k,k') z_{k'}$, which, in the case of generalized voter dynamics, Eq. (12), becomes

$$N_{\mathrm{eff}} = N \frac{\langle f(k)\rangle \langle k\rangle \left\langle \frac{kb(k)}{f(k)a(k)} \right\rangle^2}{\left\langle\!\left\langle s(k,k') kb(k) \frac{[k'b(k')]^2}{f(k')a(k')} \right\rangle\!\right\rangle}, \tag{23}$$

where we have defined $\langle\langle \cdot \rangle\rangle = \sum_{kk'} P(k)P(k')(\cdot)$.

4 Particular Cases

With the formalism developed above, we can easily recover several of the variations of the voter model proposed in the past. In order to validate the results from this formalism, we will look at some of them in this section.

4.1 Standard Voter Model and Moran Process

The standard voter model and Moran process can be recovered by setting $a(k) = b(k) = f(k) = s(k,k') = 1$ and $a(k) = k$, $b(k) = k^{-1}$ and $f(k) = s(k,k') = 1$, respectively. In this case the known results [17, 19] are recovered. Thus, for the voter model, the conserved quantity is

$$\omega = \sum_{k'} \frac{k' P(k')}{\langle k \rangle} x_{k'}(t), \tag{24}$$

the exit probability starting from a single vertex of degree k in state $+1$ is

$$E_1(k) = \frac{k}{N \langle k \rangle}, \tag{25}$$

and the consensus time takes the form

$$T_N(\omega) = -N \frac{\langle k \rangle^2}{\langle k^2 \rangle} \left[\omega \ln \omega + (1-\omega) \ln(1-\omega) \right]. \tag{26}$$

On the other hand, for the Moran process, we have

$$\omega = \frac{1}{\langle k^{-1} \rangle} \sum_k \frac{P(k)}{k} x_k \tag{27}$$

$$E_1(k) = \frac{1}{k} \frac{1}{N \langle k^{-1} \rangle} \tag{28}$$

$$T_N(\omega) = -N \langle k \rangle \langle k^{-1} \rangle \times \left[\omega \ln(\omega) + (1-\omega) \ln(1-\omega) \right]. \tag{29}$$

The difference between voter and Moran dynamics is quite evident here. By looking the expressions for the conserved quantities, Eq. (24) states that for the voter model densities are weighted with a factor $k / \langle k \rangle$ which compensates the tendency of small degree nodes to change their state, whereas in the Moran process the exact opposite occurs, being the density in Eq. (27) balanced by the factor $k^{-1} / \langle k^{-1} \rangle$. This fact translates in the different forms of the exit probability, Eqs. (25) and (28). For the voter model, a single *mutant* opinion can spread to the whole system more easily when it first starts in a vertex of large degree, due to the fact that large degree vertices are copied from with larger probability. On the other hand, a *mutant* in the Moran process is able to spread faster if it starts on a low-degree vertex, owing to the corresponding fact that low-degree vertices are invaded with low probability.

4.2 Voter Model on Weighted Networks

The extension of voter model to weighted networks [24] is motivated by those situations in which the strength of a relation can play a role in the process of opinion formation. In this sense, weights would reflect the evidence that the opinion of a given individual can be more easily influenced by a close friend rather than by a casual acquaintance. In a weighted network, the voter model is defined as follows: At each time step a vertex i is selected randomly with uniform probability; then one among the nearest neighbors of i, namely, j, is chosen with a probability proportional to the weight $w_{ij} \geq 0$ of the edge joining i and j. That is, the probability of choosing the neighbor j is

$$P_{ij} = \frac{w_{ij}}{\sum_p w_{ip}}. \tag{30}$$

Vertex i is finally updated by copying the state of vertex j. If the weights depend on the degree of the edge's endpoints, $w_{ij} = g(k_i, k_j)$, with $g(k, k')$ a symmetric multiplicative function, i.e., $g(k, k') = g_s(k)g_s(k')$ [24], voter dynamics is recovered by setting $s(k, k') = a(k) = f(k) = 1$ and $b(k) = g_s(k)\langle k\rangle / \langle k g_s(k)\rangle$, which leads to an invasion exit probability

$$E_1(k) = \frac{k g_s(k)}{N \langle k g_s(k)\rangle}, \tag{31}$$

and a consensus time

$$T_N(\omega) = -N \frac{\langle k g_s(k)\rangle^2}{\langle k^2 g_s(k)^2\rangle} \left[\omega \ln \omega + (1-\omega)\ln(1-\omega)\right], \tag{32}$$

with the conserved quantity

$$\omega = \sum_{k'} \frac{k' g_s(k') P(k')}{\langle k g_s(k)\rangle} x_{k'}(t). \tag{33}$$

In order to provide an example of weighted voter dynamics, we can consider the special case of weights scaling as a power law of the degree, $g_s(k) = k^{\theta}$ on a scale-free network with degree distribution of the form $P(k) \sim k^{-\gamma}$. The consensus time starting from homogeneous initial conditions, $x_k(0) = 1/2$, takes the form

$$T_N(1/2) = N \ln(2) \frac{\langle k^{1+\theta}\rangle^2}{\langle k^{2+2\theta}\rangle}. \tag{34}$$

From this expression, we can obtain different scalings with the network size N, depending on the characteristic exponents γ and θ. Considering only $\gamma > 2$ and using the scaling behavior of the network upper cutoff $k_c \sim N^{1/2}$ for $\gamma < 3$ and

$k_c \sim N^{1/(\gamma-1)}$ for $\gamma > 3$ [32], we are led to different regions of behavior for $T_N(1/2)$:

1. If $\theta > \gamma - 2$, both $\langle k^{1+\theta} \rangle$ and $\langle k^{2+2\theta} \rangle$ diverge. In particular, $\langle k^{1+\theta} \rangle \sim k_c^{2+\theta-\gamma}$ and $\langle k^{2+2\theta} \rangle \sim k_c^{3+2\theta-\gamma}$. Thus,

$$T_N \sim N k_c^{1-\gamma}. \tag{35}$$

   If $\gamma < 3$, $k_c \sim N^{1/2}$, and $T_N \sim N^{(3-\gamma)/2}$. If $\gamma > 3$, then $k_c \sim N^{1/(\gamma-1)}$, and $T_N \sim$ const.
2. If $\gamma - 2 > \theta > (\gamma - 3)/2$, then $\langle k^{1+\theta} \rangle$ converges and $\langle k^{2+2\theta} \rangle$ diverges. Thus,

$$T_N \sim N k_c^{\gamma-2\theta-3}. \tag{36}$$

   If $\gamma < 3$, $T_N \sim N^{(\gamma-2\theta-1)/2}$; if $\gamma > 3$, then $T_N \sim N^{2(\gamma-\theta-2)/(\gamma-1)}$.
3. If $\theta < (\gamma - 3)/2$, then both $\langle k^{1+\theta} \rangle$ and $\langle k^{2+2\theta} \rangle$ converge, and we have

$$T_N \sim N. \tag{37}$$

These scaling relations contain several interesting aspects. Among these, it is worth highlighting that, in region 1, and for $\gamma > 3$, the analytical results predict a constant scaling for the consensus time, which therefore does *not* depend on the population size. As a consequence, in the thermodynamic limit, the ordering process is instantaneous. However, this turns out to be true only at the mean-field level, i.e., on annealed networks. Simulating the process on quenched networks produces in fact different results, with a clear dependence of the consensus time on N [24]. Henceforth, this is a typical case showing the limits of the mean-field approach in predicting the behavior of phenomena occurring on true finite networks with fixed connections and disorder. Another interesting feature concerns the special value $\gamma = 3$. When $\theta > 0$, this value separates distinct scaling behavior, while as $\theta < 0$ it ceases to be a frontier, the scaling of the consensus being linear in N on both sides of it.

5 A Practical Example: Variable Opinion Strengths

The strength of the proposed generalized formalism resides in the possibility of deriving simple HMF solutions to new variations of the standard voter-like models, which were not studied in the past and whose solution would have been too hard to compute with more exact techniques. As an application of our formalism, we consider the case of voter dynamics in a society in which certain individuals are more likely to align their opinions with those of their neighbors than others [18]. In particular, the propensity of a certain individual to change opinion will depend on the strength of his/her social ties, that is, on his/her number of neighbors.

We can accomplish such a description in our formalism, by encoding this dependence in the fitness function $f(k)$ and assuming $a(k) = b(k) = s(k, k') = 1$ for the sake of simplicity. The HMF solution to such a problem will then be easy to derive. Following the steps illustrated in the previous sections, the conserved quantity reads

$$\omega = \sum_k P(k) \frac{k/f(k)}{\langle k/f(k) \rangle} x_k, \tag{38}$$

the exit probability for an individual seed in state $+1$

$$E_1(k) = \frac{k/f(k)}{N \langle k/f(k) \rangle}, \tag{39}$$

and the consensus time starting from a homogeneous state ω

$$T_N(\omega) = -N \frac{\langle f(k) \rangle \langle k/f(k) \rangle^2}{\langle k^2/f(k) \rangle} [\omega \ln \omega + (1 - \omega) \ln(1 - \omega)]. \tag{40}$$

We can provide a practical example by choosing $f(k) = k^\alpha$. If $\alpha < 0$, less connected individuals are more likely to change opinion, and connectedness can be interpreted as a measure of social self-assurance. If $\alpha > 0$, more connected individuals appear more vulnerable to opinion variability. Interestingly, the standard voter model is recovered for $\alpha = 0$, where no agent heterogeneity is postulated. The conserved quantity assumes the simple form $\omega = \langle k^{1-\alpha} x_k \rangle / \langle k^{1-\alpha} \rangle$ and the consensus time starting from an initial condition $\omega = 1/2$ is then

$$T_N(1/2) = N \ln(2) \frac{\langle k^\alpha \rangle \langle k^{1-\alpha} \rangle^2}{\langle k^{2-\alpha} \rangle}. \tag{41}$$

The size scaling behavior of the consensus time can be derived from Eq. (41), following the procedure devised in the previous section for the voter model on weighted networks. The results are illustrated in compact form in Fig. 1, where a phase diagram for the variables γ and α is shown. The phase diagram conveys great insight into the role played by the parameter α. Focusing on the region with $\gamma > 2$, which is of greater interest in the study of dynamical processes on complex networks, we observe that for $3 - \gamma < \alpha < \gamma - 1$, the simple scaling relation $T_N \sim N$ is recovered. In analogy with a similar phenomenon observed in the case of weighted networks in Sect. 4.2, the value $\gamma = 3$ does not act as frontier and the same scaling law is observed, regardless of the value of γ. Larger values of α break the balance that ensues the $T_N \sim N$ behavior and lead to nontrivial size dependence of the consensus time, with exponents that depend on the degree distribution exponent γ. If we restrict our analysis to the case of scale-free networks, corresponding to the $2 < \gamma < 3$ strip in the phase diagram, we can easily see from the results in Fig. 1 that values of α in the range $|\alpha| < \gamma - 1$ lead to either linear or sublinear size scaling of T_N, whereas for $|\alpha| > \gamma - 1$, superlinear scaling is encountered.

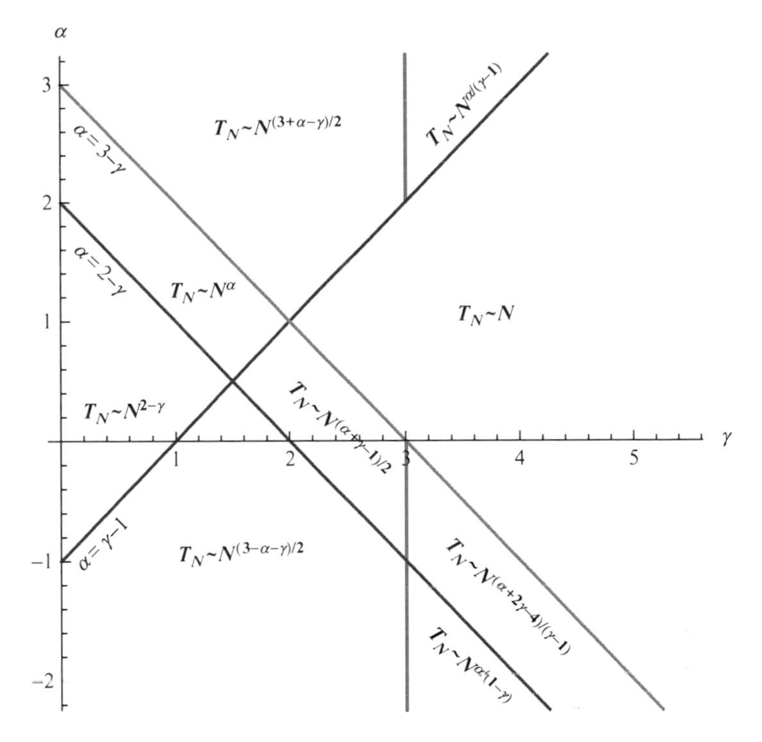

Fig. 1 Phase diagram of the heterogeneous voter model with variable opinion strengths, in which the strength of the opinion of an agent (its willingness to change opinion) is related to its degree by the relation $f(k) = k^{\alpha}$

This translates into the simple observation that in scale-free networks, *large* degree selectivity makes consensus harder to reach in larger systems, regardless of the sign of α, i.e., of whether the degree selectivity makes low-degree individuals or high-degree individuals more vulnerable to opinion change.

In order to corroborate our predictions, we have performed numerical simulations of the voter dynamics at hand in complex networks with power-law distributed degrees, generated with the Uncorrelated Model (UCM) [33]. We check the scaling behavior of the consensus time, $T_N \sim N^{\beta}$, for three pairs of (γ, α) values of the phase diagram, located in the regions where superlinear, linear, and sublinear scalings of T_N are observed respectively.

In Fig. 2 we compare the scaling exponent predicted by Eq. (41), β_{HMF}, with the exponent β_q obtained by fitting the numerical simulations run on quenched networks. The results are summarized in Fig. 2. We note that the HMF theory predicts *qualitatively* well the behavior of the voter dynamics, in the sense that the scaling of T_N with N is superlinear for $\gamma = 2$ and $\alpha = 2$, linear for $\gamma = 2.5$ and $\alpha = 1$, and sublinear for $\gamma = 3$ and $\alpha = -1$, as expected. As for the exact values of the β exponents, the HMF prediction is *quantitatively* accurate only in certain

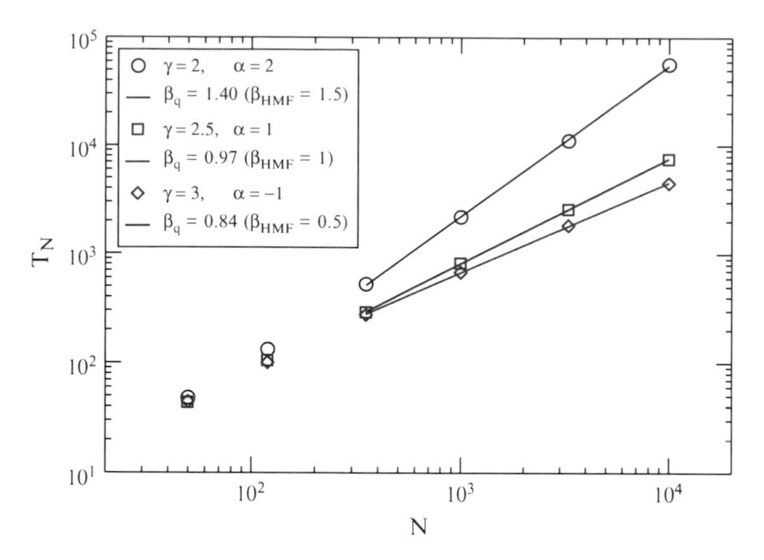

Fig. 2 Scaling behavior of the consensus time T_N as a function of the network size N for different values of γ and α

regions of the phase diagram, where the annealed-network approximation appears to hold. Elsewhere, quenched-network effects take over, and sensible deviations with respect to the HMF value of β are encountered in simulations, in analogy with results for the weighted voter model, as discussed in Sect. 4.2.

6 Conclusions

In this chapter, we have presented a generalized model of consensus formation, which is able to encompass all previous formulations of copy/invasion processes inspired by variations on the voter model and the Moran process. We considered the implementation of such generalized dynamics on a heterogeneous contact pattern, represented by a complex network, and derived the theoretical predictions for the relevant dynamical quantities, within the assumptions of the heterogeneous mean-field theory. We provided a brief review of previous results that can be recovered by our generalized formalism, and finally we considered a novel application to the case of opinion formation in a social network. In particular, we addressed the case in which the opinion strength of an individual is related to his/her degree centrality in the network. We found that in scale-free networks, strong selectivity rules (which make less connected individuals much more prone to change their opinions than more connected ones or vice versa) lead to a steeper growth of consensus time with the system size, making the ordering process slower in general. Numerical simulations on quenched networks show that the HMF theory is able to predict such behavior with reasonable accuracy. Slight deviations from the theoretical predictions are encountered in certain regions of the phase diagram, but they are due to quenched-network effects that the HMF theory is not able to capture.

Acknowledgement We acknowledge financial support from the Spanish MEC (FEDER) under Projects No. FIS2010-21781-C02-01 and FIS2009-08451, and the Junta de Andalucía, under Project No. P09-FQM4682. R.P.-S. acknowledges additional support through ICREA Academia, funded by the Generalitat de Catalunya.

References

[1] C. Castellano, S. Fortunato, V. Loreto, Statistical physics of social dynamics. Rev. Mod. Phys. **81**, 591–646 (2009)
[2] J.F. Crow, M. Kimura, *An Introduction to Population Genetics Theory* (Harper & Row, New York, 1970)
[3] F. Barth, *Ethnic Groups and Boundaries: The Social Organization of Culture Difference* (Little Brown, Boston, MA, 1969)
[4] P. Clifford, A. Sudbury, A model for spatial conflict. Biometrika **60**, 581–588 (1973)
[5] P.A.P. Moran, Random processes in genetics. Proc. Camb. Phil. Soc. **54**, 60–71 (1958)
[6] M.A. Nowak, *Evolutionary Dynamics* (Berknap/Harvard, Cambridge, 2006)
[7] T.M. Liggett, *Stochastic Interacting Particle Systems: Contact, Voter, and Exclusion Processes* (Springer, New York, 1999)
[8] L. Frachebourg, P.L. Krapivsky, Exact results for kinetics of catalytic reactions. Phys. Rev. E **53**, R3009–R3012 (1996)
[9] P.L Krapivsky, S. Redner, E. Ben-Naim, *A Kinetic View of Statistical Physics* (Cambridge University Press, Cambridge, 2010)
[10] M.E.J. Newman, *Networks: An Introduction* (Oxford University Press, Oxford, 2010)
[11] S.N. Dorogovtsev, A.V. Goltsev, J.F.F. Mendes, Critical phenomena in complex networks. Rev. Mod. Phys. **80**, 1275–1335 (2008)
[12] A. Barrat, M. Barthélemy, A. Vespignani, *Dynamical Processes on Complex Networks* (Cambridge University Press, Cambridge, 2008)
[13] K. Suchecki, V.M. Eguíluz, M. San Miguel, Conservation laws for the voter model in complex networks. Europhys. Lett. **69**(2), 228–234 (2005)
[14] C. Castellano, Effect of network topology on the ordering dynamics of voter models, in *Modeling Cooperative Behavior in the Social Sciences*, vol. 779, ed. by J. Marro, P.L. Garrido, M.A. Muñoz (AIP, Melville, NY, 2005), p. 114
[15] A.-L. Barabási, R. Albert, Emergence of scaling in random networks. Science **286**(5439), 509–512 (1999)
[16] C. Castellano, V. Loreto, A. Barrat, F. Cecconi, D. Parisi, Comparison of voter and glauber dynamics on networks. Phys. Rev. E **71**, 066107 (2005)
[17] V. Sood, S. Redner, Voter model on heterogeneous graphs. Phys. Rev. Lett. **94**(17), 178701 (2005)
[18] T. Antal, S. Redner, V. Sood, Evolutionary dynamics on degree-heterogeneous graphs. Phys. Rev. Lett. **96**, 188104 (2006)
[19] V. Sood, T. Antal, S. Redner, Voter models on heterogeneous networks. Phys. Rev. E **77**, 041121 (2008)
[20] N. Masuda, N. Gibert, S. Redner, Heterogeneous voter models. Phys. Rev. E **82**(1), 010103 (2010)
[21] C.M. Schneider-Mizell, L.M. Sander, A generalized voter model on complex networks. J. Stat. Phys. **136**, 59–71 (2009)

[22] H.-X. Yang, Z.-X. Wu, C. Zhou, T. Zhou, B.-H. Wang, Effects of social diversity on the emergence of global consensus in opinion dynamics. Phys. Rev. E **80**(4), 046108 (2009)
[23] Y.-T. Lin, H.-X. Yang, Z.-H. Rong, B.-H. Wang, Effects of heterogeneous influence of individuals on the global consensus. Int. J. Modern Phys. C **21**, 1011–1019 (2010)
[24] A. Baronchelli, C. Castellano, R. Pastor-Satorras, Voter models on weighted networks. Phys. Rev. E **83**, 066117 (2011)
[25] P. Moretti, S.Y. Liu, A. Baronchelli, R. Pastor-Satorras, Heterogenous mean-field analysis of a generalized voter-like model on networks. Eur. Phys. J. B **85**(3), 88 (2012)
[26] G.J. Baxter, R.A. Blythe, A.J. McKane, Fixation and consensus times on a network: A unified approach. Phys. Rev. Lett. **101**, 258701 (2008)
[27] R.A. Blythe, Ordering in voter models on networks: exact reduction to a single-coordinate diffusion. J. Phys. A Math. Theor. **43**(38), 385003 (2010)
[28] M. Boguñá, R. Pastor-Satorras, Epidemic spreading in correlated complex networks. Phys. Rev. E **66**, 047104 (2002)
[29] A. Baronchelli, R. Pastor-Satorras, Mean-field diffusive dynamics on weighted networks. Phys. Rev. E **82**, 011111 (2010)
[30] S.N. Dorogovtsev, J.F.F. Mendes, *Evolution of Networks: From Biological Nets to the Internet and WWW* (Oxford University Press, Oxford, 2003)
[31] F.R. Gantmacher, *The Theory of Matrices - Volume Two* (AMS Chelsea Publishing, Providence, Rhode Island, 2000)
[32] M. Boguñá, R. Pastor-Satorras, A. Vespignani, Cut-offs and finite size effects in scale-free networks. Eur. Phys. J. B **38**, 205–210 (2004)
[33] M. Catanzaro, M. Boguñá, R. Pastor-Satorras, Generation of uncorrelated random scale-free networks. Phys. Rev. E **71**, 027103 (2005)

Epidemics on a Stochastic Model of Temporal Network

Luis E.C. Rocha, Adeline Decuyper, and Vincent D. Blondel

1 Introduction

Individual contacts between people or animal movements between farms serve as pathways where infections may propagate. Characterizing and modeling these contacts are, therefore, fundamental to unveil potential infection routes, to understand the emergence of epidemics, and to control or avoid the epidemics [4]. One way to represent the patterns of contacts or movements is through networks [4, 14]. The network structure captures the heterogeneity of individual behavior and goes beyond random mixing where everyone can interact with everyone else. While static structures have been the common modeling paradigm, recent results suggest that temporal structures play different roles to regulate the spread of infections [3, 5, 16, 17, 19–21, 24, 25] (or infection-like dynamics, as in information spreading [12]) and the actual impact of evolving structures on epidemics remains unclear. This has shifted the research attention to the understanding of such changing structures [9]. On temporal networks, a particular vertex is active only at certain moments and inactive otherwise such that a contact is not continuously available [9, 12, 19]. The temporal order of contacts restricts the infection routes between vertices in comparison to nontemporal networks [15, 22]. The time between two consecutive vertex-activation events is not necessarily uniform, but in many datasets it follows heterogeneous patterns (e.g., bursts); in particular it does so on several networks of relevance to spread of infections, as, for example, sexual [18], proximity [20, 21], or communication contact networks (of interest to spread of electronic virus) [10, 12]. Previous research is mostly based on data-driven studies in which the evolution of the epidemics is contrasted between the original temporal sequence and randomized

L.E.C. Rocha (✉) • A. Decuyper • V.D. Blondel
Department of Mathematical Engineering, Université catholique de Louvain, Louvain-la-Neuve, Belgium
e-mail: Luis.Rocha@uclouvain.be; adeline.decuyper@uclouvain.be; vincent.blondel@uclouvain.be

A. Mukherjee et al. (eds.), *Dynamics On and Of Complex Networks, Volume 2*, Modeling and Simulation in Science, Engineering and Technology, DOI 10.1007/978-1-4614-6729-8_15,

versions, where the original time stamps of the edges are shuffled to destroy the temporal correlations [12, 19]. While this approach is adequate for many purposes, it is difficult to study the sole effect of a particular temporal constraint. Other topological structures (e.g., degree distribution, degree-degree correlation, or community structure) can be present and also contribute to regulate the infection spread. Some studies suggest that heterogeneous inter-event time implies a slow decay time of the prevalence during an infection dynamics [16, 23] and slowdown of the spreading dynamics in the context of communication networks [10, 12]. The opposite effect, i.e., the speedup of the infection growth due to broad inter-event activation times, is observed in a sample of sexual contacts network [19]. To improve the understanding of such contrasting behavior, in this chapter, we present a simple and intuitive stochastic model of a temporal network [16, 19] and investigate how a simulated infection coevolves with the temporal structures, focusing on the growth dynamics of the epidemics. The model assumes no underlying topological structure and is only constrained by the time between two consecutive events of vertex activation, hereafter called vertex inter-event time.

The network model consists of random activations of a vertex according to a predefined vertex inter-event time distribution. The vertices active at a given time are randomly connected in pairs during one time unit. The link is then destroyed and the vertices set to the inactive state. The infection event occurs through this link if one of the vertices is in an infective state. We study this model by using a susceptible infective dynamics with one (SI) and two (SII) infective stages. The first dynamics is motivated for being an upper limit case, where, once infected, the vertex continues infecting at every contact [7]. The second dynamics is more realistic and corresponds to a model of HIV spreading including an acute (high infectivity) and chronic (low infectivity) stages of infection with different periods [8]. If the second stage is set to zero in the SII model, we recover the susceptible-infected-recovered (SIR) dynamics [7].

To study our model, we compare the effects of two vertex inter-event time distributions on the spread of an infection. We consider the geometric[1] (corresponding to uniform probability of being active) and the power-law (to reproduce heterogeneous inter-event patterns) distributions. The main observation is that the speed of the infection spread is different for both cases but the differences depend on the stage of the epidemics. In comparison to the homogeneous scenario, the power-law case results in a faster growth in the beginning but turns out to be slower after a certain time, taking several time steps to reach the whole network.

The chapter is organized in 5 sections. We introduce the stochastic model and discuss some structural properties of the resulting network in Sect. 2 and introduce and describe the epidemics models in details in Sect. 3. The results of the SI epidemics on the networks following the different vertex inter-event time distributions are presented and discussed in Sect. 4 together with the results of the SII model. We conclude the chapter in Sect. 5 highlighting the major contributions.

---

[1]The geometric is a discrete time equivalent to the exponential distribution.

2 Network Model

In this section, we describe the stochastic model of the temporal network and discuss some key structural properties of the evolving network.

2.1 Network Evolution

The network is dynamic in the sense that vertices are active only at certain moments in time and inactive otherwise. As a consequence, edges also appear and disappear throughout the network evolution. In our model (see Table 1 for a summary of variables and parameters of the model), whenever a vertex becomes active, it is randomly connected to another active vertex [16, 19]. The corresponding edge remains available during one time step and is destroyed afterwards (see Algorithm 7).

The activation time is a fundamental property of our model and depends on the system of interest. It has been observed in different contexts that vertices are not active uniformly in time but often follow different patterns, i.e., the inter-event time ΔT_V between two vertex activations is not uniform but follows heterogeneous

---

Algorithm 7 Network model

Initialization step:
for $i = 1 \rightarrow N$ **do**
  $t_{next_i} \leftarrow$ number drawn from D_{pow}
end for
Network Evolution:
for $t = 1, 2, \ldots$ **do**
  $t_{next} \leftarrow t_{next} - 1$
  $V \leftarrow$ all nodes v such that $t_{next_v} = 0$
  if $|V|$ is odd **then**
    select one node u at random from V
    $t_{next_u} \leftarrow 1$
    $V \leftarrow V \backslash u$
  end if
  while $V \neq \emptyset$ **do**
    select two nodes at random $u, v \in V$
    make a link between u and v
    $t_{next_u} \leftarrow$ number drawn from D_{pow}
    $t_{next_v} \leftarrow$ number drawn from D_{pow}
    $V \leftarrow V \backslash \{u, v\}$
  end while
end for

Table 1 Summary list of variables, parameters, and symbols used in the text

Symbol	Description
D_{pow}	Power-law distribution
D_{geo}	Geometric distribution
D_{mf}	Mean field for the geometric case
t_{next_i}	Next time at which vertex i will be active
ΔT_{V}	Inter-event time between subsequent activations of the same vertex
ΔT_{E}	Inter-event time between subsequent activations of the same edge
N	Total number of vertices in the network
n_{active}	Number of active vertices at a given time step
$\langle \Delta T \rangle$	Average inter-event time
λ_1	Rate of the exponential cutoff for the power-law case
λ_2	Rate of the Poissonian process, i.e., $\lambda_2 = 1/\langle \Delta T \rangle$
α	Exponent of the power-law inter-event time distribution
SI	Susceptible-Infected 1 epidemics model
SII	Susceptible-Infected 1-Infected 2 epidemics model (for HIV)
β_1	Per-contact infection probability at stage 1
β_2	Per-contact infection probability at stage 2
T_1	Length of stage 1
T_2	Length of stage 2
S_{t}	Number of susceptible vertices at time t
I_{t}	Number of infected vertices at time t
$\langle \Omega(t) \rangle$	Prevalence of the infection, i.e., $\langle \Omega(t) \rangle = I_t/N$
$T_{10\%}$	Time to infect 10% of the population
$T_{90\%}$	Time to infect 90% of the population

patterns. In particular, the inter-event times on sexual and proximity contacts, e-mail, and cell phone communication patterns are reasonably well described by power-law like ΔT_{V} distributions [10, 12, 18, 20, 21]. A power-law ΔT_{V} distribution means that there are bursts, i.e., trains of activations followed by periods of inactivity of various lengths. There are different theories trying to explain such behavior in the context of communication activity. One theory is the priority queuing model [2], and the other is based on a nonhomogeneous Poissonian process including periodic activity patterns [13].

To simulate this characteristic in our model, for each active vertex at time t', we sample the next activation time t_{next} from a power-law (with cutoff) inter-event time distribution (Eq. (1), hereafter referred to as D_{pow}) such that $\Delta T_{\text{V}} = t_{\text{next}} - t'$. Note that this is equivalent to say that the probability of being active at time t_{next}, given that the vertex was active at time t', decreases as t_{next} gets larger and this decrease follows a power law. Since we have a stochastic model, the cutoff is necessary to avoid that large values of t_{next}'s to be selected during the evolution of

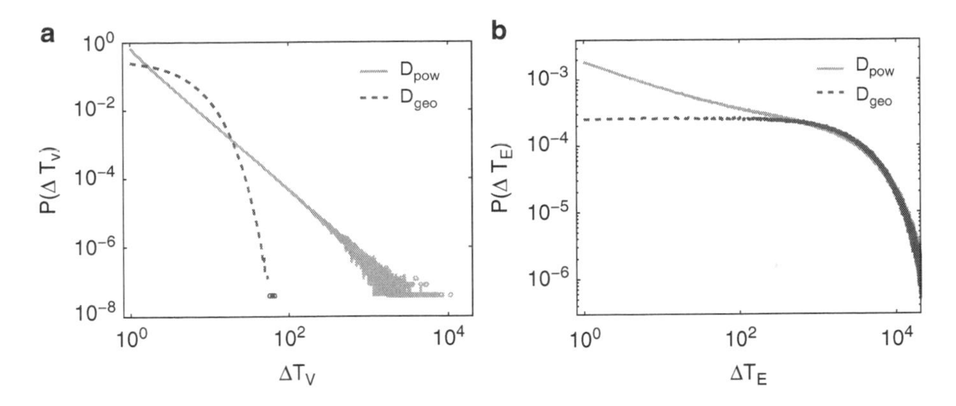

Fig. 1 Probability of ΔT, the inter-event time between subsequent activations, for the power law with cutoff (*full line*) and the geometric cases (*dashed line*), for the (**a**) vertices and (**b**) edges. The axes are in log scale

the network. Since the epidemics dynamics occurs within few time steps, we set the cutoff to a reasonable value ($\lambda_1 = 5 \cdot 10^{-4}$) so that the exponential cutoff does not affect the dynamics and the power-law regime prevails. In line with empirical observations of the inter-event time (e.g., [12, 18]), we set α to 2 (Fig. 1a). Random initial activation times t_{next} are also selected from the distribution of all vertices, and to remove transient oscillations, we let the network evolve during 1,000 time steps before the analysis (see Algorithm 7). Other distributions may be more adequate and statistically significant to model specific datasets, but the power law given by Eq. (1) represents a class of broad distributions and contains the main characteristic that we are interested to study, i.e., the presence of bursts of activities. One should expect, however, intermediate qualitative results for distributions lying between the limiting cases of the power-law and the exponential distributions of inter-event times (see below):

$$P(\Delta T) \propto \Delta T^{-\alpha} e^{-\lambda_1 \Delta T}, \tag{1}$$

We compare the effects of the heterogeneous inter-event time with a null model where the probability λ_2 of a vertex being active is constant (Poissonian process). This constant probability gives an exponential distribution of inter-event times (Eq. (2), hereafter referred to as D_{geo}) where $\lambda_2 = 1/\langle \Delta T \rangle$ ($\langle \Delta T \rangle$ is the mean inter-event time and set to be the same value as obtained from the D_{pow} case, i.e., $\lambda_2 = 4.024$ for $\alpha = 2$. To be more accurate in the simulations, we use a geometric distribution since we are working in discrete times:

$$P(\Delta T) \propto e^{-\lambda_2 \Delta T}. \tag{2}$$

Algorithm 8 Random values using a power law with exponential cutoff

while x is rejected **do**
  $r \leftarrow$ random number drawn uniformly from the interval $[0, 1)$
  $x \leftarrow \lfloor (x_{\min} - \frac{1}{2})(1-r)^{-1/(\alpha-1)} + \frac{1}{2} \rfloor$
  $\rho \leftarrow e^{-\lambda x}$, probability of acceptance of x
  $p \leftarrow$ random number drawn uniformly from the interval $[0, 1)$
  if $p < \rho$ **then**
    return x
  else
    x is rejected
  end if
end while

To generate random samples following the geometric distribution, one can use the inverse transform sampling, rounding down the value to the nearest integer [6]. This function is typically available in math libraries for most programming languages. The case of the D_{pow} distribution is a bit trickier, and we use a method similar to the one proposed in [6], shortly described by Algorithm 8.

We show on Fig. 1a numerically generated random samples using the methods described above for the two distributions D_{pow} and D_{geo}. It is visible that the exponential cutoff in the power law has little or no effect for ΔT_{V} values smaller than 300, but still the maximum values can reach about $\Delta T_{\text{V}} = 10,000$, while the maximum values of the geometric distribution are not larger than 70.

2.2 *Characteristics of the Temporal Network*

The proposed model creates evolving networks without assuming any information about the network topology. After a number of time steps, the majority of the vertices have been in contact with every other vertex at least once, which means that the vertex degree is approximately N_{V} for long times. It is, therefore, more meaningful to highlight that the distribution of the total number of contacts has a characteristic value, although the spread is larger for D_{pow} in comparison with the case of D_{geo} (e.g., see Fig. 2a for the accumulated degree distribution during 60 time steps; the number of vertices with degree equal to zero decreases for longer time steps). In other words, the vertices have a characteristic number of contacts with the same partner. The number of active vertices per time step n_{active} is similar for both cases, and for simulations using a total of $1,000$ vertices, n_{active} oscillates about a mean value of about 250 vertices (Fig. 2b).

The dynamics of the specific vertex inter-event times has different consequences on the edge inter-event time ΔT_{E}, i.e., the time between the subsequent activations of the same edge. The power-law inter-event time on the vertices results in a

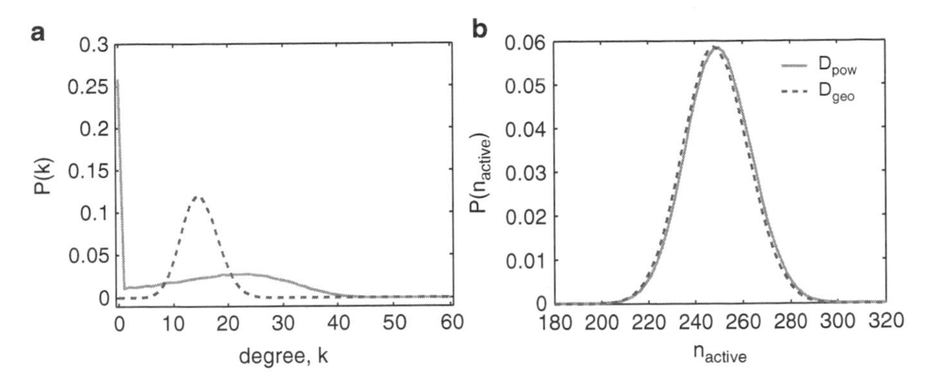

Fig. 2 (**a**) Degree distribution of the accumulated network during 60 time steps and (**b**) probability of having n_{active} active vertices per time step in a network with a total of $1,000$ vertices

distribution of ΔT_{E} following a power-law functional form during a long interval, a characteristic also observed on empirical analysis of sexual contacts [18] and cell phone communication [12]. As expected, the uniform activation probability of the D_{geo} case gives a constant probability for the ΔT_{E} distribution (Fig. 1b).

3 Epidemics Models

In this section, we describe the epidemics models and present our simulation results for the spread of infections in the evolving network structure.

3.1 SI and SII Models

To investigate the impact of the heterogeneous vertex inter-event time on the spread of infections, we simulate two complementary epidemics models on the evolving network. The first model, susceptible infective (SI), mimics a scenario where the vertices remain infective after an initial infection. Although this model is unrealistic to model actual infections, it is a good case study because it corresponds to an upper bound of infection or a worst-case scenario. To study the propagation of a hypothetical infection with a more realistic model, we run simulations for the susceptible infective model with two infective stages (SII). This model is adequate to represent the spread of HIV where an individual is highly infective during the initial months after a primary infection (acute stage), followed by a chronic stage of low infectivity.

In both epidemics models, all vertices start the dynamics at a susceptible state, except for one random vertex in the infected state. A vertex becomes infective after a contact with another infected vertex with probability β (per-contact infection probability or transmissibility). For the SI model, the states do not change after an event of infection, but for the SII, the probability of vertex A infecting vertex B is β_1 in the initial T_1 time steps after the first infection of A, and β_2 after T_1 time steps. Assuming that one time step corresponds to one day, which is a reasonable approximation for sexual contact networks, we set $T_1 = 90$ days, which is the average time of the acute stage for HIV, and following typical values, we use $\beta_1/\beta_2 = 10$ [8, 11]. We run the simulations on 100 network ensembles with 10 random infection seeds for each, and thus, we average our results over $1,000$ simulations (see Table 1 for a summary of variables and parameters of the model).

3.2 Mean-Field SI Epidemics for the Geometric Case

In this section, we discuss the case of the temporal network where the vertex activation follows a Poissonian process that can be described by a simple mean-field approximation. A discrete time SI epidemics evolving within a population of vertices homogeneously mixed can be described by the set of equations (3) [1]:

$$\begin{cases} S_{t+1} = & S_t\left(1 - \frac{\gamma\Delta t}{N} I_t\right) \\ I_{t+1} = & I_t\left(1 + \frac{\gamma\Delta t}{N} S_t\right) \end{cases} \tag{3}$$

where S_t and I_t represent respectively the number of susceptible and infected vertices in the network at time t, N is the total number of vertices in the network ($N = S_t + I_t$ is constant for all times), and γ is the average number of infectious contacts an infected individual makes per unit time interval. The initial condition is given by $I_0 > 0$ and $S_0 = N - I_0$.

The system of Eq. (3) describes the SI dynamics on our geometric case if we set $\Delta t = 1$ and $\gamma = \lambda\beta_1$, where λ is the average number of contacts per vertex per time step and β_1 is the probability of per-contact infection between a susceptible and an infected vertex. Since each active vertex makes exactly one contact during a given time step, the average number of contacts per vertex per time step is, consequently, the probability for a given vertex to be active at a given time step (4):

$$\lambda = Prob(\text{a vertex is active at time } t) = \frac{\langle n_{\text{active}} \rangle}{N} \tag{4}$$

4 Results

In this section we present the results of the coevolution of the epidemics model and the network structure and discuss the impact of the heterogeneous vertex inter-event time on the spread of infections.

4.1 *Effect of the Inter-event Time on the SI Dynamics*

To obtain a global picture of the propagation of the infection, we show in Fig. 3a the average number of infected vertices per time, $\langle \Omega(t) \rangle$. The figure contrasts the results for the different distributions of inter-event time D_{pow} and D_{geo}. For all studied parameters, the infection grows faster in the initial period for the case of D_{pow} in comparison to the case of D_{geo} but slows down afterwards, such that it takes longer to infect all vertices in the case of a power-law distribution. The difference in the growth patterns is weaker for smaller values of β. During the initial interval before the inflection point, for both inter-event cases, the infection grows exponentially but at different rates.

Figure 3b shows the evolution of $\langle \Omega(t) \rangle$ for the mean-field (D_{mf}) approximation and for the numerical simulation of the stochastic model using the geometric vertex inter-event time. For the mean-field case, we set an initial condition $I_0 = 1$ and $\langle n_{\text{active}} \rangle = 249.75$ (a value obtained from the numerical simulation; see Fig. 2) which gives $\lambda_2 = \frac{249.75}{1{,}000} \sim 0.25$ in the case of a network with $1{,}000$ vertices. The theoretical and numerical curves closely overlap for most of the interval of study, with a little mismatch when almost all the network is infected.

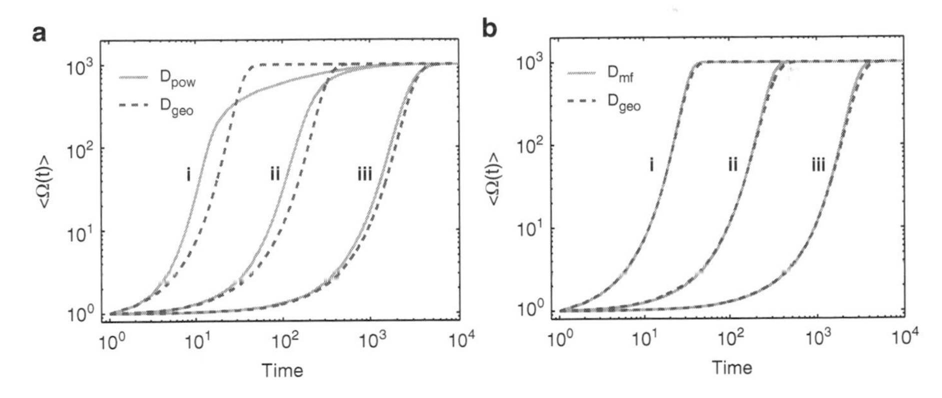

Fig. 3 The average number of infected vertices, $\langle \Omega(t) \rangle$, for 3 probabilities of infection, $\beta = 1$ (**i**), 0.1 (**ii**), and 0.01 (**iii**), for (**a**) power-law D_{pow} and geometric D_{geo} cases, and (**b**) mean-field D_{mf} and geometric D_{geo} cases. The axes are in log scale

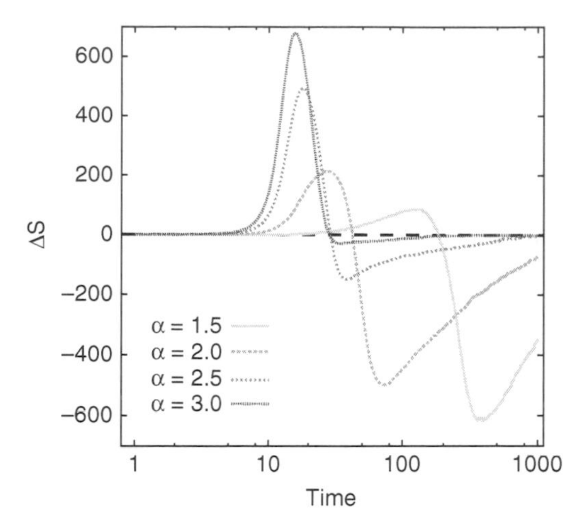

Fig. 4 Difference between the number of infected individuals if the vertex inter-event time follows a power-law distribution and the number of infected individuals if the vertex inter-event time follows a geometric distribution, $\Delta S = \langle \Omega(t)_{\mathrm{pow}} \rangle - \langle \Omega(t)_{\mathrm{geo}} \rangle$. The letter α corresponds to the exponent of the power law, and the geometric version always uses the same average inter-event time as the power law. The x-axis is in log scale

The speedup in the early stage and the slowdown in the final stage of the infection growth for the power-law case in comparison with the geometric case are observed for other values of α, i.e., the exponent of the vertex inter-event time distribution (1). Figure 4 shows the difference in the number of infected vertices in the power-law case, $\langle \Omega(t)_{\mathrm{pow}} \rangle$, and the number of infected vertices in the geometric case $\langle \Omega(t)_{\mathrm{geo}} \rangle$ during the simulation of $1,000$ time steps. Positive values indicate that the power-law case infects more vertices than the geometric one, and negative values indicate the opposite behavior. Independently of the exponent, we observe that during a significant interval, the power-law case leads to a faster infection growth, and this interval increases with decreasing α.

To better understand and characterize the differences in the growth in the early and final stages, we plot the number of time steps necessary to reach 10% ($T_{10\%}$, Fig. 5a) and 90% ($T_{90\%}$, Fig. 5b) of the vertices. We compare how the mean and the median values of this measure vary with β for each vertex inter-event case. The median is a better statistics for nonsymmetric broad distributions, whereas both the median and the mean are adequate statistics for symmetric distributions. Since the distribution of $T_{10\%}$ has no characteristic value for the power-law case (see Fig. 5c and discussion below), we use both statistics to characterize the distributions and to compare the diverging effects. Figure 5a shows that the mean and median for both D_{pow} and D_{geo} converge to similar values for small β, and the median of $T_{10\%}$ for D_{pow} is always smaller than for the case of D_{geo}, indicating that the infection takes less time to reach 10% of the vertices in the D_{pow} case. Conversely, due to the shape of the distribution (Fig. 5c), the mean value gives a biased value and suggests

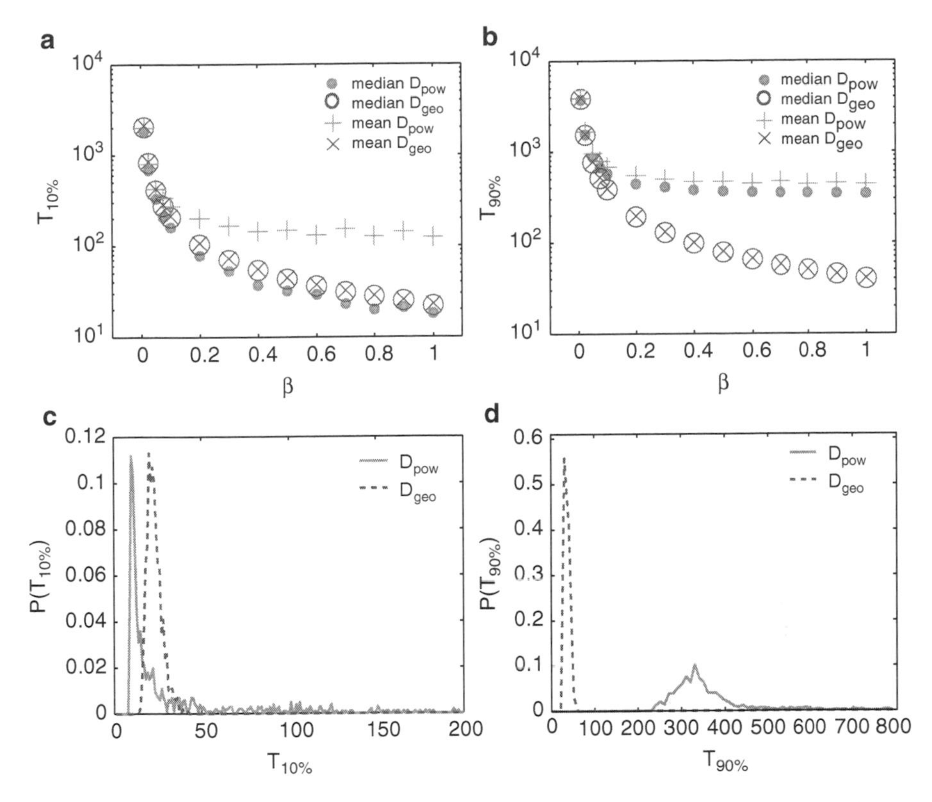

Fig. 5 The per-contact infection probability β versus the mean and median time for the infection to reach (**a**) 10% and (**b**) 90% of the network for both D_{pow} and D_{geo} cases. In case of $\beta = 1$, the probability of (**c**) 10% and (**d**) 90% of the vertices being infected after, respectively, $T_{10\%}$ and $T_{90\%}$ time steps. The y-axes in (**a**) and (**b**) are in log scale

the opposite effect (Fig. 5a). To reach 90% of the vertices, however, the situation is different and the two statistics (the median and mean) are similar for both the case of D_{pow} and for D_{geo} (Fig. 5b). At this later stage of infection, the behavior of the epidemics is more homogeneous, and we identify a characteristic time when 90% of the vertices are infected (Fig. 5d). These results indicate a slowdown of the infection growth in case of a broad inter-event time distribution (Fig. 5b), a result that is in agreement with theoretical studies [10, 16, 23].

As discussed above, the D_{pow} case results in a broad distribution of times for early infections ($T_{10\%}$), meaning that although there is a peak in Fig. 5c, there are several possible scenarios of outbreaks, and these possible scenarios are mainly because of the heterogeneous vertex activity. In fact, out of 1,000 trials, 145 result in times above 200 time steps, and the largest one is 4,453 time steps. As a comparison, the largest time observed in the D_{geo} case is 42 time steps. This narrow distribution of the D_{geo} cases shows that the choice of the initial condition has little influence on the propagation of the infection. The large $T_{10\%}$ values observed in the

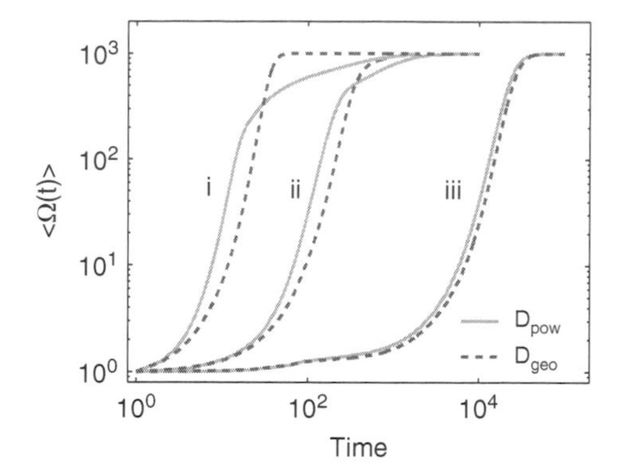

Fig. 6 Average number of infected vertices $\langle \Omega(t) \rangle$ in case of SII dynamics for different combinations of per-contact infection probabilities β_1 and β_2; (**i**) $\beta_1 = 1$ and $\beta_2 = 0.1$, (**ii**) $\beta_1 = 0.1$ and $\beta_2 = 0.01$, and (**iii**) $\beta_1 = 0.01$ and $\beta_2 = 0.001$. The axes are in log scale

D_{pow} case are responsible for the divergence between the mean and the median. A similar effect, i.e., the distribution of inter-event times for the D_{pow} case being broader than for the D_{geo} case, is observed when 90% of the vertices are infected. This is not that strong however because for $T_{90\%}$ the distribution is more symmetric in case of D_{pow} (Fig. 5d). Out of 1,000 trials, 71 result in times above 800 time steps, and 4,843 is the largest value. As a comparison, the largest value for the D_{geo} case is 60 time steps.

4.2 *Effects of the Inter-event Time on the SII Dynamics*

In the case of the SII dynamics, vertices go through two infective stages where the individual contribution to the global epidemics varies at each stage. Since we set the period of the first infective stage to $T_1 = 90$, the growth curve for $\beta_1 = 1$ and $\beta_2 = 0.1$ (Fig. 6) is not much affected in the early stage because most of the network is infected before the period T_1. More specifically, according to Fig. 5b, the mean value to reach 90% of the network is at about 400 time steps. After the inflection point, however, we see that the slowdown interval starts earlier in the SII epidemics. For smaller β, the high and low infective regimes affect the infection growth significantly, especially, for $\beta_1 = 0.01$ and $\beta_2 = 0.001$. This effect is particularly interesting because it corresponds to the range of per-contact infection probability of HIV on several societies [11]. For these values, the epidemics waits roughly 10 times longer to take off in comparison to a one infective stage scenario ($t \sim 100$ for SI and $t \sim 1,000$ for SII; see also Fig. 3a).

5 Conclusion

Recent research has suggested that temporal constraints on network structures affect the spread of infections. In particular, there is evidence that the time between two consecutive events of vertex activation, observed in some empirical networks, speeds up (or slows down in case of SI-like models of information spreading) the spread of infections in comparison to homogeneous vertex inter-event times. These results are based on contrasting the infection growth on the original empirical network and on null models of networks where the original time stamps are randomized by a reshuffling procedure. To investigate the changes in the shape of the infection growth more carefully, we propose in this chapter a simple and intuitive stochastic model of a temporal network where the topology is left completely random but the vertex inter-event time is explicitly controlled by using a power-law (with cutoff) and a geometric inter-event time distributions.

By simulating a SI epidemics model coevolving with the network structure, we observe that the inter-event time affects the infection growth differently according to the stage of the epidemics. In the early stages, in comparison to the homogeneous scenario (Poisson-like dynamics), the heterogeneous (power-law) inter-event time results in a faster infection growth, while at later stages, the infection growth is slower. In our model, these differences result from the inter-event time and do not depend on the network topology (e.g., degree distribution, community structure, degree–degree correlations). We also observe that the heterogeneous inter-event time creates a diversity of outbreaks and highlights the influence of the initial conditions on the infection spread. In other words, while the homogeneous inter-event time results in characteristic times to reach a certain number of vertices, the heterogeneous case produces several scenarios where the infection may take long times before affecting the network significantly.

Acknowledgement LECR is beneficiary of a FSR incoming postdoctoral fellowship of the Académie universitaire Louvain, co-funded by the Marie Curie Actions of the European Commission. AD is a research fellow with the Fonds National de la Recherche Scientifique (FRS-FNRS). Computational resources have been provided by the supercomputing facilities of the Université catholique de Louvain (CISM/UCL) and the Consortium des Équipements de Calcul Intensif en Fédération Wallonie Bruxelles (CECI) funded by FRS-FNRS.

References

[1] L.J.S. Allen, Some discrete-time SI, SIR and SIS epidemic models. Math. Biosci. **124**, 83–105 (1994)
[2] A.-L. Barabási, The origin of bursts and heavy tails in human dynamics. Nature **435**, 207–210 (2005)
[3] P. Bajardi, A. Barrat, F. Natale, L. Savini, et al., Dynamical patterns of cattle trade movements. PLoS ONE **6**, 5 e19869 (2011)

[4] A. Barrat, M. Barthélemy, A. Vespignani, *Dynamical Processes on Complex Networks* (Cambridge University Press, Cambridge, 2008)
[5] S. Bansal, J. Read, B. Pourbohloul, L.A. Meyers, The dynamic nature of contact networks in infectious disease epidemiology. J. Biol. Dyn. **4**, 5 478–489 (2010)
[6] A. Clauset, C.R. Shalizi, M.E.J. Newman, Power law distributions in empirical data. SIAM Rev. **51**(4), 661–703 (2009)
[7] H.W. Hethcote, The mathematics of infectious diseases. SIAM Rev. **42**(4), 599–653 (2000)
[8] T.D. Hollingsworth, R.M. Anderson, C. Fraser, HIV-1 transmission, by stage of infection. JID **198**, 687–693 (2008)
[9] P. Holme, J. Saramäki, Temporal networks. To appear in Phys. Rep. **519**(3), 97–125 (2012)
[10] J.L. Iribarren, E. Moro, Impact of human activity patterns on the dynamics of information diffusion. PRL **103**, 038702 (2009)
[11] J.S. Koopman, J.A. Jacquez, G.W. Welch, et al., The role of early HIV infection in the spread of HIV through populations. JAIDS **14**(3), 249–258 (1997)
[12] M. Karsai, M. Kivelä, R.K. Pan, K. Kaski, et al., Small but slow world: How network topology and burstiness slow down spreading. PRE **83**, 025102 (2011)
[13] R.D. Malmgren, D.B. Stouffer, A.E. Motter, L.A.N. Amaral, A poissonian explanation for heavy tails in e-mail communication. PNAS **105**(47), 18153–18158 (2008)
[14] M. Newman, *Networks: An Introduction* (Oxford University Press, Oxford, 2010)
[15] R.K. Pan, J. Saramäki, Path lengths, correlations, and centrality in temporal networks. PRE **84**, 016105 (2011)
[16] L.E.C. Rocha, V.D. Blondel, Bursts of vertex activation and epidemics in evolving networks. To appear in PLoS Comput. Bio. arXiv:1206.6036 (2013)
[17] J.M. Read, K.T.D. Eames, W.J. Edmunds, Dynamic social networks and the implications for the spread of infectious disease. J. R. Soc. Interface **5**, 26 1001–1007 (2008)
[18] L.E.C. Rocha, F. Liljeros, P. Holme, Information dynamics shape the sexual networks of internet-mediated prostitution. PNAS **107**(13), 5706–5711 (2010)
[19] L.E.C. Rocha, F. Liljeros, P. Holme, Simulated epidemics in an empirical spatiotemporal network of 50,185 sexual contacts. PLoS Comput. Biol. **7**(3), e1001109 (2011)
[20] M. Salathé, M. Kazandjieva, J.W. Leeb, et al., A high-resolution human contact network for infectious disease transmission. PNAS **107**(51), 22020–22025 (2010)
[21] J. Stehlé, N. Voirin, A. Barrat, C. Cattuto, et al., Simulation of a SEIR infectious disease model on the dynamic contact network of conference attendees. BMC Med. **9**(87), 1–15 (2011)
[22] J. Tang, S. Scellato, M. Musolesi, et al., Small-world behavior in time-varying graphs. PRE **81**, 055101(R) (2010)
[23] A. Vazquez, B. Rácz, A. Lukács, A.-L. Barabási, Impact of non-Poisson activity patterns on spreading processes. PRL **98**, 158702 (2007)
[24] M. Vernon, M.J. Keeling, Representing the UK cattle herd as static and dynamic networks. Proc. R. Soc. Lond. B Bio. **276**, 469–476 (2009)
[25] E. Volz, L.A. Meyers, Susceptible-infected-recovered epidemics in dynamic contact networks. Proc. R. Soc. B **274** 2925–2933 (2007)

Network-Based Information Filtering Algorithms: Ranking and Recommendation

Matúš Medo

1 Introduction

After the Internet and the World Wide Web have become popular and widely available, the electronically stored online interactions of individuals have fast emerged as a challenge for researchers and, perhaps even faster, as a source of valuable information for entrepreneurs. We now have detailed records of informal friendship relations in social networks, purchases on e-commerce sites, various sorts of information being sent from one user to another, online collections of web bookmarks, and many other data sets that allow us to pose questions that are of interest from both academical and commercial point of view. For example, which other users of a social network you might want to be friend with? Which other items you might be interested to purchase? Who are the most influential users in a network? Which web page you might want to visit next? All these questions are not only interesting per se, but the answers to them may help entrepreneurs provide better service to their customers and, ultimately, increase their profits.

All the questions posed above have many different ways to be approached that belong to the field of information filtering [1]. The goal of information filtering is to eliminate the redundant or unsuitable information and thus overcome the information overload. In our case, information filtering helps users to choose from an abundant number of possibilities (available products, potential friends, etc.), those that are most likely to be of interest or use for them. Common approaches to this task are based on mathematical statistics, machine learning, and artificial intelligence [2, 3]. They formulate a parametric mathematical model which is calibrated using the readily available data and then use to predict unknown user opinions.

M. Medo (✉)
Physics Department, University of Fribourg, CH-1700 Fribourg, Switzerland
e-mail: matus.medo@unifr.ch

A. Mukherjee et al. (eds.), *Dynamics On and Of Complex Networks, Volume 2*,
Modeling and Simulation in Science, Engineering and Technology,
DOI 10.1007/978-1-4614-6729-8_16,

In this chapter we discuss a different class of algorithms that all make use of a network representation of the data. The current classical example of such an algorithm is PageRank which, while having a far-reaching history [4], has been reinvented and popularized by the founders of Google where it serves up to now as the key element of their Internet search engine [5]. As we shall see below, this algorithm is closely related to random walks that play an important role in physics. (In the case of PageRank, of course, we do not face a random walk in physical space but a random walk on a network consisting of web pages and directed links among them.) These network-based methods can be used alone or in combination with other information filtering techniques, giving rise to hybrid methods [6].

We focus here on two important information filtering tasks—ranking and recommendation. By ranking we mean producing a general list of available items (users or objects) that captures some inherent quality of them. Finding influential users or exceptional web pages belongs to this. By recommendation we mean preparing a specific "recommendation list" for each individual user, each list containing items that are likely to be appreciated by the given user. Finding potential friends or items to purchase belongs here. In addition to traditional unipartite networks where only nodes of one kind are present (such as the network of web sites connected by hyperlinks or a network of users connected by friendship relations), we will often make use of bipartite networks where nodes of two kinds are present. For example, a network connecting users with the items that they have purchased is bipartite because every link connects a user with an item while links between users or between items are entirely absent. For a review of networks and network analysis that do not directly contribute to ranking and recommendation yet they can help to understand the structure of the data in hand, see the survey of complex networks measurements in [7]. For a general overview of dynamical processes on complex networks, see [8].

2 Ranking

When we want to rank nodes of a network, there are obviously many approaches, each of them suiting a different purpose. The simplest possible ranking is by node degree (or, in the case of a directed network, node in-degree) which is based on the assumption that "important" nodes are those that are referred by many other nodes. Many other measures of node importance exist, based either on local or global properties of the given network [9]. In this section, we discuss the importance rankings where score of a node is directly computed by random walk or where score spreads among the nodes in a manner akin to random walk.

2.1 PageRank

When given a directed unipartite network, PageRank [5] is arguably the most famous method to produce a general ranking of the network's nodes. The method is based on the circular idea *"A node is important if it is pointed by other important nodes"* which can be applied to many different situations, including ranking of web sites (an important site is referred by important sites), scientific journals (articles from an important journal are cited by articles from important journals), and people (an important person is referred/trusted by important people). For a review of past research in this direction and the use of this circular idea in various disciplines, see [4].

We begin with a general exposition of the approach, denoting the importance/score of node i as h_i and the nonnegative strength of the link pointing from node i to node j as w_{ij} ($i = 1, \dots, N$ where N is the number of nodes in the network). The above circular thesis can now be formalized as

$$h_j = \sum_i \frac{w_{ij}}{\sum_k w_{ik}} h_i \tag{1}$$

where the division with $\sum_k w_{ik}$ ensures that the importance of node i is distributed among the nodes pointed by it with each node receiving part proportional to w_{ij}. To simplify our notation, we introduce normalized weights $P_{ij} := w_{ij} / \sum_j w_{ij}$. Now we can write $h_j = \sum_i P_{ij} h_i$ which can be further simplified by matrix notation to get

$$\boldsymbol{h} = \mathsf{P}^{\mathrm{T}} \boldsymbol{h}. \tag{2}$$

This matrix form shows that the sought-for vector $\boldsymbol{h}$ is the right eigenvector of P^{T} associated with eigenvalue 1. Since P^{T} is now a column-normalized matrix (also called stochastic matrix), the Frobenius-Perron theorem applies and states that 1 is its largest eigenvalue. A solution to Eq. (2) thus always exists, and when matrix P is irreducible, this solution is unique. (A matrix is irreducible if and only if in the directed graph that the matrix represents there exists a directed path between any two vertices.) The uniqueness is of course up to multiplication of $\boldsymbol{h}$ by a constant factor which allows us to impose the normalization condition $\sum_i h_i = 1$. Note that Eq. (2) is similar to the eigenvector centrality measures that are common in the analysis of social networks [10, 11]. In that case, however, one does not employ a normalized matrix P but the network's adjacency matrix itself and searches for a vector x satisfying $\mathsf{A}^{\mathrm{T}} x = \lambda x$ where λ is a number.

In addition to the redistribution point of view described above, a random-walk view can often provide useful insights. The normalized weights P_{ij} can be interpreted as probabilities of moving from node i to node j and, consequently, h_i as the probability of being at node i. An initial probability distribution $\boldsymbol{h}^{(0)}$ transforms gradually by $\boldsymbol{h}^{(n+1)} = \mathsf{P}^{\mathrm{T}} \boldsymbol{h}^{(n)}$ until a stationary probability distribution corresponding to the largest eigenvalue of P^{T} is established. If this eigenvalue is degenerated, the stationary solution is not unique. The rate of convergence of this iterative method is determined by the magnitude of the second-largest eigenvalue of P^{T}.

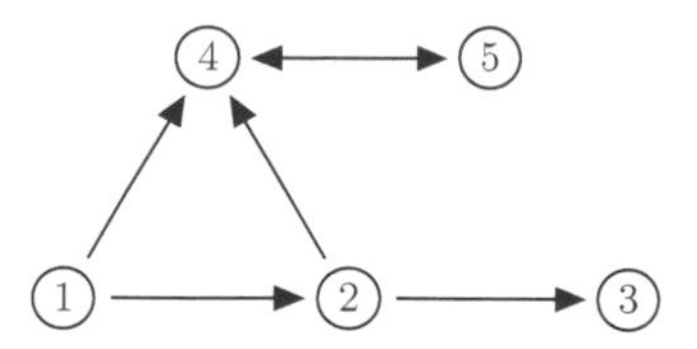

Fig. 1 PageRank computation for a toy network. When $\alpha = 1$ (no random teleportation), scores of nodes 1, 2, and 3 go with iterations to zero, but no stationary distribution exists because one of the eigenvalues is -1 and causes ceaseless alternations of score of nodes 4 and 5. When $\alpha = 0.85$ (the usual value adopted for web site ranking), the resulting score vector is $\boldsymbol{h} = (0.04, 0.06, 0.07, 0.43, 0.40)$. When $\alpha = 0$ (no link-following), all nodes have the same score $1/5$

Our treatment up to now was fully general and applies to any redistribution of h_i values over a weighted network given by weights w_{ij}. Depending on the nature of the problem and the input data, one needs to choose the weights so that the resulting importance vector $\boldsymbol{h}$ contains the information that we are interested in. In the case of PageRank, which was designed to produce the importance score for web sites, the input data consists of a directed network of web sites where a hyperlink from site A to site B can be interpreted as a sign of approval of site B by site A. Since no additional strength information is attached to hyperlinks, the network of hyperlinks is represented by its adjacency matrix A where $A_{ij} = 1$ if there is a link pointing from node i to node j (the network is directed and hence this matrix is not symmetric in general). Weights P_{ij} thus should be the same for all sites j referred by a given node i which, respecting the weight normalization condition, leads to $P_{ij} = A_{ij}/k_i^o$ where k_i^o is the out-degree of node i. Since this is ill defined for nodes with no out-going links ("dangling nodes"), one usually assumes that if $k_i^o = 0$, $P_{ij} = 1/N$ for all j.

One can easily see that even when the problem of nodes with zero out-degree is solved, the resulting solution can easily be pathological in some sense. If the network contains a component without out-going links (so-called bucket; see nodes 4 and 5 in Fig. 1), this part of the network would act as a trap for the random-walk process. Would concentrate there, and it would thus give us little useful information. The inventors of PageRank overcame the problem by postulating that links leading from a node are followed only with certain probability α [5]. With the complementary probability $1 - \alpha$, teleportation (jump) occurs, ending at a randomly chosen node of the network. The corresponding transition matrix (also called Google matrix) is

$$\mathsf{G} = \alpha \mathsf{P}^{\mathrm{T}} + (1 - \alpha)\mathsf{T} \tag{3}$$

where T is the teleportation matrix with all elements equal $1/N$. The parameter α (also called damping) and $1 - \alpha$ determines the weight given to link-following and teleportation, respectively. Since α is the probability of following an out-going link, one can easily compute that the average number of links followed in a row is $\sum_{k=0}^{\infty} k\alpha^k(1-\alpha) = \alpha/(1-\alpha)$. In the original PageRank paper, α was proposed to be set around 0.85 which corresponds to following five or six hyperlinks in a row and then jumping to a random page [5]. The value of α is closely related to the convergence rate of the iterative PageRank computation (the lower the value, the faster the convergence; see [12] for more details). While PageRank was originally devised for directed networks, one can apply it also to undirected networks [13, 14]. The teleportation parameter then plays a crucial role—without it, PageRank score on an undirected network reduces to node degree.

Alternatively, one can replace the uniform teleportation matrix with $\mathbf{1}_N \boldsymbol{v}^{\mathrm{T}}$ where $\mathbf{1}_N$ is an N-dimensional vector of ones and $\boldsymbol{v}$ is a normalized N-dimensional vector which allows us to give preference to some nodes. This provides an important additional degree of freedom and allows one to, for example, devise a topic-specific ranking as described in [15]. A complementary point of view is presented, for example, in [16] where an inverse problem of finding matrix elements G_{ij} from some partial knowledge of node-pair preferences ("we want the score of node i to be higher than that of node j") is studied.

Using the definition of G given in Eq. (3), the PageRank equation $\mathsf{G}\boldsymbol{h} = \boldsymbol{h}$ can be written as $\alpha \mathsf{P}^{\mathrm{T}} \boldsymbol{h} + (1-\alpha)\mathbf{1}_N/N = \boldsymbol{h}$, leading to

$$\boldsymbol{h} = \frac{1-\alpha}{N}(\mathsf{I}_N - \alpha \mathsf{P}^{\mathrm{T}})^{-1}\mathbf{1}_N = \frac{1-\alpha}{N}\sum_{k=0}^{\infty}(\alpha \mathsf{P}^{\mathrm{T}})^k \mathbf{1}_N \tag{4}$$

where I_N is an $N \times N$ identity matrix and $\mathbf{1}_N$ is an N-dimensional vector of ones. Here both the inverse and the series expansion exist as long as $\alpha < 1$. While these formulas for computing $\boldsymbol{h}$ can be easily applied for small systems, a critical advantage of PageRank lies in the fact that the above-mentioned iterative method for finding $\boldsymbol{h}$ is in practice very effective even for very large systems. Thanks to that, PageRank serves as an important input for the Google's ranking of web sites where scores are computed for several billions of pages (for more information on the data mining for the WWW, see [17, 18]). Even for the enormous size of the WWW, only a few tens of iterations are sufficient to compute PageRank to a required precision [19]. The iterative method is also easy to parallelize and, in addition, one can write $\boldsymbol{h}^{(n+1)} = \alpha \mathsf{P}^{\mathrm{T}} \boldsymbol{h}^{(n)} + (1-\alpha)\mathbf{1}_N/N$ and thus benefit from the sparsity of P. In comparison with that, directly multiplying $\mathsf{G}\boldsymbol{h}^{(n)}$ is computationally much more expensive because G has no zero entries.

Another advantage of PageRank is that it is robust to spamming and malicious behavior. This robustness is rooted in the inability of web site administrators to create new hyperlinks pointing to their sites. If they simply create fake new web sites pointing to the site whose status they want to enhance, the artificially created web sites themselves have low scores (because no one points at them) and contribute little to the score of the target site. Of course, various sophisticated methods of manipulating the PageRank still exist [20]. The stability of node rankings obtained with PageRank is the central point in [21] where the authors show that PageRank is particularly prone to noisy data when the network is random (and thus the degree distribution, which is crucial for the ranking's stability, decays exponentially). By contrast, a small number of super-stable nodes whose ranking is particularly resistant to perturbations emerge in scale-free networks.

2.2 Variants of PageRank

From the conceptual point of view, an interesting generalization of PageRank has been proposed in [22] where spreading of the score was separated into branching (due to out-degree) and damping (due to the damping parameter α). In the case of PageRank, damping is exponential because with each propagation step, another multiplication with α is added. The authors show that a power-law damping of the form $1/[(t+1)(t+2)]$ where t denotes the number of steps is equivalent to a so-called TotalRank which is obtained simply by integrating the α-dependent PageRank score over α. Importantly, a linear damping can produce results very close to those obtained with PageRank while requiring fewer iterations to be computed. An important variant of PageRank, EigenTrust, has been proposed to compute trust values in distributed peer-to-peer systems [23]. EigenTrust, which replaces uniform teleportation matrix with random jumps to a set of pre-trusted peers, can be easily computed in a distributed way and is thus suitable for deployment in distributed P2P systems. A very different perspective was adopted in [24] where a class of quantum PageRank algorithms was proposed based on quantized Markov chains.

Almost at the same time as PageRank, another important algorithm based on random walks and circular reasoning was developed independently. It is called HITS (Hypertext-Induced Topic Search), and by contrast to PageRank, it assigns two distinct scores to each node—authority score x_i and hub score y_i [25]. The basic thesis is that a good hub is pointed to by good authorities and vice versa. In mathematical terms, this can be written as

$$\boldsymbol{x}^{(n+1)} = \mathsf{A}^{\mathrm{T}} \boldsymbol{y}^{(n)}, \quad \boldsymbol{y}^{(n+1)} = \mathsf{A} \boldsymbol{x}^{(n+1)} \tag{5}$$

Consequently, one can write $\boldsymbol{x}^{(n+1)} = \mathsf{A}^{\mathrm{T}}\mathsf{A}\boldsymbol{x}^{(n)}$ and $\boldsymbol{y}^{(n+1)} = \mathsf{A}\mathsf{A}^{\mathrm{T}}\boldsymbol{y}^{(n)}$, showing that the stationary authority and hub vectors are the dominant eigenvectors of $\mathsf{A}^{\mathrm{T}}\mathsf{A}$ and $\mathsf{A}\mathsf{A}^{\mathrm{T}}$, respectively. Since these two matrices are not stochastic matrices as it was the case for PageRank, when finding them by iterations, one has to implement additional normalization of the score vectors. In [26], HITS has been generalized to bipartite graphs with the goal to weaken the score reinforcement among the connected nodes and which improve the algorithm's robustness to noisy links. See an extensive review of eigenvector methods for web information retrieval in [27].

In [28], PageRank has been applied to citations among scientific papers (which naturally constitute a directed unweighted network) to assess the relative importance of papers. The authors argued that readers of scientific papers typically follow paths of length two, corresponding to the damping parameter $\alpha = 0.5$ much lower than the original value of 0.85. Albeit the PageRank score of papers was found to be highly correlated with the number of citations (similarly as the PageRank score of web sites is correlated with the number of incoming hyperlinks), significant outliers from this trend were found and identified as seminal publications. This is because the PageRank score redistribution allows a paper with moderate citation count score high thanks to high citation counts of the papers that cite it. As later argued in [29], time decay is of crucial importance in the analysis of citation networks because, unlike hyperlinks in the WWW, citations basically cannot be updated after a paper is published. There is also an increasing evidence that time plays an eminent role in the growth of citation networks—see [30] for a recent account. See also [31] for a general overview of our knowledge about citation networks.

The effect of aging of publications is included in the CiteRank algorithm [32] where the uniform teleportation matrix is replaced with $\mathbf{1}_N \boldsymbol{\varrho}^{\mathrm{T}}$ where $\varrho_i = \exp[-t_i/\tau]$, t_i is the age of paper i, and τ is a characteristic decay time. Interestingly, when the correlation between the CiteRank score and the number of recently gained citations is investigated, the optimal damping parameter α is found to be close to the value of 0.5 which was before only hypothesized on the basis of reading habits of researchers. The authors consequently show that apart from selecting papers that contribute most to the current research, CiteRank is particularly successful in selecting papers of long-lasting interest.

Similarly, the network of scientific journals with links weighted by the number of times an article from journal i cites an article from journal j is again suitable for PageRank-like computation of journal status [33]. Albeit the number of citations does not directly enter here, the resulting ranking of journals is similar to that obtained with the so-called impact factor (which is essentially the average number of citations of recent papers in a given journal). The observed differences in these two measures allowed the authors to introduce the categories of popular journals (which have high impact factors but their citations come from lesser journals, hence the resulting PageRank score of the popular journals is comparatively small) and prestigious journals (which have moderate impact factor but their citations come from journals with high PageRank score, thus allowing the prestigious journals to score high too). A publicly available web site SJR runs a slightly different algorithm based on citations among journals to rank scientific value of journals and countries (see www.scimagojr.com) [34].

What is perhaps of even a greater interest to researchers than rankings of papers and journals are rankings of the researchers themselves. The simplest approach to achieve that would be to create a directed networks of authors where links are created according to who cites whom and weight these links according to the citation frequency for a given pair of authors. To better represent the diffusion of scientific credit in such a network, the authors in [35] propose additional weights reflecting the number of authors of the citing and cited paper, respectively. If the citing paper A was authored by n_A authors and the cited paper B was authored by n_B authors, $n_A n_B$ independent links pointing from an author of paper A to an author of paper B are created, each with weight $1/(n_A n_B)$. The credit of individuals is then redistributed over the weighted author–author network in a usual twofold way: part $1-q$ of i's credit goes to the authors cited by i and part q of i's credit is distributed to all authors according to their productivity. For authors with zero out-strength, it is their whole credit what is distributed to all authors in the network. It is then observed how the resulting ranking of authors changes in time and significant correlations are found between highly ranked authors and important scientific prizes being given to them. A very similar algorithm has been used to rank professional tennis players [36].

Another possible approach to the ranking of researchers is by running a PageRank variant on a so-called coauthorship network which is an undirected network where researchers are connected if they have authored a paper together (it is again natural to weight the connection by the number of papers authored together) [37]. Co-citation networks where authors are connected if they were cited together by a paper were also used as input for a PageRank-based algorithm to obtain a ranking of authors [14].

PageRank has been used also to measure the importance of species in the network of ecological relationships where the loss of a single species can trigger a cascade of extinctions [38]. Upon a minor modification of the input network by introducing a root node which is pointed to by each species and which points back to all "primary producers" (species that do not rely on any other species and produce biomass from inorganic compounds) and setting the damping parameter to one (because nutrients cannot randomly jump among nodes in a food web), the authors were able to use the standard PageRank formula. The obtained importance ranking of species was shown to be very effective in choosing nodes leading to the fastest collapse of the food web, outperforming rankings by betweenness and closeness centrality.

A root node pointed by and pointing to all nodes was used also later in [39] where the PageRank algorithm was used to quantify user influence in a directed social network. It is useful to realize that such a root node in fact serves as a teleportation probability: it leads from a given node to the root node and then in the next step to a randomly chosen normal node. This teleportation probability is node dependent: jump to the ground node occurs with a 50% probability for a node with only one original out-going link, but the probability is only 1% for a node with 99 original out-going links. In addition, this root node causes the transition matrix to be irreducible and primitive which guarantees existence and uniqueness of a stationary solution. Based on the tests on data obtained from the social bookmarking service "Delicious.com," the authors of [39] argue that their variant of PageRank is particularly suitable for social networks as it better detects influential users and it is more resistant to manipulations and noisy data.

2.3 Random Walks with Sources and Sinks

As we have seen above, PageRank is built on a process where the initial node occupancy distribution $\boldsymbol{h}^{(0)}$ is gradually washed away by the random walk and an equilibrium distribution $\boldsymbol{h}^{(\infty)}$ emerges. In some cases, there exist nodes that act as sources or sinks—they constantly emit or absorb, respectively, "particles" that are transported over the network [40]. To allow for termination of the random walk, it is assumed that sources not only emit new particles but also absorb particles arriving in them. Denoting the set of source/sink nodes as S and the set of remaining (transient) nodes as T where $|T| := M$ and thus $|S| = N - M$, we can write the transition matrix in the form

$$\mathsf{P} = \begin{pmatrix} \mathsf{P}_{SS} & \mathsf{P}_{ST} \\ \mathsf{P}_{TS} & \mathsf{P}_{TT} \end{pmatrix} \tag{6}$$

where we have sorted the nodes so that the first $N - M$ nodes are from S and the next M nodes are from T. If S is the set of sinks, then $\mathsf{P}_{ST} = 0$ and $\mathsf{P}_{SS} = \mathsf{I}_{N-M}$. We can now ask what is the probability $F_{ij}(t)$ that a particle originating at $i \in T$ gets absorbed in $j \in S$ in t steps or less, avoiding all other sink nodes on its path. This absorption can either occur in one step, with the probability P_{ij}, or the particle can first go to another transient node k and then be absorbed from there in $t - 1$ steps or less. Together we have

$$F_{ij}(t) = P_{ij} + \sum_{k \in T} P_{ik} F_{kj}(t-1) \tag{7}$$

where, of course, $F_{kj}(0) = 0$ for all k and j. This formula can be written in a matrix form as $\mathsf{F}(t) = \mathsf{P}_{TS} + \mathsf{P}_{TT}\mathsf{F}(t-1)$ where $\mathsf{F}(t)$ is an $M \times (N - M)$ matrix of absorption probabilities. The stationary solution F thus fulfills $\mathsf{F} = \mathsf{P}_{TS} + \mathsf{P}_{TT}\mathsf{F}$, and one can express it as

$$\mathsf{F} = (\mathsf{I}_M - \mathsf{P}_{TT})^{-1}\mathsf{P}_{TS} \tag{8}$$

In the simplest case when $\mathsf{P}_{TT} = 0$ (all links from transient nodes lead directly to sink nodes), we obtain $\mathsf{F} = \mathsf{P}_{TS}$ as expected. One can show that the inverse $(\mathsf{I}_N - \mathsf{P}_{TT})^{-1}$ exists if for every $i \in T$ and $j \in S$, there is a directed path from i to j [40].

The dual problem of particle diffusion from sources can be solved analogously, leading to the average number of times, $H_{ij}(t)$, that a particle originating at a source node i visits a transient node j in t steps or less, without being absorbed in a source node. The final result reads

$$\mathsf{H} = \mathsf{P}_{ST}(\mathsf{I}_M - \mathsf{P}_{TT})^{-1}. \tag{9}$$

Unlike F, a particle can visit a transient node j repeatedly and therefore H_{ij} can be greater than one. The described picture can be generalized to include the possibility of particle dissipation also in transient nodes [40]. There is a close relation between random walks with sinks/sources and currents in electric networks—for details, see [41,42].

PageRank augmented with sinks was shown to increase the diversity of top ranked items [43]. After the top ranked object is found by ordinary PageRank computation, it is turned into a sink and the second object is selected from the remaining transient nodes as the one that has the longest time to absorption. The selected node is then turned into a sink too, and the third object is again found by the absorption time criterion. Since the expected number of visits of node j when starting with node i is $V_{ij} = [(\mathsf{I}_M - \mathsf{P}_{TT})^{-1}]_{ij}$, the expected absorption time of node i is $t_i = \sum_j V_{ij} = (\mathsf{V}\mathbf{1}_M)_i$. The absorption time maximization leads to the preference for nodes that are far away in the given network from the nodes already selected for the top of the ranking, which provides a stimulus to the diversity of results.

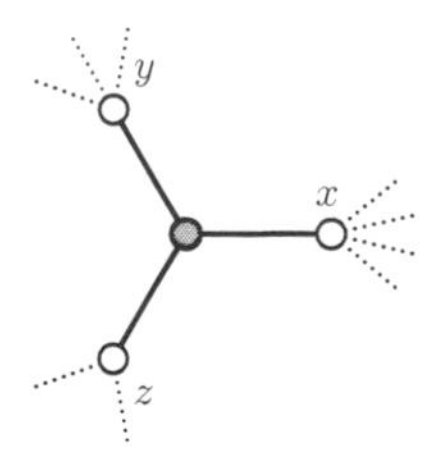

Fig. 2 In random walk, the occupancy probability of the central node in the next time step is $\frac{x}{5} + \frac{y}{4} + \frac{z}{3}$ (where x, y, z are the current occupancy probabilities of the neighboring nodes, respectively). In heat diffusion, the temperature of the central node in the next time step is $\frac{x}{3} + \frac{y}{3} + \frac{z}{3}$ (where x, y, z are the current temperatures of the neighboring nodes, respectively)

We finally note a close connection between random walk and heat diffusion. In random walk, the occupancy probability of a node in the next time step depends on the current occupancy probabilities and degrees of its neighbors. By contrast, in heat diffusion, the temperature of a node in the next time step depends on the current temperatures of its neighbors and the degree of the given node (see Fig. 2 for an illustration). In mathematical terms, while the transition matrix of random walk reads $P_{ij} = A_{ij}/k_i$ and thus P^T is column normalized, the matrix converting the current vector of temperature values into a next time step vector reads $O_{ij} = A_{ij}/k_j$ and thus O^T is row normalized.

Further connections can be found by studying the emission and absorption processes described above. If we fix a sink node j, the probabilities of absorption in j for particles starting in node i, F_{ij}, satisfy the discrete heat equation on the network. This is easy to see on an unweighted undirected network—given a transient node i and its set of neighbors $\mathcal{N}_i$, we can write similarly as in Eq. (7)

$$F_{ij} = \frac{1}{k_i} \sum_{k \in \mathcal{N}_i} F_{kj}$$

That is, the probability of being absorbed in node j when starting in node i is simply the average over these absorption probabilities when starting in neighbors of node i. The boundary condition is given by the sure absorption in j when starting in j and impossible absorption in j when starting in another sink node (corresponding to the boundary probability values 1 and 0, respectively). Generalization to a weighted or undirected network is straightforward. This duality is illustrated on a toy network in Fig. 3.

2.4 Other Algorithms

Node betweenness in a network is calculated as the fraction of the shortest paths between node pairs that pass through a selected node. If the node lies on many shortest paths, it is assumed to be important for information spreading over the

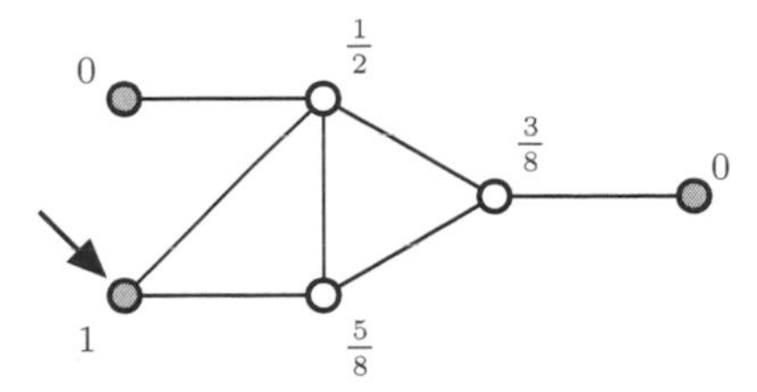

Fig. 3 Random walk with absorption in sink nodes (*shaded with gray*): the probability of being absorbed in the arrow-marked node is shown for each node. These probabilities solve the heat equation with the boundary condition given by the temperature of sink nodes fixed at one (for the *arrow-marked node*) and zero (for all other sink nodes), respectively. For example, the absorption probability $5/8$ for one of the transient nodes can be obtained by averaging the absorption probabilities 1, $1/2$, and $3/8$ of the neighboring nodes

network (e.g., it connects extensive clusters). However, node betweenness considers only the shortest paths and thus neglects a significant part of the network's topology. Random-walk betweenness improves this by considering paths of essentially all lengths, albeit still giving more weight to short ones [42]. It is based on a simple assumption—if random walk starts in node i and ends (gets absorbed) in node j, its contribution to the betweenness of node k is given by the average "net" number of visits of this node during the random walk, $n_k^{(ij)}$. The net number of visits means that passing through a node and then passing through it again from an opposite direction cancel out. Also, if various realizations of random walk are equally likely to pass through a node in opposite directions, these two directions cancel. The resulting betweenness of k is then obtained by averaging the number of visits over start-end node pairs (i, j)

$$b_k = \frac{\sum_{i<j} n_k^{(ij)}}{\frac{1}{2}N(N-1)} \tag{10}$$

where N is the number of nodes in the network. Alternatively, one can obtain the same result building on the electric current injected and removed in a node pair with the contribution to betweenness of node k given by the current passing through this node. The further development is similar to that presented in Sect. 2.3 and ultimately allows to find betweenness values for all nodes in time $O((E + N)N^2)$. This betweenness measure is shown to outperform not only the shortest-path betweenness but also the flow betweenness [42]. With a similar goal, several network flows were typologized and studied by simulations in [44].

A very recent second-order centrality also makes use of random walks but with three distinctions [45]. Firstly, it can be computed in a distributed manner with nodes having only information of who are their neighbors. Secondly, it relies on "unbiased" random walk where the stationary occupancy probability is equal for all nodes regardless of their degree (this is achieved by a Metropolis-Hastings algorithm where step from node i to a neighboring node j is accepted with the probability k_i/k_j for $k_j > k_i$ and always for $k_j \leq k_i$). Finally, it is based on the standard

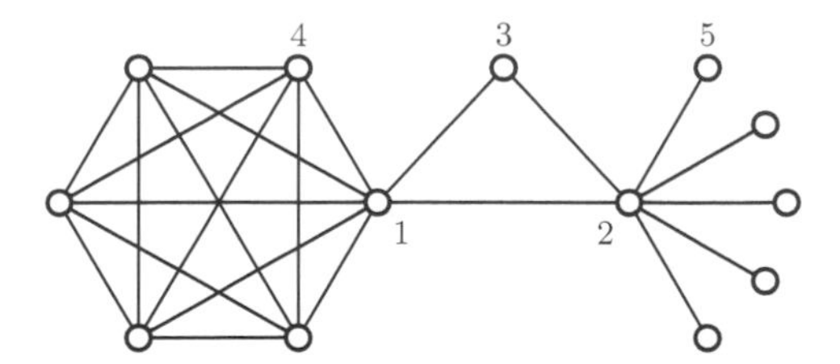

Fig. 4 A toy network for the computation of node centrality (see results in Table 1)

deviation σ_i of the return times to a given node i. The basic idea is that a node with a central position in the network is visited more regularly than peripheral nodes (those are visited in "clusters" with closely grouped subsequent visits interrupted by longer periods when the random walk is in a different part of the network). In addition to numerical stochastic computation of this centrality, various analytical results can be derived and used to better calibrate the numerical implementation.

The network of citations among scientific papers has the special property of being directed and acyclic (the acyclicity is due to citations pointing from a newer paper to older ones). This acyclicity allows one to use the probability of passing through a given node instead of the more traditional occupancy probability. In [46], the probability of passing through node i when the random walk starts in node j, G_{ij}, was proposed to quantify the influence of node i on node j. By summing over j, one consequently obtains the aggregate impact of node i which may be in turn used to rank the nodes. Since aggregate impact of node i correlates with the i's progeny size (by i's progeny we mean the set of nodes from which i can be reached by random walk respecting directions of links), one can better distinguish outstanding nodes by comparing the two characteristics. This passing probability framework has been also used to introduce a new node similarity which is based on the assumption that two nodes are similar if they are both influenced by the same nodes.

To better illustrate performance of the presented methods, we use them to compute node centrality in the network shown in Fig. 4. (Unlabeled nodes have standing identical with that of node 4 or 5.) For the shortest path centrality (also called betweenness centrality), we count also shortest paths where a node lies on the path's beginning or the path's end. For the PageRank score, we use the usual damping value $\alpha = 0.85$. For the random-walk centrality, we follow the prescription given in [42]. For the second-order centrality, we convert the standard deviation of return times σ_i into a centrality value $1/\sigma_i$ (recall that small σ_i is expected for centrally placed nodes). The results summarized in Table 1 show that there are considerable differences between respective centrality measures. While measures agree on a high centrality value of node 1 and a low centrality value of node 5, respectively, big differences exist in assessment of nodes 2, 3, and 4. In particular, eigenvector centrality puts emphasis on the tightly connected part of the network (represented by the complete 6-graph in our toy network) and values little node with low-degree neighbors (in our case, node 2). Random-walk centrality awards the central position of node 3 more than other tested measures which is a direct

Table 1 Centrality values for the network shown in Fig. 4. Values are normalized so that the average centrality is one in all cases

	Node				
Measure	1	2	3	4	5
Degree	1.98	1.98	0.57	1.41	0.28
Shortest path	2.59	3.14	0.66	0.66	0.66
Eigenvector	2.03	0.62	0.52	1.84	0.12
PageRank	1.71	2.65	0.68	1.12	0.47
Random walk	2.31	2.69	1.09	0.84	0.55
Second order	2.23	2.23	0.87	1.17	0.36

consequence of including not only the shortest paths in computation. One can note that degree centrality and second-order centrality rank nodes identically—the value difference between nodes 3 and 4 is however smaller in the case of second-order centrality which is again due to its random-walk origin being able to appreciate the central location of node 3.

3 Recommendation

The task of recommender systems is to utilize past evaluations of items by users to select further items that could be appreciated by the users. We often speak about personalized recommendations because a good recommender system should be able to recognize preferences of individuals and select the object to be recommended accordingly. Thanks to the availability of large-scale data on user behavior and the ever-increasing power of computers at our disposal, the field of recommendation grows rapidly. Nowadays, one can hardly imagine a successful e-commerce site without a sophisticated recommender system (think of Amazon.com) and companies organize competitions aiming to improve their recommendation methods (as it was prominently done by Netflix by their NetflixPrize) [47]. Approaches used to produce recommendations range from variants of the simple thesis "recommend to a user what was already appreciated by similar users" to complicated mathematical models and machine learning techniques [48–50]. The problem of link prediction is closely related to recommendation with the task being to identify possible missing or future links in a given network [51].

In this section, we aim to discuss the use of random walks in recommendation. First of all, similarity measures based on random walks can be used in similarity-based (sometimes called memory-based) collaborative filtering algorithms. Denoting the rating of object α given by user i as $r_{i\alpha}$ and the average rating of user i as μ_i, the generic form of collaborative filtering using user similarity is

$$\tilde{r}_{i\alpha} = \mu_i + \frac{\sum_{j \in R_\alpha} s_{ij}(r_{j\alpha} - \mu_j)}{\sum_{j \in R_\alpha} |s_{ij}|} \tag{11}$$

where $\tilde{r}_{i\alpha}$ is the expected (predicted) rating of object α by user i and R_α is the set of users who have already rated object α. User similarity s_{ij} (or object similarity $s_{\alpha\beta}$ for an item-based variant of collaborative filtering) is usually computed using the standard Pearson similarity or cosine similarity. Our interest now is in random-walk-based similarity measures that can be used instead of traditional ones.

Assuming that random walk starts in node i, one can introduce the average first passage time for node j, $T(j|i)$. The symmetrized quantity $C(i,j) := T(j|i) + T(i|j)$, the average commute time, was shown to act as a distance on the graph and can be further transformed into $\sqrt{C(i,j)}$, a so-called Euclidean Commute Time Distance [52]. In addition, both $C(i,j)$ and $\sqrt{C(i,j)}$ can serve as node similarity measures and in turn effectively used for collaborative filtering. While one can compute $C(i,j)$ on a node-by-node basis using the sink-node machinery described in Sect. 2.3, it is computationally more efficient to employ the formula

$$C(i,j) = 2E\left(l_{ii}^{+} + l_{jj}^{+} - 2l_{ij}^{+}\right) \tag{12}$$

where l_{ij}^{+} is an element of the Moore-Penrose pseudoinverse L^{+} of the network's Laplacian matrix $\mathsf{L} = \mathsf{D} - \mathsf{A}$ (here D is the degree matrix with elements $d_{ij} = k_i \delta_{ij}$) [52]. Pseudoinverse is applied because L cannot be inverted (zero is one of its eigenvalues) and can be computed as $\mathsf{L}^{+} = (\mathsf{L} - \mathbf{1}_N \mathbf{1}_N^{\mathrm{T}}/N)^{-1} + \mathbf{1}_N \mathbf{1}_N^{\mathrm{T}}/N$.

A simple node similarity measure based on local random walk was proposed in [53]. Denoting the probability that a random walker starting at node i is located at node j after t time steps as $\pi_{ij}(t)$, the similarity of nodes i and j was proposed in the form

$$s_{ij}^{LRW}(t) = \frac{1}{2E}\left(k_i \pi_{ij}(t) + k_j \pi_{ji}(t)\right) \tag{13}$$

where E is the total number of edges in the graph. Multiplication with node degree (k_i and k_j, respectively) gives more weight to nodes with high degree and compensates for the increased dispersion of random walk at those nodes (if many links lead from x, π_{xy} can be low). The obtained quantity can be summed over different t, leading to "superposed" similarity $s_{ij}^{SRW}(t) = \sum_{\theta=1}^{t} s_{ij}^{RW}(\theta)$. Numerical evaluation on five distinct real networks showed that s^{LRW} and s^{SRW} in most cases outperform traditional similarity metrics in accuracy and are less computationally demanding than other well-performing methods [53]. A method for random-walk computation of paper similarity was proposed specifically for scientific citation data [54]. When computing similarity of papers i and j, two two-step random walks are combined. One aims "downstream" to papers cited by both i and j, thus reflecting the opinion of the authors of i and j. The other aims "upstream" to papers citing both i and j, thus reflecting the opinion of the readers of i and j. It is then shown that this novel similarity measure is able to identify the backbone of the citation network, leading to accurate characterization of hierarchical structure of the scientific development and its classification into fields and subfields.

Due to sparsity of the input data, traditional similarity measures based on overlapping neighborhoods can fail to accurately assess node similarity. To alleviate this problem, it was suggested to transform the similarity matrix into a PageRank-

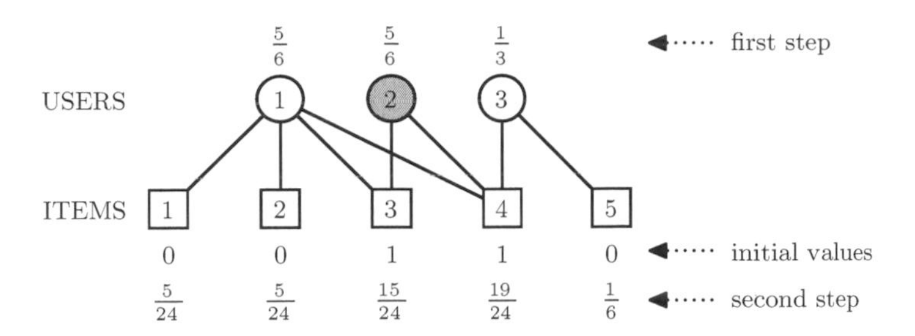

Fig. 5 Illustration of random-walk recommendation for user 2. Items collected by user 2 are initially assigned unit resource which then spreads uniformly to users connected with these items and finally back to the item side. Items with the highest resulting resource amount are then recommended to the given user. In this case, items 1 and 2 score best (items 3 and 4 have higher resulting values but are ignored as they have been already collected by user 2)

like form P by normalization and addition of random jumps and then use $\mathsf{P}(1-\alpha\mathsf{P})^{-1}$ as a new similarity matrix where similarity values are assigned also to item pairs that have not been evaluated by any users [55]. Here $\alpha \in (0,1]$ is the probability of continuing the random walk, and thus $1/\alpha$ is the characteristic number of steps over which similarity is transferred.

Apart from using random walks to quantify node similarity, there are also recommendation methods that are directly based on random walks. In [56], the authors consider the bipartite user-item network where links connect users with the items they collected or appreciated. Note that explicit ratings given by users to items play no role here—the method only requires the knowledge of items that have been collected/favored by individual users. Assuming that each item collected by a given user i is assigned a unit initial resource, this resource is spread uniformly from the collected items to the users connected with them and then in the second step back to items connected with those users (see Fig. 5 for an illustration). The final amount of resource on respective items is then interpreted as their recommendation score and items with the highest score are then recommended to user i (already collected items are of course excluded). The reasoning behind this spreading process is that it selects items that have been collected by users who share some interests with the given user i. At the same time, if user i has collected an extremely popular item α or if a collected item has been co-collected by an extremely active user j, the information signal is weak because the overlap between i and j as well as between i and α is rather small in those cases. The random-walk-based even spreading of the resource is thus a reasonable approach to quantify the resulting recommendation scores.

The transition matrices from objects to users and vice versa have the form $U_{i\alpha} = A_{i\alpha}/k_\alpha$ and $V_{\alpha i} = A_{i\alpha}/k_i$, respectively, where k_α is the degree of item α (the number of users who collected it) and k_i is the degree of user i (the number of items collected by this user). The vector with object recommendation scores for user i then reads $\tilde{\boldsymbol{h}}_i = \mathsf{VU}\boldsymbol{h}_i$ where $(\boldsymbol{h}_i)_\alpha = A_{i\alpha}$ encodes which items have been actually collected by user i. One can introduce $\mathsf{W}^P := \mathsf{VU}$ and show that

$$W^P_{\alpha\beta} = \frac{1}{k_\beta} \sum_{i=1}^{U} \frac{A_{i\alpha} A_{i\beta}}{k_i} \tag{14}$$

where indices α and β are used to enumerate items, i enumerates users, and U is the total number of user nodes. One can also spread the initial resource over $2n$ steps in the bipartite network by $(\mathsf{W}^{\mathsf{P}})^n \boldsymbol{h}_i$, but the result converges fast to a vector whose elements are proportional to object degree k_α and hence conveys little information for personalized recommendation.

This basic method has been subsequently generalized in multiple ways. For example, it was proposed to assign the initial amount of resource to items not uniformly but depending on the item degree as k_α^θ [57]. Best results were achieved with $\theta \approx -1$ when the produced recommendations were both more accurate and more personalized. To better answer the need for diversity in recommendations, a hybrid algorithm was proposed which combines the random-walk algorithm with heat spreading [58]. As we have already seen, heat diffusion differs from random walk in normalization of their matrices and thus the matrix of heat diffusion reads $W^H_{\alpha\beta} = (1/k_\alpha) \sum_{i=1}^{U} A_{i\alpha} A_{i\beta} / k_i$. The best performing hybrid of the two has the form

$$W^{P+H}_{\alpha\beta} = \frac{1}{k_\alpha^{1-\lambda} k_\beta^{\lambda}} \sum_{i=1}^{U} \frac{A_{i\alpha} A_{i\beta}}{k_i} \tag{15}$$

where the parameter λ controls the balance between the contribution of random walk and heat spreading. This method was shown to simultaneously increase accuracy and diversity of recommendations.

A combination of random walk and heat diffusion for data with explicit ratings was presented in [59] where recommendation scores obtained by each respective process are multiplied to obtain the final recommendation score. In addition, the employed random walk is self-avoiding, i.e., there is no possibility to return to the initial item node after two steps. If user evaluations are given in an integer scale (a very typical case nowadays), a multichannel spreading can be employed where the states of the random walk are represented not only by the current item but also by the rating given to this item [60]. If, for example, a five-level rating scale is used, 5×5 connections are created between any two items. However, this approach suffers from aggravating the data sparsity problem (the same amount of data is used to construct many more connections between (item, rating) pairs) which limits its performance.

Spreading over a bipartite network is considered also in [61] where the bipartite user-item network is augmented with social links among users (this kind of data is often produced in online gaming). The random walk starting at the user for which recommendations are being made follows a social link to another user with probability α or a link to an item with probability $1 - \alpha$ where it is absorbed. Items are then ranked according to the fraction of random walks absorbed in them. A different mechanism of heat diffusion on an item–item network was used to produce recommendations by representing items liked and disliked by a given user as nodes with fixed temperature 1 and 0, respectively [62]. From the remaining nodes, those with the highest resulting temperature are then chosen to be recommended to the given user. See [50, Chap. 6] for other related works and more detailed information.

4 Conclusion

We attempted here to give an overview of applications of random walks to information filtering, focusing on the tasks of ranking and recommendation in particular. Despite the amount of work done in these two directions, multiple important research challenges still remain open. Due to the massive amounts of available data, scalability of algorithms is of critical importance. Even when full computation is possible, one can think of potential approaches to update the output gradually when new data arrives. To achieve that, one can use or learn from perturbation theory which is a well-known tool in physics. We have seen that results based on random walks often correlate strongly with mere popularity (represented by degree) of nodes in the network. Such bias toward popularity may be beneficial for an algorithm's accuracy, but it may also narrow our view of the given system and perhaps create a self-reinforcing loop further boosting popularity of already popular nodes. We thus need information filtering algorithms that converge less to the center of the given network. Random walks biased by node centrality or time information about nodes and links could provide a solution to this problem. As a beneficial side effect, this line of research could yield algorithms pointing us to fresh and promising content instead of highlighting old victors over and over again.

Acknowledgement This work was partially supported by the Swiss National Science Foundation Grant No. 200020-132253. I wish to thank a number of wonderful friends and colleagues who helped to shape many of the ideas presented here.

References

[1] U. Hanani, B. Shapira, P. Shoval, Information filtering: Overview of issues, research and systems. User Model. User Adapted Interact. **11**, 203–259 (2001)
[2] T. Hastie, R. Tibshirani, J. Friedman, *The Elements of Statistical Learning: Data Mining, Inference, and Prediction*, 2nd edn. (Springer, New York, 2001)
[3] I.H. Witten, E. Frank, M.A. Hall, *Data Mining: Practical Machine Learning Tools and Techniques*, 3rd edn. (Morgan Kaufmann/Elsevier, San Francisco, 2011)
[4] M. Franceschet, PageRank: Standing on the shoulders of giants. Comm. ACM **54**, 92–101 (2011)
[5] S. Brin, L. Page, The anatomy of a large-scale hypertextual web search engine. Comput. Network ISDN Syst. **30**, 107–117 (1998)
[6] R. Burke, Hybrid web recommender systems, in *The Adaptive Web: Methods and Strategies of Web Personalization*, ed. by P. Brusilovsky, A. Kobsa, W. Nejdl (Springer, Heidelberg, 2007)

[7] L. Costa, F. da, F.A. Rodrigues, G. Travieso, P.R. Villas Boas, Characterization of complex networks: A survey of measurements. Adv. Phys. **56**, 167–242 (2007)
[8] A. Barrat, M. Barthelemy, A. Vespignani, *Dynamical Processes on Complex Networks* (Cambridge University Press, New York, 2008)
[9] S. Boccaletti, V. Latora, Y. Moreno, M. Chavez, D.-U. Huang, Complex networks: structure and dynamics. Phys. Rep. **424**, 175–308 (2006)
[10] S. Wasserman, K. Faust, *Social Network Analysis: Methods and Applications* (Cambridge University Press, Cambridge, 1994)
[11] P. Bonacich, P. Lloyd, Eigenvector-like measures of centrality for asymmetric relations. Soc. Network **23**, 191–201 (2001)
[12] P. Berkhin, A survey on PageRank computing. Internet Math. **2**, 73–120 (2005)
[13] N. Perra, S. Fortunato, Spectral centrality measures in complex networks. Phys. Rev. E **78**, 036107 (2008)
[14] Y. Ding, E. Yan, A. Frazho, J. Caverlee, PageRank for ranking authors in co-citation networks. J. Am. Soc. Inform. Sci. Tech. **60**, 2229–2243 (2009)
[15] A. Hotho, R. Jäschke, C. Schmitz, G. Stumme, Information retrieval in folksonomies: search and ranking, in *Lecture Notes in Computer Science*, vol. 4011, ed. by Y. Sure, J. Domingue, pp. 84–95 (2006)
[16] A. Agarwal, S. Chakrabarti, S. Aggarwal, Learning to rank networked entities, in *Proceedings of the 12th ACM SIGKDD International Conference on Knowledge Discovery and Data Mining (KDD'06)* (ACM, New York, 14–23 2006)
[17] A.N. Langville, C.D. Meyer, *Google's PageRank and Beyond: The Science of Search Engine Rankings* (Princeton University Press, Princeton, 2006)
[18] L. Bing, *Web Data Mining: Exploring Hyperlinks, Contents and Usage Data* (Springer, Heidelberg, 2007)
[19] T.H. Haveliwala, Efficient computation of PageRank. Technical Report, Stanford University Database Group, http://ilpubs.stanford.edu:8090/386/ (1999)
[20] A. Cheng, E. Friedman, Manipulability of PageRank under Sybil strategies, in *Proceedings of the First Workshop on the Economics of Networked Systems (NetEcon06)*, Ann Arbor, 2006
[21] G. Ghoshal, A.-L. Barabsi, Ranking stability and super-stable nodes in complex networks. Nat. Comm. **2**, 394 (2011)
[22] R. Baeza-Yates, P. Boldi, C. Castillo, Generalizing PageRank: damping functions for link-based ranking algorithms, in *Proceedings of the 29th Annual International ACM SIGIR Conference on Research and Development in Information Retrieval (SIGIR'06)* (ACM, New York, 2006)
[23] S.D. Kamvar, M.T. Schlosser, H. Garcia-Molina, The eigentrust algorithm for reputation management in P2P networks, in *Proceedings of the 12th International Conference on World Wide Web (WWW'03)* (ACM, New York, 2003)
[24] G.D. Paparo, M.A. Martin-Delgado, Google in a quantum network. Sci. Rep. **2**(444), arXiv.org/abs/1112.2079. http://www.nature.com/srep/2012/120608/srep00444/full/srep00444.html?WT.mc_id=FBK_SciReports (2012)
[25] J.M. Kleinberg, Authoritative sources in a hyperlinked environment. J. ACM **46**, 604–632 (1999)
[26] H. Deng, M.R. Lyu, I. King, A generalized Co-HITS algorithm and its application to bipartite graphs, in *Proceedings of the 15th ACM SIGKDD International Conference on Knowledge Discovery and Data Mining (KDD'09)* (ACM, New York, 2009)
[27] A.N. Langville, C.D. Meyer, A survey of eigenvector methods for web information retrieval. SIAM Rev. **47**, 135–161 (2005)
[28] P. Chen, H. Xie, S. Maslov, S. Redner, Finding scientific gems with Google's PageRank algorithm. J. Informetrics **1**, 8–15 (2007)
[29] S. Maslov, S. Redner, Promise and pitfalls of extending Google's PageRank algorithm to citation networks. J. Neurosci. **28**, 11103–11105 (2008)
[30] M. Medo, G. Cimini, S. Gualdi, Temporal effects in the growth of networks. Phys. Rev. Lett. **107**, 238701 (2011)

[31] F. Radicchi, S. Fortunato, A. Vespignani, Citation networks, in *Models of Science Dynamics, Understanding Complex Systems*, ed. by A. Scharnhorst, et al. (Springer, Berlin Heidelberg, 2012)
[32] D. Walker, H. Xie, K.K. Yan, S. Maslov, Ranking scientific publications using a model of network traffic. J. Stat. Mech. **6**, P06010 (2007)
[33] J. Bollen, M.A. Rodriguez, H. Van de Sompel, J. Status. Scientometrics **69**, 669–687 (2006)
[34] B. Gonzlez-Pereiraa, V.P. Guerrero-Bote, F. Moya-Anegn, A new approach to the metric of journals scientific prestige: The SJR indicator. J. Informetrics **4**, 379–391 (2010)
[35] F. Radicchi, S. Fortunato, B. Markines, A. Vespignani, Diffusion of scientific credits and the ranking of scientists. Phys. Rev. E **80**, 056103 (2009)
[36] F. Radicchi, Who is the best player ever? A complex network analysis of the history of professional tennis. PLoS ONE **6**, e17249 (2011)
[37] E. Yan, Y. Ding, Discovering author impact: A PageRank perspective. Inform. Process. Manag. **47**, 125–134 (2011)
[38] S. Allesina, M. Pascual, Googling food webs: can an eigenvector measure species' importance for coextinctions? PLoS Comput. Biol. **5**, e1000494 (2009)
[39] L. Lü, Y.-C. Zhang, C.H. Yeung, T. Zhou, Leaders in social networks, the delicious case. PLoS ONE **6**, e21202 (2011)
[40] A. Stojmirović, Y.-K. Yu, Information flow in interaction networks. J. Comput. Biol. **14**, 1115–1143 (2007)
[41] P.G. Doyle, J.L. Snell, Random walks and electric networks. *Carus Mathematical Monographs*, vol. 22 (Mathematical Association of America, Washington, 1984)
[42] M.E.J. Newman, A measure of betweenness centrality based on random walks. Soc. Network **27**, 39–54 (2005)
[43] G.-L. Lin, H. Peng, Q.-L. Ma, J. Wei, J.-W. Qin, Improving diversity in Web search results re-ranking using absorbing random walks, in *Proceedings of the International Conference on Machine Learning and Cybernetics (ICMLC'10)* (IEEE, 2116–2421 2010)
[44] S.P. Borgatti, Centrality and network flow. Soc. Network **27**, 55–71 (2005)
[45] A.-M. Kermarrec, E. Le Merrer, B. Sericola, G. Trdan, Second order centrality: Distributed assessment of nodes criticity in complex networks. Comput. Comm. **34**, 619–628 (2011)
[46] S. Gualdi, M. Medo, Y.-C. Zhang, Influence, originality and similarity in directed acyclic graphs. EPL **96**, 18004 (2011)
[47] J.B. Schafer, J.A. Konstan, J. Riedl, E-commerce recommendation applications. Data Min. Knowl. Discov. **5**, 115–153 (2001)
[48] G. Adomavicius, A. Tuzhilin, Toward the next generation of recommender systems: a survey of the state-of-the-art and possible extensions. IEEE Trans. Knowl. Data Eng. **17**, 734–749 (2005)
[49] F. Ricci, L. Rokach, B. Shapira, P.B. Kantor (eds.), *Recommender Systems Handbook* (Springer, New York, 2011)
[50] L. Lü, M. Medo, C.H. Yeung, Y.-C. Zhang, Z.-K. Zhang, T. Zhou, Recommender systems. Phys. Rep. **519**, 1–49. arXiv.org/abs/1202.1112 (2012)
[51] L. Lü, T. Zhou, Link prediction in complex networks: a survey. Phys. A **390**, 1150–1170 (2011)
[52] F. Fouss, A. Pirotte, J.-M. Renders, M. Saerens, Random-walk computation of similarities between nodes of a graph with application to collaborative recommendation. IEEE Trans. Knowl. Data Eng. **19**, 355–369 (2007)
[53] W. Liu, L. Lü, Link prediction based on local random walk. EPL **89**, 58007 (2010)
[54] S. Gualdi, C.H. Yeung, Y.-C. Zhang, Tracing the evolution of physics on the backbone of citation networks. Phys. Rev. E **84**, 046104 (2011)
[55] H. Yildirim, M.S. Krishnamoorthy, A random walk method for alleviating the sparsity problem in collaborative filtering, in *Proceedings of the 2008 ACM Conference on Recommender Systems (RecSys'08)* (ACM, New York, 2008)
[56] T. Zhou, J. Ren, M. Medo, Y.-C. Zhang, Bipartite network projection and personal recommendation. Phys. Rev. E **76**, 046115 (2007)

[57] T. Zhou, L.-L. Jinag, R.-Q. Su, Y.-C. Zhang, Effect of initial configuration on network-based recommendation. EPL **81**, 58004 (2008)
[58] T. Zhou, Z. Kuscsik, J.-G. Liu, M. Medo, J.R. Wakeling, Y.-C. Zhang, Solving the apparent diversity-accuracy dilemma of recommender systems. Proc. Natl. Acad. Sci. USA **107**, 4511–4515 (2010)
[59] M. Blattner, B-rank: A top N recommendation algorithm, in *Proceedings of the 1st International Multi-Conference on Complexity, Informatics and Cybernetics*, pp. 336–341, 2010
[60] Y.-C. Zhang, M. Medo, J. Ren, T. Zhou, T. Li, F. Yang, Recommendation model based on opinion diffusion. EPL **80**, 68003 (2007)
[61] A.P. Singh, A. Gunawardana, C. Meek, A.C. Surendran, Recommendations using absorbing random walks, in *Proceedings of the North East Student Colloquium on Artificial Intelligence*, 2007
[62] Y.-C. Zhang, M. Blattner, Y.-K. Yu, Heat conduction process on community networks as a recommendation model. Phys. Rev. Lett. **99**, 154301 (2007)

Index

A. Mukherjee et al. (eds.), *Dynamics On and Of Complex Networks, Volume 2*,
Modeling and Simulation in Science, Engineering and Technology,
DOI 10.1007/978-1-4614-6729-8, © Springer Science+Business Media New York 2013

C

Printed by Publishers' Graphics LLC